Organic Metal and Metalloid Species in the Environment

Springer-Verlag Berlin Heidelberg GmbH

Alfred V. Hirner • Hendrik Emons (Eds.)

Organic Metal and Metalloid Species in the Environment

Analysis, Distribution, Processes and Toxicological Evaluation

With 85 Figures

Springer

Editors

Prof. Dr. Alfred V. Hirner
University of Duisburg-Essen
Institute of Environmental Analytics
45117 Essen, Germany

Prof. Dr. Hendrik Emons
European Commission
Joint Research Centre
Institute for Reference Materials and Measurements
Retieseweg
2440 Geel, Belgium

ISBN 978-3-642-05873-8 ISBN 978-3-662-09135-7 (eBook)
DOI 10.1007/978-3-662-09135-7

Library of Congress Cataloging-in-Publication Data Applied For

A catalog record for this book is available from the Library of Congress.
Bibliographic information published by Die Deutsche Bibliothek
Die Deutsche Bibliothek lists this publication in die Deutsche Nationalbibliographie; detailed bibliographic data is available in the Internet at <http://dnb.ddb.de>.

springeronline.com

Originally published by Springer-Verlag Berlin Heidelberg New York in 2004
Softcover reprint of the hardcover 1st edition 2004

Cover Design: Erich Kirchner, Heidelberg

Typesetting: Camera-ready by the editors

Printed on acid free paper 30/3141/LT – 5 4 3 2 1 0

Preface

In recent years, a number of books devoted to various aspects of trace metal speciation and analysis have been published. For organometal(loid) and alkylated metal(loid) compounds in particular, the rapid progress in the field of speciation becomes evident when the second edition of Craig's *Organometallic Compounds in the Environment* (J. Wiley & Sons, 2003) is compared with the first edition published in 1986. The growing number of publications in the environmental and life sciences concerned with the biogeochemical cycling and the toxicity of metal(loid)s indicates the increasing awareness that reliable data on the occurrence, reactions, and fate of such species are needed.

In the course of the international *Workshop on Organometallics in the Environment* at the University of Essen in 2002, it was proposed to publish the workshop proceedings supplemented with additional data in form of a book. The wide range of articles on the current state and future perspectives of speciation presented in this book by chemists, biologists, toxicologists, and physicians reflects the multidisciplinary nature of that workshop. The interdisciplinary exchange of views particularly revealed that, in contrast to the progress in analytical techniques made in recent years, data on the toxic effects of organometal(loid) compounds and the underlying mechanisms are still sparse. A more intensive transfer of information from fields such as geosciences or marine chemistry to those of microbiology, neurochemistry, and genomics is required to assess the role of organometal(loid) species in living systems.

To pursue this aim in an interdisciplinary effort, several authors of this book work closely together in a joint project funded by the German Research Foundation (Deutsche Forschungsgemeinschaft, Grant No. FOR 415). While the majority of the articles represent original research papers, a number of review-type papers give short overviews on the environmental and analytical chemistry of organometal(loid) species (Ch. 1 and 2) as well as on the geno- and neurotoxicity of these compounds (Ch. 11, 14 and 15). The series of articles begins with an introduction to the environmental chemistry of organometallic compounds (Ch. 1), followed by an overview on instrumental analytical techniques used in metal(loid) speciation (Ch. 2). Particular emphasis is laid on arsenic speciation in environmental systems (Ch. 3) and in biomonitoring studies (Ch. 4). Biotransformation processes in microorganisms (Ch. 5, 7, and 8) and in man (Ch. 6 and 10) are described, and a chemical modelling study concerning organometallics/DNA interactions is presented (Ch. 9). Aspects of genotoxicity (Ch. 11 to 13) and neurotoxicity of organometal(loid)s (Ch. 14 to 16) are discussed in detail. The last two chapters of the book comprise the results of panel discussions on current analytical (Ch. 17) and toxicological issues (Ch. 18) concerning organometal(loid) compounds.

The editors are confident that newcomers as well as experienced scientists will find useful information, new ideas, and stimulating hypotheses in many of the following chapters.

Last but not least, we would like to thank staff members of the Institute of Environmental Analytical Chemistry, in particular Rita Lehmann, for substantial help in preparing and editing this book.

Essen, October 2003

Alfred V. Hirner
Hendrik Emons

Contents

1 Organometallic compounds in the environment: An overview 1
P. J. Craig and R. O. Jenkins

General points 1
Biomethylation 5
Arsenic 6
Antimony 8
Mercury 8
Microbial demethylation/dealkylation 11
Literature 12

2 Advances in analytical methods for speciation of trace elements in the environment 17
J. Szpunar, B. Bouyssiere, R. Lobinski

Introduction 17
Organometall(oid) Species in the Environment 18
Hyphenated Techniques in Speciation Analysis 18
Advances in GC/ICP-MS 20
Advances in Gas Chromatography prior to ICP-MS 21
ICP-MS Detection in Gas Chromatography 22
GC/ICP-MS Studies Using Stable Isotopes 24
Advances in Sample Preparation 27
Multidimensional LC with ICP-MS Detection 29
The Coupling HPLC/ICP-MS 29
Electrospray Mass Spectrometry for the Species Identification 30
Multidimensional Analytical Strategy for the Identification of Organoarsenic Species in Marine Biota 34
Conclusions 35
Literature 35

3 Analytical strategies for arsenic speciation in environmental and biological samples 41
J. Feldmann, S. Devalla, A. Raab, H. R. Hansen

Arsenic species in the environment and in biological samples 41
Analytical strategies 45
Sample preparation 45
Separation methods 47

Identification 49
Quantification 51
Analytical aspects of selected arsenic speciation studies 52
Gas chromatography coupled to MS or ICP-MS 52
Liquid chromatography coupled to ESI-MS and ICP-MS 57
Conclusions 67
Acknowledgements 68
Literature 68

4 Occurrence and speciation of arsenic, antimony and tin in specimens used for environmental biomonitoring of limnic ecosystems 71
H. Emons, Z. Sebesvari, K. Falk, M. Krachler

Environmental biomonitoring 71
Occurrence of As, Sb and Sn in freshwater ecosystems 72
Arsenic 73
Antimony 74
Tin 75
Speciation analysis of As, Sb and Sn in limnic samples 76
Arsenic species 80
Antimony species 82
Tin species 84
Quality assurance 84
Speciation of As, Sb, Sn in limnic bioindicators – first results 86
Arsenic 86
Antimony 88
Tin 88
Outlook and future challenges 91
Acknowledgment 92
Literature 92

5 Methylated metal(loid) species in biological waste treatment 97
R.A. Diaz-Bone, B. Menzel, A. Barrenstein, A.V. Hirner

Introduction 97
Biological waste treatment 98
Experimental 100
Derivatisation of ionic species 100
Sampling of gaseous species 102
Metal analysis 102
Set-up of garden composting experiments 103
Results and discussion 103
Ionic organometal(loid) species in compost from waste treatment facilities 103
Metal(loid) species in different aged compost 105

Biomethylation in garden compost 108
Conclusions 110

6 Volatile mercury species in environmental gases and biological samples 113
J. Hippler, J. Kresimon, A. V. Hirner

Introduction 113
Analytical methods 114
Instrumental techniques 114
Results 119
Environmental samples 119
Biological samples 122
Validation of data 127
Discussion 129
Dimethylmercury in environmental gases 129
Mercury species generated in metabolic processes 130
Conclusions 131
Acknowledgements 132
Literature 132

7 Biovolatilisation of metal(loid)s by microorganisms 137
K. Michalke & R. Hensel

Introduction 137
Analytical setup for identifying and quantifying metal(loid)-organic derivatives 138
Volatile metal(loid) compounds in the environment 138
Towards an understanding of metal(loid) biotransformation processes 140
Direct monitoring of volatile metal(loid) derivatives in the environment 140
Monitoring of metal(loid) volatilisation by microbial populations enriched from original environments 142
Monitoring of metal(loid) volatilisation by isolated strains 147
Conclusion and perspectives 149
Literature 150

8 The Effect of Phosphate on the Bioaccumulation and Biotransformation of Arsenic(V) by the Marine Alga *Fucus gardneri* 155
S. C. R. Granchinho, W. R. Cullen, E. Polishchuk, and K. J. Reimer

Introduction ... 155
Experimental ... 156
Reagents and Chemicals ... 156
Medium and Antibiotics ... 156
Fucus gardneri Samples ... 157
Sample Preparation and Analysis ... 157
Results and discussion ... 158
Acknowledgments ... 165
Literature ... 165

9 Molecular modeling studies of specific interactions between organometallic compounds and DNA ... 167
R. Yonchev, H. Rehage, H. Kuhn

Introduction ... 167
Theory and methods ... 167
Extensible Systematic Force Field (ESFF) ... 168
Ligand aligning algorithm ... 172
Results and discussion ... 175
Conclusions ... 180
Future work ... 180
Literature ... 180

10 Organometal(loid) compounds associated with human metabolism ... 181
A.V. Hirner, L. M. Hartmann, J. Hippler, J. Kresimon, J. Koesters, K. Michalke, M. Sulkowski, A. W. Rettenmeier

Introduction ... 181
Experimental ... 183
Ingestion experiment ... 183
Analytical methods ... 184
Results ... 185
Gaseous samples ... 185
Saliva ... 186
Urine ... 187
Blood ... 192
Faeces ... 196
Discussion ... 198
Conclusions ... 201
Acknowledgements ... 201
Literature ... 201

11 Genotoxicity of organometallic species ... 205
A.-M. Florea, E. Dopp, G. Obe, A.W. Rettenmeier

Human exposure to organometallic species 205
Genotoxicity of organometallic species 206
Genotoxic effects of organomercury compounds 206
Genotoxic effects of organoarsenic compounds 208
Genotoxic effects of organotin compounds 211
In vitro genotoxicity of different organometal(loid) compounds in CHO cells 214
Summary 215
Literature 216

12 Current aspects on the genotoxicity of arsenite and its methylated metabolites: Oxidative stress and interactions with the cellular response to DNA damage 221
A. Hartwig, T. Schwerdtle, I. Walter

Introduction 221
Oxidative DNA damage 223
DNA repair systems and interactions by arsenite 225
Poly(ADP-ribosyl)ation as sensitive intracellular target for arsenite 226
Conclusions and perspectives 228
Acknowledgements 230
Literature 230

13 Cytogenetic investigations in employees from waste industries 235
Helga Fender, Giesela Wolf

Introduction 235
Materials and methods 236
Subjects 236
Smoking Status 237
Cytogenetic methods 238
Results 239
Discussion 243
Acknowledgements 244
Literature 244

14 Neurotoxicity of metals 247
G. Stoltenburg-Didinger

Introduction 247
Mercury 248
Lead 251
Literature 255

15 Actions of metals on membrane channels, calcium homeostasis and synaptic plasticity 259
D. Büsselberg

Relevance of metals 259
Lead 259
Aluminium 261
Zinc 262
Mercury 262
Interference of metals with cell functions 263
Metals on Membrane Channels 264
Other channels involved in lead toxicity 268
Metals and calcium homeostasis 269
Lead and aluminium actions on synaptic plasticity 269
Relevance of the data 270
General Implications 271
Literature 272

16 Effects of organometal(loid) compounds on neuronal ion channels: possible sites for neurotoxicity 283
J. Gruner, K. Krüger, N. Binding, M. Madeja, U. Mußhoff

Toxicity of organometal(loid)s 283
Neurotoxic aspects of organometal(loid)s 283
Neurotoxicity of inorganic and organic arsenicals 285
Target structures of neurotoxicity 286
Function of ion channels in the nervous system 287
Prediction of neurotoxic potency of hazardous substances with an in vivo expression system: *Xenopus laevis* oocytes 288
The use of *Xenopus laevis* oocytes for electrophysiological studies of ion channels 288
Testing the Xenopus expression system for neurotoxicological investigations: Effects of lead on voltage- and transmitter-operated ion channels 291
Effects of arsenicals on transmitter-operated ion channels 293
Transmembraneous ion currents through the different glutamate-operated ion channels 293
Electrophysiological techniques 294
Reproducibility of the receptor-mediated ion currents (control experiments) 295
Application of the arsenicals 296
Effects of inorganic arsenite 297
Effects of monomethylarsonic acid ($MeAsO(OH)_2$) 299
Effects of dimethylarsinic acid (Me_2AsOOH) 301
Effects of arsenicals on voltage-operated potassium channels 303
Transmembraneous ion currents through different voltage-operated potassium channels 304
Electrophysiological techniques 304

Neuronal ion channels as targets for organic arsenicals 306
Perspectives and limitation of Xenopus oocytes 307
Effects of arsenicals on glutamate-operated ion channels 307
Effects of arsenicals on voltage-operated potassium channels 309
Significance in arsenic poisoning 310
Literature 310

17 Panel discussion: Analytical aspects
Discussion Session on “Speciation Analysis of Environmental Samples” 317
Convener: H. Emons

18 Panel discussion: Toxicological Aspects
Discussion Session on "Toxicological Aspects of Alkylated Metal(loid) Species" 319
Convener: A. W. Rettenmeier

Resumé 323

Subject Index 325

List of contributors

Dr. Axel Barrenstein, Landesumweltamt NRW, Postfach 10 23 63, 45023 Essen, Germany

Dr. Norbert Binding, Institut für Arbeitsmedizin, Westfälische Wilhelms-Universität Münster, Robert-Koch-Str. 51, 48149 Münster, Germany

Brice Bouyssiere, GKSS-Research Centre, Institute for Coastal Research / Physical and Chemical Analysis, Max-Planck-Straße, 21502 Geesthacht, Germany

Prof. Dr. Dietrich Büsselberg, Universitätsklinikum Essen, Institut für Physiologie, 45122 Essen, Germany

Prof. Dr. Peter J. Craig, School of Molecular Sciences, De Montfort University, The Gateway, Leicester LE1 9BH, UK

Prof. Dr. William R. Cullen, Environmental Chemistry Group, Chemistry Department, University of British Columbia, Vancouver, B.C., Canada V6T 1Z1

Dr. Sandhya Devalla, University of Aberdeen, Department of Chemistry, Meston Walk, Old Aberdeen AB24 3UE, UK

Roland Diaz-Bone, Institut für Umweltanalytik, Universität Duisburg-Essen, Universitätsstr. 3 – 5, 45141 Essen, Germany

Dr. Elke Dopp, Institut für Hygiene und Arbeitsmedizin, Universitätsklinikum Essen,
Hufelandstraße 55, 45122 Essen, Germany

Prof. Dr. Hendrik Emons, European Commission, Joint Research Centre, Institute for Reference Materials and Measurements, Retieseweg, 2440 Geel, Belgium

Dr. Kirsten Falk, Institut für Phytosphäre, Forschungszentrum Jülich, D-52425 Jülich, Germany

Dr. Jörg Feldmann, University of Aberdeen, Department of Chemistry, Meston Walk, Old Aberdeen AB24 3UE, UK

Dipl.-Biol. Ana-Maria Florea, Institut für Hygiene und Arbeitsmedizin, Universitätsklinikum Essen, Hufelandstraße 55, 45122 Essen, Germany

Dr. Helga Fender, Robert Koch-Institute Berlin, 13302 Berlin, P.O. Box 65 02 80, Germany

Sophia C. R. Granchinho, Environmental Chemistry Group, Chemistry Department, University of British Columbia, Vancouver, B.C., Canada V6T 1Z1

Janina Gruner, Institut für Physiologie, Universität Münster, Robert-Koch-Str. 51, 48149 Münster, Germany

Helle Hansen, University of Aberdeen, Department of Chemistry, Meston Walk, Old Aberdeen AB24 3UE, UK

Dr. Louise M. Hartmann, Institut für Umweltanalytik, Universität Essen, Universitätsstr. 3-5, 45141 Essen, Germany

Prof. Dr. Andrea Hartwig, Institut für Lebensmittelchemie und Toxikologie Universität Karlsruhe, Kaiserstr.12, Postfach 6980, D-76128 Karlsruhe, Germany

Prof. Dr. Reinhard Hensel, Mikrobiologie, Universität Essen, Universitätsstr. 3-5, 45141 Essen, Germany

Jörg Hippler, Institut für Umweltanalytik, Universität Essen, Universitätsstr. 3-5, 45141 Essen, Germany

Prof. Dr. Alfred V. Hirner, Institut für Umweltanalytik, Universität Essen, Universitätsstr. 3-5, 45141 Essen, Germany

Dr. Richard O. Jenkins, School of Molecular Sciences, De Montfort University, The Gateway, Leicester LE1 9BH, UK

Jan Kösters, Institut für Umweltanalytik, Universität Essen, Universitätsstr. 3-5, 45141 Essen, Germany

Dr. Michael Krachler, Universität Heidelberg, Institut für Umwelt-Geochemie, Im Neuenheimer Feld 236, 6900 Heidelberg, Germany

Dr. Jutta Kresimon, Institut für Umweltanalytik, Universität Essen, Universitätsstr. 3-5, 45141 Essen, Germany

Dr. Hubert Kuhn, Institut für Physikalische Chemie, Universität Essen, Schützenbahn 70, 45117 Essen, Germany

Dr. Katharina Krüger, Institut für Physiologie, Universität Münster, Robert-Koch-Str. 27a, 48149 Münster, Germany

Prof. Dr. Ryszard Lobinski, CNRS, Pau, France

Bernd Menzel, Institut für Umweltanalytik, Universität Essen, Universitätsstr. 3-5, 45141 Essen, Germany

Dr. Klaus Michalke, Mikrobiologie, Universität Essen, Universitätsstr. 5, 45141 Essen, Germany

Prof. Dr. Mußhoff, Institut für Physiologie, Universität Münster, Robert-Koch-Str. 51, 48149 Münster, Germany

Prof. Dr. Günter Obe, Genetik, Universität Essen, Universitätsstr. 3-5, 45141 Essen, Germany

Dr. Elena Polishchuk, Environmental Chemistry Group, Chemistry Department, University of British Columbia, Vancouver, B.C., Canada V6T 1Z1

Dr. Andrea Raab, University of Aberdeen, Department of Chemistry, Meston Walk, Old Aberdeen AB24 3UE, UK

Prof. Dr. Heinz Rehage, Institut f. Physikalische Chemie, Universität Essen, Universitätsstr. 3-5, 45141 Essen, Germany

Dr. Kenneth J. Reimer, Environmental Sciences Group, Royal Military College, Kingston, Ontario, Canada K7K 5LO

Prof. Dr. Albert W. Rettenmeier, Institut f. Hygiene u. Arbeitsmedizin, Klinikum Essen, Hufelandstr. 55, 45122 Essen, Germany

Dr. Tanja Schwerdtle, Institut für Lebensmittelchemie und Toxikologie, Universität Karlsruhe, Kaiserstr.12, Postfach 6980, D-76128 Karlsruhe, Germany

Zita Sebesvari, Institut für Phytosphäre, Forschungszentrum Jülich, 52425 Jülich, Germany

Prof. Dr. Gisela Stoltenburg-Didinger, Department of Neuropathology, Freie Universität Berlin, Universitätsklinikum Benjamin Franklin, 12200 Berlin, Germany

Dr. Margareta Sulkowski, Institut für Umweltanalytik, Universität Essen, Universitätsstr. 3-5, 45141 Essen, Germany

Dr. Joanna Szpunar CNRS, Pau, France

Ingo Walter, Institut für Lebensmittelchemie und Toxikologie, Universität Karlsruhe, Kaiserstr.12, Postfach 6980, D-76128 Karlsruhe, Germany

Dr. Gisela Wolf, Berliner Wissenschaftliche Gesellschaft, Robert-Koch-Platz 7, 10115 Berlin, Germany

Raycho Yonchev, Institut für Physikalische Chemie, Universität Essen, Schützenbahn 70, 45117 Essen, Germany.

Chapter 1

Organometallic compounds in the environment: An overview

P. J. Craig and R. O. Jenkins

General points

In this chapter organometallic compounds are defined as those which have a carbon to metal single sigma bond polarized $M^{\delta+}$ - $C^{\delta-}$ (some useful sources are listed in the References Craig 2003; Abel et al. 1995; Bennett et al. 1994, Crompton 2002; Ebdon et al: 2001; Elschenbroich and Salzer 1992; Sigel and Sigel 1993; Hock 2001; Ure and Davidson 1995). The metals of interest are usually main group metals in environmental matters (They are shown in Tables 1 and 2). For useful information to be derived, these substances need to be analysed. In most cases a full speciation analysis is not possible; the organometallic fragment is usually bound to a complex environmental moiety which may not be identifiable. Nevertheless much progress in speciation analysis has been made in recent years (see references above). For speciation analysis in the environment to be possible the organometallic fragment has to be separated from its environmental binding and then measured.

(i) *Methods of Separation*

a. Gas chromatography
b. Thermal desorption methods
c. High performance liquid chromatography
d. Flow injection methods
e. Ion exchange chromatography
f. Ion chromatography

(ii) *Methods of Detection*

a. Atomic absorption spectroscopy
b. Atomic fluorescence spectroscopy (sometimes with hydride generation)
c. Atomic emission spectroscopy
d. Voltammetry
e. Mass spectrometry
f. X-ray and neutron methods

Table 1. Stability of methylmetals to oxygen [a]

Stable	Unstable [b]
Me_2Hg	$MePbX_3$
Me_4Si, $[Me_2SiO]_n$, $(Me)_nSi^{(4-n)+}$, Me_6Si_2	$MeTl^+$
Me_4Ge,$Me_4Ge^{(4\ n)+}$, Me_6Ge_2	Me_2Zn, $MeZn^+$
Me_4Sn	Me_2Cd, $MeCd^+$
Me_4Pb§	Me_3B
MeHgX (Ph and Et also stable)	Me_3Al
$Me_{4\ n}SnX_n$	Me_3Ga
Me_3PbX	Me_3In
Me_2PbX_2	Me_3Tl
$Me(C_5H_4)Mn(CO)_3$ [c]	Me_3As [e]
$MeM_n(CO)_4L$ [d]	Me_3Sb [e]
$Me_2AsO(OH)$	Me_3Bi [e]
$MeAsO(OH)_2$	Me_2AsH
Me_2S	$MeAsX_2$
Me_2Se	$MeSbX_2$
MeHgSeMe	$Me_{4n}SnH_n$ [e]
$MeCoB_{12}$ (solid state)	Me_6Sn_2 (At RT gives $[Me_3Sn]_2O$)
Me_3SbO	Me_6Pb_2 (to methyl lead products)
$Me_2SbO(OH)$	Me_3Sb
$MeSbO(OH)_2$	Me_3AsO
Me_2Tl^+, Me_2Ga^+	Me_3P
Me_3S^+	Me_4SiH_{4-n}
Me_3Se^+	Me_4GeH_{4-n}
Me_3PO	

a At room temperature in bulk as against rapid (seconds, minutes) oxidation. Assume similar but lesser environmental stability for ethyls.

b Variously unstable because of empty low lying orbitals on the metal, polar metal-carbon bonds and/or lone electron pairs on the metal.

c Gasoline additive.

d To exemplify ligand-complexed transition metal organometallics. Many of these synthetic compounds are oxygen-stable but none have been found in the natural environment.

e But stable in dilute form and detected in the environment

Reproduced in part from Craig (2003) with permission

Table 2 Stability of organometallic species to water

Organometallic	Stability, comments
R_2Hg, R_4Sn, R_4Pb	Only slightly soluble, stable, diffuse to atmosphere. Higher alkyls less stable and less volatile. Species generally hydrophobic and variously volatile
MeHgX	Stable, slightly soluble depending on X
$(Me)_nSn^{(4-n)+}$	Soluble, methyltin units stable but made hexa- and penta-coordinate by H_2O, OH^-. Species are solvated, partly hydrolysed to various hydroxo species. At high pH polynuclear bridged hydroxo species form for $(Me)_2Sn^{2+}$
Me_3Pb^+	Soluble, hydrolysis as methyltins above. Also dismutates to Me_4Pb and Me_2Pb^{2+} at 20 °C
Me_2Pb^{2+}	Soluble as for Me_3Pb^+ above. Disproportionates to Me_3Pb^+ and CH^+_3 slowly. These reactions cause eventual total loss of Me_3Pb^+ and Me_2Pb^{2+} from water.
Me_2As^+	Hydrolyses to Me_2AsOH then to slightly soluble $[Me_2As]_2O$
$MeAs^{2+}$	Hydrolyses to $MeAs(OH)_2$ then to soluble $(MeAsO)_n$
$Me_2AsO(OH)$	Stable and soluble (330 g dm^{-3}). Acidic pKa = 6.27. i.e. cacodylic acid, dimethylarsinic acid. Detected in oceans
$MeAsO(OH)_2$	Stable and soluble. Strong acid pK_1 = 3.6, pK_2 = 8.3 – methylarsonic acid. Detected in oceans
Me_3S^+, Me_3Se^+	Stable and slightly soluble
Me_nSICl_{4-n}	Hydrolyse and condense, but methylsilicon groups retained
$Me_nGe^{(4-n)+}$	Stable, soluble, has been discovered in oceans. Hydrolyses but Me_nGe moiety preserved
Me_2Tl^+	Very stable, soluble, but not been detected as a natural environment product
Me_3AsO, Me_3SbO	Stable and soluble
MeAsH, CH_3AsH_2	Insoluble, diffuses to atmosphere, air unstable
MeSbO(OH)	Stable and soluble. Detected in oceans
$CH_3SbO(OH)_2$	Stable and soluble. Detected in oceans

Solubility refers to air-free distilled water, no complexing ligands. Range of solubility is from mg dm^{-3} to g dm^{-3}.

Other species

Stable and insoluble – R_4Si, $(R_2SiO)_n$,$H_3HgSeCH$, most PhHg derivatives, Me_2S, Me_2Se, Me_4Ge, Me_3B

Unstable – $MePb^+$ (has not been detected in the environment), R_2Zn, R_2Cd, R_3Al, R_3Ga, Me_6Sn_2, Me_6Pb_2, $MeTl^{2+}$, $MeCd^+$, Me_2Cd, Me_2Sb^+, $MeSb^{2+}$.

Reproduced in part from Craig (2003) with permission

Separation and detection of organometallics in the environment presupposes that the species are stable. Stability is not discussed here but it is discussed in detail in a recent work (Craig 2003). Toxicology is also covered elsewhere in this work.

As many organometallic moieties are tightly bound in the environment, they may not be volatile and susceptible to separation as above. For such cases separation by derivatization is usually carried out.

This is generally achieved by (formal) SN_2 attack by hydride (from $NaBH_4$), ethyl (NaBEt) or other alkyl group (e.g. from a Gignard reagent), e.g. (Eqn. 1)

$$Bu_3SnX \xrightarrow{NaBH_4} Bu_3SnH \qquad [1]$$

X = environmental counter ion

Organometallic compounds may be found in the natural environment because they are *formed* or because they are *introduced.* The chemistry of the latter group is better known, and their environmental impact has been widely discussed. Organometallic compounds introduced to the environment directly may enter *via* use as products whose properties relate to the environment (e.g. biocides) or they may enter additionally to a separate, main function (e.g. gasoline additives, polymer stabilizers). Compounds of arsenic, mercury, tin and lead have important environmental roles as organometallic compounds.

Where organometallics are formed in the environment it is usually as a result of a methylation process, "biomethylation". This process is discussed below.

As mentioned above, stabilities will be little discussed here (but are shown in Tables 1 and 2), but it should be pointed out that the metal carbon bond is not necessarily weak (i.e. leading to decay for thermodynamic reasons), but there are often low energy routes for metal-carbon decomposition (i.e. kinetic decay). In the latter case the activation energy for decay is low.

In a similar way, all organometallic compounds are thermodynamically unstable to oxidation because of the lower free energies of the products of oxidation. However, some do not oxidise (or are not inflammable) for kinetic reasons. Compounds which in bulk may e.g. ignite are sometimes stable at high attenuation. Most organometallics are thermodynamically unstable to hydrolysis to the metal hydroxide and hydrocarbon. Again many are kinetically stable e.g. if nucleophilic attack by water on the metal cannot take place because lack of suitable orbitals on the metal or if the attack is physically blocked by ligands.

Stability to light for organometallics is most important for those volatile species which enter the atmosphere, and less relevant for those coordinated to biological or environmental ligands in organisms or water. Chemistry, decay and toxicity of organometallic species in biological systems is discussed elsewhere (see above) and in the present work.

Biomethylation

Biomethylation is the process whereby living organisms produce a direct linkage of a methyl group to a metal or metalloid, thus forming metal-carbon bonds. Methylation has been extensively studied and biomethylation activity has been found in soil, but mainly occurs in sediments in e.g. estuaries, harbours, rivers, lakes and oceans. The addition of a methyl group to a metall(oid) changes the chemical and physical properties of that element, and this then influences toxicity. The organisms responsible for metall(oid) biomethylation are nearly all microorganisms. Anaerobic bacteria are believed to be the main agents of biomethylation in sediments and other anoxic environments. Some aerobic and facultatively anaerobic bacteria, as well as certain fungi and lower algae, may also methylate metals. With the exception of vitamin B_{12} higher organisms, containing a methyl-cobalt bond, do not seem to be able to methylate genuine metals. The situation is different for the metalloids arsenic, selenium and tellurium: many higher organisms have been shown to form methyl derivatives of these elements, e.g. methylarsenicals are formed in a wide range of organisms, including marine biota and mammals (and man).

Volatile methyl and hydride derivatives of metal(loids) can often be found in gases released from natural and anthropogenic environments (Feldmann and Hirner 1999; Feldmann et al. 1999; Hirner et al. 1998; Feldmann et al. 1994), e.g. geothermal gases, sewage treatment plants, marine sediments, landfill deposits), and biological production of volatile compounds is believed to be an important part of the biogeological cycles of metal(loids)s, such as arsenic, mercury, selenium and tin. With the exception of arsenic and selenium (and perhaps antimony), biomethylation increases toxicity, because methyl derivates are more lipophilic and therefore more biologically active.

Biomethylation of organic molecules – e.g. such as proteins, nucleic acid bases, polysaccharides and fatty acids – occurs in all living cells and is an essential part of normal intracellular metabolism. The three main biological methylating agents for organic molecules – S-adenosylmethionine (SAM), methylcobalamine and N-methyltetrahydrofolate – have all been shown to be capable of involvement in the biomethylation of metall(oids). SAM is a ubiquitous methylating agent and is synthesised by the transfer of an adenosyl group from ATP to the sulphur atom of methionine.

A positive charge on the sulphur atom activates the methyl group of methionine, making SAM a methyl carbonium ion donor. The methyl group in biochemistry is transferred as a radical ($CH_3•$) or as a carbonium ion (CH_3^+) and the atom receiving the methyl group must be nucleophilic, which requires a lone pair of electrons in the metal valency shell; i.e. oxidative addition, where there is alternation of M^{n+} and $M^{(n+2)+}$ oxidation states. The reaction (involving carbonium ions) is usually referred to as the Challenger mechanism, devised originally to account for the methylation of arsenic (Challenger et al. 1933; Brinckman and Bellama 1978) (Figure 1). SAM has subsequently been shown also to be the methylating agent for selenium, tellurium, phosphorus (Fatoki 1997) and antimony (Andrewes et al. 1999a), all of which have lone paid of electrons. N-Methyltetrahydrofolate like SAM, is thought to transfer methyl groups as carbonium ions or an intermediate radical, but its transfer potential is not as high as SAM. Conversely, for methylcobalamin in the environment – a derivative of vitamin B_{12} – the methyl group is transferred as a carbanion (CH_3^-) and the recipient atoms must be electrophilic e.g. mercury (II). Methylcobalamin is well established as a methylating agent for mercury and can also be involved in (in vitro) methylation for lead, tin, palladium, platinum, gold and thallium (Fatoki 1997). Although methylcobalamin appears to be the only carbanion methylating agent, in the natural environment, a carbanion may also be transferred to metals from other organometallic species which may be present, e.g. Me_3Pb^+, Me_3Sn^+. Regardless of the methylating agent, only one methyl group transfers to the metal(loids) in each step, although further groups may then be transferred to the same receiving atom. Methylation of some individual elements is now discussed.

Arsenic

Many microorganisms can methylate arsenic and methylarsenic compounds are produced, under both aerobic and anaerobic conditions (Cullen and Reimer 1989). Fungal methylation of arsenic was reported in the nineteenth century, when several poisoning incidents in England and Germany were associated with arsenic containing wall papers. Gosio in 1901, reported that a garlic-smelling, methylated arsenic compound was released from moulds growing in the presence of inorganic arsenic. Challenger et al. (1933) identified this as trimethylarsine (Me_3As). Several fungi have been shown to methylate arsenic under aerobic conditions, including *Scopulariopsis brevicaulis, Penicillium* sp., *Gliocladium roseum* and the yeast *Cryptococcus humicola* (Brinckman and Bellama 1978; Thayer 1984; Andreae 1986). The mechanism of arsenic methylation in fungi was established by Challenger et al 1933 and involves a series of reductions and oxidative methylations, using SAM (Figure 1.) as a methyl donor.

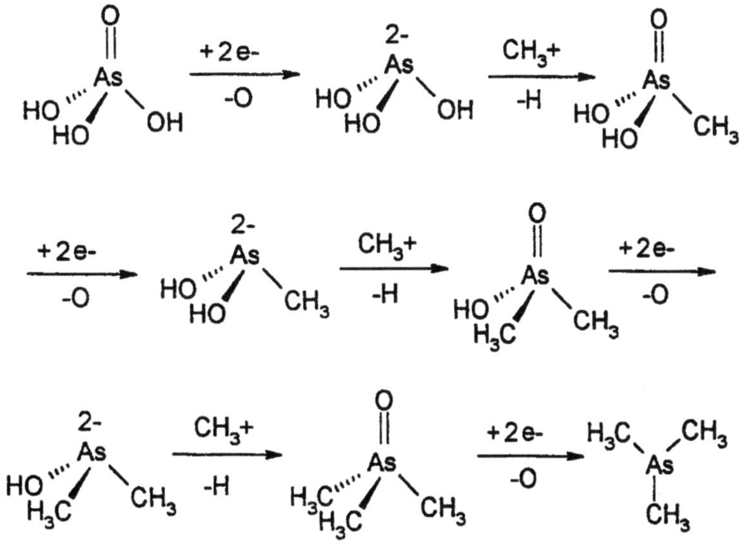

Fig. 1. Arsenic methylation according to the Challenger mechanism

Certain bacteria have also been shown to methylate arsenic under aerobic conditions, including *Flavobacterium* sp., *Escherichia coli* and *Aeromonas* sp. (Thayer 1984).

Arsenic biomethylation can also be catalysed by obligate anaerobic bacteria, e.g. *Methanobacterium* sp. can reduce AsO_3^- to AsO_2^-, with subsequent methylation to methylarsonic acid, dimethylarsenic acid and dimethylarsine (Takahashi et al. 1990). Methylcobalamin is the methyl donor for arsenic methylation by methanogenic archaea (McBride and Wolfe 1971). Michalke et al. (2000) have reported trimethylarsine in the gas phase above anaerobic cultures of *Methanobacterium formicicum, Clostridium collagenovorans* and two *Desulfovibrio* spp. For *C. collagenovorans* and *D, vulgaris*, trimethylarsine was the only volatile arsenic species detected. Conversely, *M. formicicum* volatilised arsenic as mono- di- and trimethylarsine and as arsine (AsH_3), while *Methanobacterium thermoautotrophicum* produced only arsine. This data serves to illustrate the key role of the organism for metal(loid) microbial transformations.

There is an extensive marine chemistry for arsenic with many methylarsenic riboside species identified and analogous compounds are now being found in terrestrial situations.

Antimony

Mono- and dimethylantimony species have been found to be formed in both marine and terrestrial natural waters (Andreae 1981; Andreae 1984). Freshwater plants from two lakes contaminated by mine effluent, and plant material from an abandoned antimony mine have also been shown to contain methylantimony species (Craig et al. 1999; Dodd et al. 1996).

Biomethylation of inorganic antimony by the aerobic fungus *Scopulariopsis brevicaulis* is documented (Craig et al. 1998; Jenkins et al. 1998a; Jenkins et al. 1998b; Andrewes et al. 1998; Andrewes et al. 1999b; Andrewes et al. 2000) and is thought to involve methyl transfer *via* SAM (Andrewes et al. 1999a). Recently, the fungus *Phaeolus schweinitzii* has also been shown able to biomethylate antimony (Andrewes et al. 2001). Mixed cultures of bacteria growing under anaerobic conditions have been shown to generate volatile trimethylantimony as the sole volatilised antimony species (Gates et al. 1997, Gurleyuk et al 1997, Jenkins et al 1998c). Volatilisation of antimony from environmental sediments and municipal waste sites also suggests that certain anaerobic and/or facultatively anaerobic bacteria can biomethylate (Hirner et al. 1994; Feldmann et al. 1994) antimony. Michalke et al. (2000) reported biomethylation of inorganic antimony by pure cultures of anaerobic bacteria. *C. collagenovorans*, *D. vulgaris* and three species of methanogenic archaea, were shown to produce trimethylantimony in culture headspace gases. *M. formicium* was shown to produce SbH_3 together with $Me_{3-n}SbH_n$ (n = 0,72) into culture headspace gases.

Mercury

Biomethylation of mercury has been studied extensively. Methylation of mercury increases lipid solubility and thus enhances bioaccumulation in living organisms. This can lead to mercury entering the food chain and to mercury poisoning incidents, e.g. in Japan and Iraq (Craig 2003). Many different bacteria in aerobic and anaerobic environments are known to cause mercury biomethylation, although anaerobic sediments are the main sites of environmental methylmercury formation (Fatoki 1997). Sulphur-reducing bacteria, such as *Desulfovibrio desulfuricans*, are thought to be the most important methylators of mercury (Compeau and Bartha 1985; Compeau and Bartha 1987; Kerry et al. 1991). Both low pH and high sulfate concentrations promote mercury biomethylation activity within environmental sediments (Winfrey and Rudd 1990; Craig and Moreton 1984). Methylcobalamin is the methyl donor, giving rise to monomethylmercury and (in a slower step) dimethylmercury (volatile) products in a transfer to electrophilic mercury (II) of a methyl carbonium ion (Eqn 2) viz.

$$CH_3CoB_{12} + Hg^{2+} \xrightarrow{H_2O} CH_3Hg^+ + H_2OCoB_{12} \qquad [2]$$

Further reaction of methylmercury can occur *via* various biotic mechanisms (e.g. disproportionation reaction involving H_2S) also giving rise to dimethylmercury (Eqn 3) (Craig and Moreton 1984).

$$2CH_3Hg^+ + S^{2-} \rightarrow (CH_3Hg)_2S \rightarrow (CH_3)_2Hg + HgS \quad [3]$$

An important factor governing the concentration of mercury in biota is the concentration of methyl mercury in environmental waters, which is determined by the relative efficiencies of methylation and demethylation processes.

Several other metals and metalloids can undergo methylation and these systems have been discussed in detail in a recent work (Craig 2003). The elements involved include tin, bismuth, selenium and others. An example of a biogeochemical cycle for tin is shown in Figure 2.

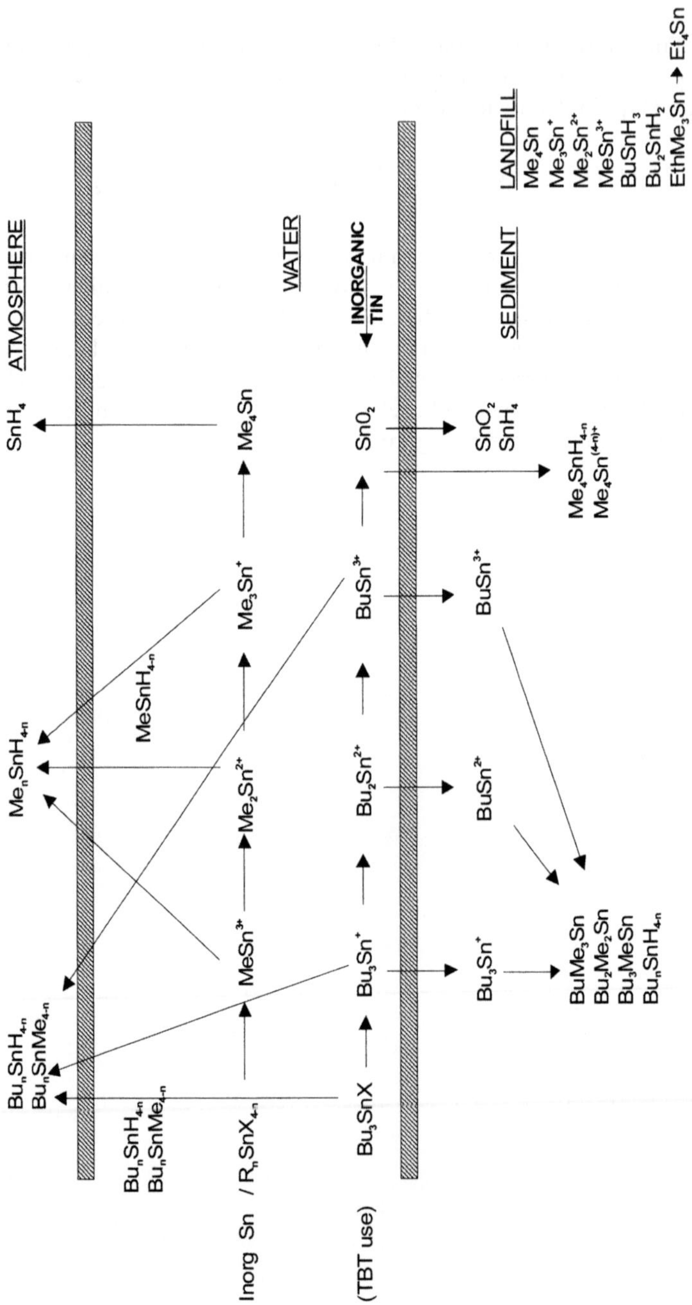

Fig. 2. The biogeochemical cycle of tin

Microbial demethylation/dealkylation

For some metals, such as mercury and tin, microbial demethylation or dealkylation of organometallic forms are important in detoxication mechanisms. Bacterial mercury (II) resistance, for example, involves reduction of Hg(II) to Hg(O) *via* mercuric reductase (MR). The reduced form of the metal is less toxic, more volatile and is rapidly removed to the atmospheric. Detoxification of organomercurials proceeds *via* organomercurial lyase (OL), a product of the *Mer* B gene, that enzymatically cleaves the Hg-C bond to form Hg(II), which is then removed *via* mercuric reductase (Gadd 1993; Losi and Frankenberger 1997) (Eqns 4 and 5). For example

$$CH_3Hg^+ \xrightarrow{OL} CH_4 + Hg^{2+} \xrightarrow{MR} Hg^0 \qquad [4]$$

$$C_2H_5Hg^+ \xrightarrow{OL} C_2H_6 + Hg^{2+} \xrightarrow{MR} Hg^0 \qquad [5]$$

Oxidative demethylation of methylmercury, involving bacterial liberation of carbon dioxide, also occurs. The mechanism is thought to involve enzymes associated with bacterial metabolism of C-1 compounds (Oremland et al. 1991) and is believed to occur widely in freshwater and aqueous environments, under both aerobic and anaerobic conditions.

Degradation of organotin compounds has been shown for a wide range of microorganisms, including bacteria, fungi and algae (Gadd 2000). Organotin degradation (Eqn.6) is thought to involve sequential cleavage of tin-carbon bonds, resulting in removal of organic groups and a reduction in toxicity (Cooney and Wuertz 1989; Crompton 1998; Cooney 1988a; Cooney et al. 1988b). Cleavage is initiated by hydroxo attack (Eqn. 6).

$$R_4Sn \rightarrow R_3SnX \rightarrow R_2SnX_2 \rightarrow RSnX_3 \rightarrow SnX_4 \qquad [6]$$

Tributyltin oxide (TBTO) and tributyltin napthenate (TBTN) have been shown to be degraded by fungal action to di- and monobutyltins. Certain gram-negative bacteria and the green alga *Ankistrodesmuc falcatus* can also dealkylate tributyltin, giving rise to dibutyl-, monobutyl- and inorganic tin products; the alga was able to metabolise around 50% of the accumulated tributyl over a four-week period (Maguire 1984). Similar end-products are formed by the action of soil microorganisms on triphenyltin acetate (Barnes et al. 1973). Tin-carbon bonds are also cleaved abiotically, for example by UV, and it has been difficult to establish the relative importance of abiotic and biotic mechanisms of degradation in the natural environment (Gadd 2000). In some circumstances, environmental conditions (e.g. pH or redox potential), established by microbial activity, strongly pH or redox potential), established by microbial activity, strongly influence the extent of abiotic degradation of organotins (Gadd 2000). This discussion relating to abiotic and biotic organotin degradation also apply to other organometal(loids).

Bacterial demethylation of methylarsenicals is known to occur in aerobic aqueous and terrestrial environments, giving CO_2 and arsenate (Andreae et al. 1986). Several soil bacteria, such as *Achromobacter, Flavobacterium* and *Pseudomonas*, have been shown to possess organoarsenical demethylation ability. Bacterial demethylation of methylarsenic acids excreted by marine algae is an important part of the biogeochemical cycling of arsenic.

Dimethylselenide has been shown to be demethylated in anaerobic environments by methanogenic and suphate reducing bacteria, producing CO_2 and CH_4 (Newman et al. 1997). Several bacterial isolates form aerobic soils, - including members of the genera *Pseudomonas*, *Xanthmononas* and *Corynebacterium* – have been shown to use methylselenides as sole source of carbon (Doran and Alexander 1977).

Literature

Abel EW, Stone FGA, Wilkinson G (1995) Comprehensive organometallic chemistry II, Vols 1-14, Pergamon (Elsevier), Oxford, UK

Andreae MO (1986) Arsenic Compounds in the Environment, in Craig, PJ, Organometallic Compounds in the Environment, Longman, Harlow, UK, 198-228 pp

Andreae MO, Asmode J-F, Foster P, van't Dack L (1981) Determination of antimony(III), antimony (V), and methylantimony species in natural waters by atomic absorption spectrometry with hydride generation. Anal Chem 53: 1766-1771

Andreae MO, Froelich P N (1984) Arsenic, antimony, and germanium biogeochemistry in the Baltic Sea. Tellus 36b:101-117

Andrewes P, Cullen WR, Feldmann J, Koch I, Polishchuk E, Reimer K J (1998) The production of methylated organoantimony compounds by *Scopulariopsis brevicaulis*. Appl Organometal Chem 12:827-842

Andrewes P, Cullen WR, Feldmann J, Koch I, Polishchuk E (1999a) Methylantimony compound formation in the medium of *Scopulariopsis brevicaulis* cultures: $^{13}CD_3$-L-methionine as a source of the methyl group.Applied Organometallic Chemistry 13:681-687

Andrewes P, Cullen WR, Polishchuk E (1999b) Confirmation of the aerobic production of trimethylstibine by Scopulariopsis brevicaulis. Appl Organometal Chem13:659-664

Andrewes P, Cullen WR, Polishchuk E (2000) Arsenic and antimony biomethylation by *S.brevicaulis*: Interactions of arsenic and antimony compounds. Environ Sci Technol 34:2249-2253

Andrewes P, Cullen WR, Polishchuk E, Reimer KJ (2001) Antimony biomethylation by the wood rotting fungus *Phaeolus schweinitzii*. Appl Organometal Chem 6:473-480

Barnes RD, Bull AT, Poller RC (1973) Persistence of the organotin fungicide fentin acetate (triphenyltin acetate) in the soil and on surfaces exposed to light. Pesticide Sci 4:305-317

Bennet S (1994) The Open University, Science, a Third Level Course, S343, Inorganic Chemistry, Block 6, Open University Press, second edition

Brinckman F, Bellama JM (1978) Organometals and organometalloids; occurrence and fate in the environment, American Chemical Society, Washington DC

Challenger F, Higginbottom C, Ellis L (1933) The formation of organometalloidal compounds by microorganisms. Part 1: Trimethylarsine. J Chem Soc Transact 32:95-101

Compeau GC, Bartha, R (1985) Sulfate-reducing bacteria: principal methylators of mercury in anoxic estuarine sediment. Appl Environ Microbiol 50:498-502

Compeau GC, Bartha R (1987) of salinity on mercury-methylating activity of sulfate-reducing bacteria in estuarine sediments. Appl Environ Microbiol 53:261-265

Cooney JJ (1988a) Ineractions between microorganisms and tin compounds. In: Craig P J, Glockling F (eds) The biological alkylation of heavy elements. Royal Society of Chemistry, London, UK, 92-104pp

Cooney JJ (1988b) Microbial transformations of tin and tin compounds. J Ind Microbiol 3:195-204

Cooney JJ, Wuertz S (1989) effects of tin compounds on microorganisms. J Ind Microbiol 4:375-402

Craig PJ, Moreton PA (1984) Total mercury, methyl mercury and sulphide in River Carron sediments. Marine Poll Bull 15:406-408

Craig PJ, Forster SN, Jenkins RO, Miller DP (1999) An analytical method for the detection of methylantimony species in environmental matrices: methylantimony levels in some UK plant material. Analyst 124:1243-1248

Craig PJ (2003) (ed), Organometallic compounds in the environment, second edition, John Wiley, Chichester, UK

Crompton TR (1998) Occurrence and analysis of organometallic compounds in the environment, John Wiley, Chichester, UK

Crompton TR (2002) Determination of metals in natural and treated waters, Spon Press (Taylor and Francis), London, UK

Cullen WR, Reimer KJ (1989) Arsenic speciation in the environment. Chem Rev 89:731-764

Dodd MS, Pergantis S, Cullen WR, Li H, Eigendorf GK, Reimer KJ (1996) Antimony speciation in freshwater plant extracts by using hydride generation-gas chromatography-mass spectrometry. Analyst 121:223-228

Doran JW, Alexander M (1977) Microbial transformations of selenium. Appl Environ Microbiol 33:31-37

Ebdon L, Pitts L, Cornelis R, Crews H, Donard OFX, Quevauviller Ph, (eds) (2001) Trace Elements Speciation for Environment, Food and Health, Royal Society of Chemistry, London, UK

Elschenbroich CH, Salzer A (1992) Organometallics, a concise introduction, second edition, VCH Press, Weinheim, Germany

Fatoki OS (1997) Biomethylation in the natural environment: A review. S Afr J Sci 93:366-370

Feldmann J, Gruemping R, Hirner AV (1994) Determination of volatile metal and metalloid compounds in gases from domestic waste deposits with GC/ICP-MS. Fresenius J Anal Chem, 350:228-234

Feldmann J, Hirner AV (1995) Occurence of volatile metal and metalloid species in landfill and sewage gases. Int J Environ Anal Chem 60:339-359

Feldmann J, Krupp E, Glindemann D, Hirner AV, Cullen W R (1999) Methylated bismuth in the environment. Appl Organometal Chem 13:739-748

Gadd GM (2000) Microbial interactions with tributyltin compounds: detoxification, accumulation, and environmental fate. Sci Total Environ 258:119-127

Gates PN, Harrop HA, Pridham JB, Smethurst B (1997) Can microorganisms convert methylated antimony trioxide or potassium antimonyl tartrate to methylated stibines? Sci Total Environ 205:215-222

Gurleyuk H, Van Fleet-Stadler V, Chasteen TG (1997) Confirmation of the biomethylation of antimony compounds. Appl Organometal Chem11:471-483

Hirner AV, Feldmann J, Goguel R, Rapsomanikis S, Fischer R, Andreae M (1994) Volatile metal and metalloid species in gases from municipal waste deposits. Appl Organometal Chem 8:65-69

Hirner AV, Feldmann J, Krupp E, Gruemping R, Goguel R, Cullen WR (1999) Metal(loid)organic compounds in geothermal gases and waters. Org Geochem 29:1765-1778

Hock CB (2001) Thermochemical proceses, principles and models, Butterworth-Heinemann, Oxford, UK

Jenkins RO, Craig PJ, Goessler W, Irgolic KJ (1998a) Biovolatilization of antimony and sudden infant death syndrome (SIDS). Human Exp Toxicol 17:231-238

Jenkins RO, Craig PJ, Goessler W, Miller D, Ostah N, Irgolic KJ (1998b) Biomethylation of inorganic antimony compounds by an aerobic fungus: *Scopulariopsis brevicaulis*. Environ Sci Technol 32:882-885

Jenkins RO, Craig PJ, Miller DP, Stoop LCAM, Ostah N, Morris T-A (1998c) Antimony biomethylation by mixed cultures of micro-organisms under anaerobic conditions. Appl Organometal Chem 12:449-455

Kerry A, Welbourn PM, Prucha B, Mierle G (1991) Mercury methylation sulphate- reducing bacteria from sediments of an acid stressed lake. Water Air Soil Pollut 56:565-575

Losi ME, Frankenburger WT (1997) Bioremediation of selenium in soil and water. Soil Sci 162:692-702

McBride BC, Wolfe RS (1971) Biosynthesis of dimethylarsine by Methanobacterium. Biochemistry 10:4312-4317

Maguire RJ (1984) Butyltin compounds and inorganic tin in sediments in Ontario. Environ Sci Technol 18:291-294

Michalke K, Wickenheiser A , Mehring A, Hirner AV, Hensel R (2000) Production of volatile derivatives of metal(loid)s by microflora involved in anaerobic digestion of sewage sludge. Appl Environ Microbiol 66:2791-2796

Newman DK, Kennedy EK, Coates JD, Ahmann D, Ellis DJ, Lovley DR, Morel FMM (1997) Dissimilatory arsenate and sulfate reduction in *Desulfotomaculum auripigmentum* sp. nov. Arch Microbiol 168:380-388

Oremland RS, Culbertson CW, Winfrey MR (1991) *In situ* bacterial selenate reduction in the agricultural drainage systems of western Nevada. Appl Environ Microbiol 578:130-137

Sigel H, Sigel A, (eds) (1993) Metal ions in biological systems, Vol 29, Biological properties of metal alkyl derivatives, Marcel Dekker, New York

Takahashi K, Yamauchi H, Mashiko M, Yamamura Y (1990) Effect of S-adenosylmethionine on methylation of inorganic arsenic. *Nippon Eiseigaku Zasshi* 45:613-618

Thayer JS (1984) Organometallic compounds and living organisms, Academic Press, New York

Ure AM, Davidson CM, (eds) (1995) Chemical speciation in the environment, Blackie Academic and Professional, Glasgow

Winfrey MR, Rudd JMW (1990) Environmental factors affecting the formation of methylmercury in low pH lakes. Environ Toxicol Chem 9:853-869

Chapter 2

Advances in analytical methods for speciation of trace elements in the environment

J. Szpunar, B. Bouyssiere, R. Lobinski

Introduction

The recognition of the fact that, in environmental chemistry, occupational health, nutrition and medicine, the chemical, biological and toxicological properties of an element are critically dependent on the form in which the element occurs in the sample has spurred a rapid development of an area of analytical chemistry referred to as speciation analysis (Templeton et al. 2000). IUPAC defines a chemical species as a specific and unique molecular, electronic, or nuclear structure of an element (Templeton et al. 2000). Speciation of an individual element refers to its occurrence in or distribution among different species. Speciation analysis is the analytical activity of identifying and quantifying one or more chemical species of an element present in a sample (Templeton et al. 2000). The combination of a chromatographic separation technique, that ensures that the analyte compound leaves the column unaccompanied by other species of the analyte element, with atomic spectrometry, permitting a sensitive and specific detection of the target element, has become a fundamental tool for speciation analysis, as discussed in many review publications (Caruso et al. 2000; Cornelis et al. 2003; Lobinski 1997; Szpunar 2000; Tomlinson et al. 1995; Zoorob et al. 1998).

In the original concept (Lobinski 1997), speciation analysis targeted well-defined analytes, usually anthropogenic organometallic compounds, such as alkyllead, butyl- and phenyltin compounds, and simple organoarsenic and organoselenium species, and products of their environmental degradation. Calibration standards were either available or could be readily synthesized. The presence of a metal(loid)-carbon covalent bond assured a reasonable stability of the analyte(s) during sample preparation. The volatility of the species allowed the use of gas chromatography with its inherent advantages, such as the high separation efficiency and the absence of the condensed mobile phase, that enabled a sensitive (down to the femtogram levels) element-specific detection by atomic spectroscopy (Bouyssiere et al. 2002; Lobinski and Adams 1997).

Metalloids, such as notably arsenic and selenium, are known to be metabolised by living organisms in a way that leads to the formation of a covalent bond between the heteroatom and the carbon incorporated in a larger structure (e.g. arsenosugars, selenoproteins). The resulting compounds are difficult to be converted into volatile species that prevents the use of GC for their separation. Moreover, standards for most of these species are unavailable since many of the naturally synthesized compounds have not been identified and characterized yet. The analytical challenges spur therefore the use of separation techniques in the liquid phase, such as HPLC and CZE with ICP-MS detection, and the use of molecule specific techniques, notably electrospray tandem mass spectrometry (ES-MS/MS).

This paper reviews recent developments in the areas of gas chromatography with ICP-MS detection for the determination of anthropogenic organometallic contaminants and the advances in multidimensional LC with the parallel ICP-MS and electrospray MS/MS detection for the detection and identification of natural metabolites of metalloids in the environment.

Organometall(oid) Species in the Environment

The species of interest, containing a covalent carbon-metal(loid) bond, in environmental speciation analysis can be divided into the following classes:

(i) products of environmental methylation of mercury, selenium, arsenic, tin, bismuth, or carbonylation of molybdenum and tungsten ($Mo(CO)_6$, $W(CO)_6$),
(ii) organometallic anthropogenic contaminants and products of their degradation or environmental transformation. This group includes tetraalkylated lead (Et_xMe_yPb, $x+y = 4$), an antiknock additive to gasoline, that is degraded to trialkyl or dialkyl species, ingredients of antifouling paints, such as butyl-, octyl and phenyl tin species, released into the aquatic environment, and products of their degradation or biomethylation,
(iii) products of the metabolism of arsenic by marine biota leading to the formation of the carbon-arsenic bond, as e.g. in arsenobetaine or arsenosugars,
(iv) selenoaminoacids, -peptides and proteins, biosynthesized by bacteria, fungi and plants.

Hyphenated Techniques in Speciation Analysis

A suitable analytical technique for speciation analysis should address three issues:

(i) selectivity of the separation technique allowing the target analyte species to arrive at the detector well separated from potential matrix interferents and from each other,
(ii) sensitivity of the element or molecular selective detection technique since the already low concentrations of trace elements in environmental samples are usually distributed among several species,

(iii) species identification. Retention time matching usually employed requires the availability of standards. When standards are not available, the use of a molecule-specific detection technique is mandatory.

The above challenges can be addressed by a hyphenated technique of which the choice available is schematically shown in Fig. 1. In the most frequent case a separation technique: chromatography (gas or liquid), electrochromatography or gel electrophoresis is combined with ICP-MS. The coupling is realized directly (for GC), *via* a nebulizer (for column liquid separation techniques) or by laser ablation (for planar techniques).

The separation component of the coupled system becomes of particular concern when the targeted species have similar physicochemical properties. Gas chromatography should be chosen wherever possible because of the high separation efficiency and the very low achievable detection limits because of the absence of the condensed mobile phase. For non-volatile species column liquid phase separation techniques, such as HPLC and CE, are the usual choice because of the ease of on-line coupling and the variety of separation mechanisms and mobile phases available allowing the preservation of the species identity. The two-dimensional gel electrophoresis is indispensable in seleno- and phosphoroproteomics because of its impressive peak capacity.

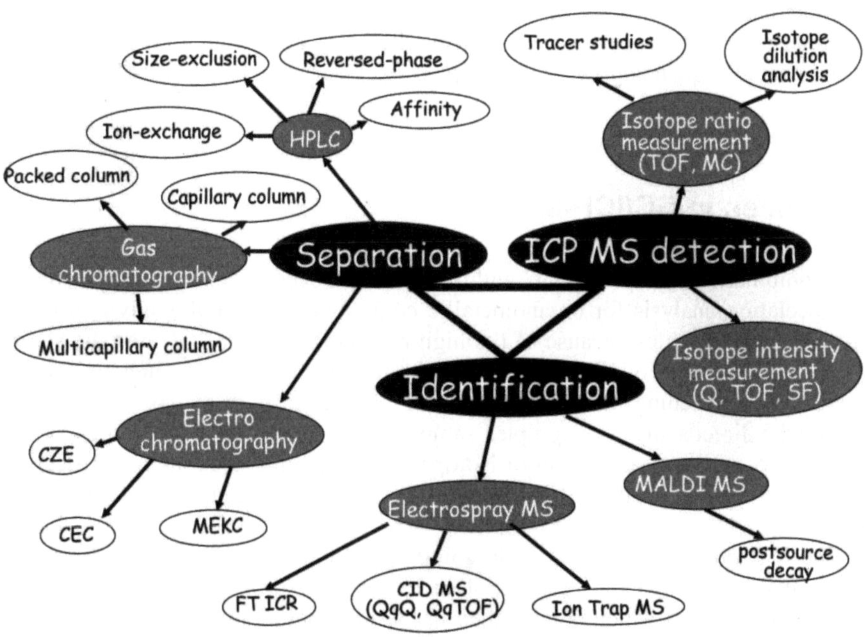

Fig. 1. Hyphenated techniques in speciation analysis.

For element-specific detection in gas chromatography, a number of dedicated spectrometric detection techniques can be used, e.g. quartz furnace atomic absorption or atomic fluorescence for mercury, microwave induced plasma atomic emission for lead or tin, but it is ICP-MS that has been establishing its position as the versatile detector of choice. ICP-MS is virtually the only technique capable of coping, in on-line mode, with the trace element concentrations in LC and CE effluents. The femtogram level absolute detection limits may turn out to be insufficient if an element present at the ng/ml level is split into a number of species, or when the actual sample amount analysed is limited to several nanolitres as in the case of CE. The isotope specificity of ICP-MS offers a still underexploited potential for tracer studies and for improved accuracy in quantification *via* the use of isotope dilution techniques.

The third important component of an analytical strategy is identification and characterization of metallospecies, either newly discovered, or those for which standards are unavailable. This can be achieved by electrospray MS or MALDI MS for column or planar separation techniques, respectively. The use of TOF-MS is recommended; the 5-10 ppm accuracy of the M_r (average molecular mass) measurement allows the determination of the empiric formula for metallospecies with $M_r < 500$. Structural information can be acquired by collision induced dissociation (CID) of an ion selected by a quadrupole mass filter followed by a product ion scan using a quadrupole or a TOF mass analyser. Despite recent considerable advantages in instrumentation, an enzymic digestion of a selenopolypeptide prior to aminoacid sequencing by ES-MS/MS, is necessary to complete the characterization.

Advances in GC/ICP-MS

The combination of capillary GC with ICP-MS has become an ideal methodology for speciation analysis for organometallic compounds in complex environmental and industrial samples because of the high resolving power of GC and the sensitivity and specificity of ICP-MS. Indeed, the features of ICP-MS such as low detection limits reaching the one femtogram (1 fg) level, high matrix tolerance allowing the direct analysis of complex samples, such as e.g. gas condensates, or the capabibility of the measurement of isotope ratios enabling accurate quantification by isotope dilution position ICP-MS at the lead of the GC element specific detectors.

The ICP quadrupole MS is undergoing a constant improvement leading to a wider availability of more sensitive, less interference prone, smaller in size and cheaper instruments which favours their use as chromatographic detectors. The introduction of ICP-TOF-MS increased the speed of data acquisition allowing multiisotope measurement of millisecond-wide chromatographic peaks and improving precision of isotope ratios determination (Haas et al. 2001; Heisterkamp and Adams 2001; Leach et al. 2000; Pack et al. 1998; Vanhaecke and Moens 1999). An even better precision was recently reported for magnetic sector multicollector instruments used as on-line GC detectors (Krupp et al. 2001a, 2001b). These instrumental developments go in parallel with the miniaturization of GC hardware allowing the time-resolved introduction of gaseous analytes into an ICP, e.g. based on microcolumn multicapillary GC, and sample preparation methods including microwave-assisted, solid phase microextraction or purge and capillary trap automated sample introduction systems (Lobinski et al. 1998). A GC/ICP-MS interface is commercially available having proven the recognition of the maturity of this coupling by analytical instrumentation industry.

Advances in Gas Chromatography prior to ICP-MS

Packed column GC used in the early studies on GC/ICP-MS coupling (Chong and Houk 1987; Vanloon et al. 1986) has practically given way to capillary GC; the coupling of the latter to ICP-MS was first described by Kim et al. (Kim et al. 1992a, 1992b). Packed columns can, by design, handle high flow rates and large sample sizes, but the efficiency and resolution properties are compromised because of the high dispersion of the analytes on the column. The large column volume negatively affects the sensitivity in the peak height mode and the detection limits. The packing itself may be chemically active toward many organometallic species, which makes silanization necessary and worsens the reliability of results. It should be noted, however, that still a considerable number of works, especially those using hydride generation purge and trap are carried out with packed column chromatography because of easier handling of highly volatile species at temperatures below -100 °C (Amouroux et al. 1998; Pecheyran et al. 1998).

Capillary GC offers improved resolving power over packed column GC/ICP-MS which is of importance for the separation of complex mixtures of organometallic compounds found in many environmental samples. Capillary GC allows one to cope with the co-elution of the solvent and the volatile compounds, such as Me_4Sn or Me_2Hg, and thus to avoid or to minimize the plasma quenching. The reduced sample size and the high dilution factor with the detector's makeup gas necessary to match the spectrometer's optimum flow rate result in a loss of sensitivity.

Recently, a number of papers appeared on rapid (flash) GC employing columns that consist of a bundle of 900 - 2000 capillaries of a small (20-40 µm) internal diameter, referred to as multicapillary columns (for a review see Lobinski et al. 1999). Such a bundle of capillaries allows the elimination of the deficiencies associated with the use of capillary and packed columns while the advantages of both are preserved. Multicapillary (MC)GC features high flow rates which minimize the dilution factor and facilitate the transport of the analytes to the plasma. The coupling between MC GC and ICP-MS using a non-heated interface offered 0.08 pg detection limits for Hg speciation (Slaets et al. 1999).

An interesting feature is the use multicapillary microcolumns for sample introduction into an ICP. Rodriguez et al. (Rodriguez et al. 1999) showed isothermal separations of organometallic species using a 50-mm long microcolumn which opens the way to the miniaturisation of GC sample introduction units, possibly making the classical GC oven redundant.

ICP-MS Detection in Gas Chromatography

Quadrupole mass analysers have predominantly been used, their sensitivity has improved a factor of 10 during the past decade. Tao et al. (Tao et al. 1999) reported a fabulous instrumental detection limit of 0.7 fg by operating the shield torch at normal plasma conditions using the HP 4500 instrument. Of other types of analysers, TOF-MS has been extensively studied as a GC detector during the last few years (Baena et al. 2001; Haas et al. 2001; Heisterkamp and Adams 2001; Leach et al. 2000). Applications of sector-field analysers (Krupp et al. 2001a, 2001b; Rodriguez-Fernandez et al. 2001), also with multicollectors (Krupp et al. 2001a, 2001b), are slowly appearing.

GC/ICP Time-Of-Flight MS

The measurement of a time-dependent, transient signal by a sequential scanning using a quadrupole or a sector-field (single collector) mass spectrometer results in two major types of difficulties. The first one is the limited number of isotope intensity measurements that can be carried out within the time-span of a chromatographic (especially from capillary or multicapillary GC) peak. The other one is the quantification error known as spectral skew (Holland et al. 1983), which arises during the measurement of adjacent mass-spectral peaks at different times along a transient signal. Alleviating these difficulties requires increasing the number of measurement points per time unit and the simultaneous measurement of the isotopes of which the ratio is investigated.The ability to produce complete mass spectra at a high frequency (typically > 20 000 s^{-1}) makes TOF-MS nearly ideal for the detection of transient signals produced by high speed chromatographic techniques. The simultaneous extraction of all *m/z* ions for mass analysis in TOF-MS eliminates the quantification errors of spectral skew, reduces multiplicative noise and makes TOF-MS a valuable tool for determining multiple transient isotopic ratios (Baena et al. 2001; Haas et al. 2001; Heisterkamp and Adams 2001; Leach et al. 2000). Whereas some of GC/ICP-TOF-MS applications reported so far (Baena et al. 2001; Heisterkamp and Adams 2001; Leach et al. 2000) could have readily been carried with a quadrupole mass analyzer, the emerging potential of ICP-TOF-MS is evident in three areas: (i) as a diagnostic tool for tracer experiments (ii) for studying isotope fractionation reactions in the biovolatilization processes of metals, (iii) for truly multielemental (above 3 elements) screening for volatile metal(loids) in a sample.

The practical problems to be addressed in GC/ICP-TOF-MS include:

(i) the need for the removal of C^+ ions originating from the solvent that would otherwise overload the detector. This is realised by use of the transverse rejection ion pulse option. Ions are deflected by means of a high voltage pulse applied perpendicularly at a time appropriate for the ion being rejected (Heisterkamp and Adams 2001),

(ii) the theoretically achievably number of acquired mass spectra per second is impressive but the tremendous volume of data collected at this speed makes it necessary to be limited. Data acquisition speed of 200 individual spectra per second was found sufficient for data acquistion every 10 ms which allows the measurement of all but fastest transient signals (peak widths <50 ms) (Leach et al. 2000),

(iii) the lower sensitivity in the monoelemental mode in comparison with the last generation of ICP quadrupole mass spectrometers. A minimum of approx. 500 pg of each species is necessary for the measurement of isotope ratios with a precision better than 0.5 % [14]. The limitations of the pulse counting system are clearly seen, with peak heights of more than 2000 counts reaching saturation (for an integration time of 100 ms) (Haas et al. 2001). On the other hand, Heisterkamp et al. reported a DL of 10-15 fg for alkyllead compounds, a value which is comparable with ICP-MS (Heisterkamp and Adams 2001). It should be emphasized that the loss of sensitivity in the monoelemental mode is compensated by the fact that the number of isotopes determined during one chromatographic run is no longer limited by peak definition (like in ICP-MS employing a quadrupole analyser), because the number of data points per chromatographic peak is independent of the number of measured isotopes.

GC/ICP-MS Using Sector Field Mass Analysers

The hitherto works have been scarce (Krupp et al. 2001a, 2001b; Rodriguez-Fernandez et al. 2001). Prohaska et al. used an ICP sector-field double-focussing mass spectrometer for the analysis of arsenic hydrides and organoarsenic compound in gaseous emissions from a microcosm, after their separation on a capillary column (Prohaska et al. 1999). Neither instrumental detection limits were reported nor chromatograms were, however, shown. Sanz-Medel's group proposed the same technique for the detection of ^{32}S in GC analysis of volatile sulfur compounds in bad breath and reported, when a guard electrode was used under cold plasma conditions, absolute DLs in the low nanogram range (Rodriguez-Fernandez et al. 2001).

The other two reports concerned the precise measurement of isotopic ratios by GC/ICP-MS with multicollector detection factility (Krupp et al. 2001a, 2001b). Four lead, ^{203}Tl, ^{205}Tl (for the correction of the mass bias) and ^{202}Hg (for the correction of the ^{204}Pb isobaric overlap) were monitored simultaneosly. A double-focussing instrument applied for this purpose allowed a detection limit of 1 pg for ^{207}Pb (introduced as Et_4Pb) which is poorer than values reported with other ICP mass spectrometers (Krupp et al. 2001b). The minimum time resolution was limited by the Axiom software to 50 ms; 60 points could define a full peak width of 3 s (Krupp et al. 2001b). Much lower detection limits were obtained by GC coupled to a single magnetic sector instrument equipped with a hexapole collision cell. A value of 2.9 fg was reported for the most abundant ^{208}Pb isotope (Krupp et al. 2001a); the values cited for the other isotopes were proportional to the isotopic abundance.

GC/ICP-MS Studies Using Stable Isotopes

The use of enriched isotopes with ICP-MS detectors has been of benefit for the development of speciation methodology. The isotopic specificity of ICP-MS opens the way to the use of stable isotopes or stable isotope enriched species for studies of transformations and of artefact formation during extraction and derivatization processes and to the wider implementation of the isotope dilution quantification. The latter had until recently been limited by the non-availability of organometallic species with the isotopically enriched element. However, standards for the isotopically enriched $Me^{201}Hg$ (Demuth and Heumann 2001), BuSn, Bu_2Sn and Bu_3Sn (Encinar et al. 2001) have recently been synthesized and applications are being developed. The prerequisite of the use of stable dilution techniques is the precise and accurate measurement of the isotopic ratios. The to date applications to real-world samples have been exclusively carried out with ICP-MS but precision and accuracy values for the measurement of isotope ratios in standard compounds by ICP-TOF-MS (Leach et al. 2000) and by sector-field multicollector (Krupp et al. 2001a, 2001b) instruments has been reported. Tracer experiments can be performed with ICP-MS but for the natural fractionation studies it may not be sufficient.

Isotope Ratio Measurements

In GC/ICP-MS the isotope ratio determinations are more precise if the intensities of the isotopes are integrated over the whole chromatographic peak instead of only measuring the isotope ratio at a single point of the peak (Heumann et al. 1998). A precision of 1% was reported for the Hg isotope ratios determined for MeEtHg eluted from a packed column by GC/ICP-MS (Hintelmann et al. 1995). Heumann et al. reported a 0.5% precision for Se (derivatized as piazselenol) in GC Q ICP-MS (Heumann et al. 1998).

A tin isotope-ratio measurement accuracy of 0.28% and a precision of 2.88% was calculated for a 1-s wide GC peak of Me_4Sn (Leach et al. 2000). Haas et al. reported that a minimum of 0.5 ng of an organometallic species was necessary for the measurement of isotope ratios with a precision better than 0.5%, the best value (0.34%) was attained for Me_2SnH_2 (Haas et al. 2001). ICP-TOF-MS should give an order of magnitude better precision than ICP-MS in comparable conditions when natural gas samples were analysed (Haas et al. 2001). The precision of the determination of isotopic ratios is improved by the use of simultaneous multi-elemental ion extraction (Heisterkamp and Adams 2001).

The precision of the isotope ratio measurements can be improved by using a sector field instrument with a multicollector facility. The precision values reported for the measurement of major Pb isotope ratio with a double-focusing instrument were better than 0.07% (for a 3s transient signal) corresponding to a 0.35% accuracy (Krupp et al. 2001a). When a single magnetic sector instrument (with a hexapole collision cell and multicollector detection) was used the precision was in the range 0.02 - 0.07% for ratios of high-abundance isotopes and injections of 5-50 pg (Krupp et al. 2001b). After mass bias correction the accuracy was within 0.02-0.15% (Krupp et al. 2001b).

For accurate determinations by the isotope dilution technique the mass discrimination effect must be taken into account. Mass bias was about 0.5% per mass unit (Lobinski et al. 1998). The ways to measure and to correct for it included the sequential measurement of the isotope ratio in the sample and the standard (Gallus and Heumann 1996) or the addition of an internal standard, such as e.g. Cd (Haas et al. 2001; Heisterkamp and Adams 2001) or Tl (Krupp et al. 2001a, 2001b) and the simultaneous measurement of the $^{111}Cd/^{113}Cd$ or $^{203}Tl/^{205}Tl$ isotopic ratios, respectively. From the practical point of the latter system requires the simultaneous delivery of the analyte and of the internal standard to the ICP-MS which can be done only, *via* a spray-chamber interface in view of the nonvolatility of Cd and Tl species.

In order to enable an alternate measurement of the isotope ratio of the analyte element in a standard, a diffusion cell containing pure chemical of the element to be determined for calibration of the measured isotope ratios was proposed (Gallus and Heumann 1996). It consisted of a glass vial covered by a membrane which allowed diffusion of the volatile calibrant species into the flow cell. If the isotope ratios of the element in the calibration compound are known, the measured isotope ratio of the separated species in the sample can be corrected (Gallus and Heumann 1996).

Stable Isotopes in the Monitoring of Artefacts During Sample Preparation Procedure

Enriched isotopes and isotope-enriched species provide an important diagnostic tool for the development of new analytical methods in which isotopes can be used as tracers. In a landmark paper Hintelmann et al. (1997) added isotopically enriched Hg^{2+} to prove the formation of $MeHg^{+}$ during a water vapor distillation procedure (Quevauviller et al. 1999) that still raised unanswered questions about the accuracy of methylmercury determinations in sediment samples. The mercury methylation in sediments was extensively studied using the method developed (Hintelmann et al. 1995). Stable isotope labelled mercury species allowed a GC/ICP-MS study of the simultaneous Hg^{2+} biomethylation and CH_3Hg^{+} demethylation at ambient trace levels with the sensitivity superior to traditional tracer or radiotracer techniques (Hintelmann et al. 1997).

Using a similar approach and a $Me^{201}Hg$ spike, Demuth et al. showed the transformation of $MeHg^{+}$ into elemental mercury (Hg^{0}) in the presence of chloride and bromide during the derivatization by $NaBEt_4$ but not by $NaBPr_4$ (Demuth and Heumann 2001). Isotope enriched Hg^{2+} allowed the observation of an artifactual formation of methylmercury when the water vapour distillation was applied for aqueous rain samples containing visible particles (Holz et al. 1999). No artefact formation of methylmercury during sample preparation was observed following the addition of a $^{201}Hg^{2+}$ isotope standard (Tu et al. 2000).

Speciated Isotope Dilution Analysis

Isotope dilution (ID) MS is a method of proven high accuracy. The sources of systematic errors are well understood and can be controlled which makes ID-MS accepted as a powerful method of analysis. Fundamentals of ID-GC/ICP-MS for species-specific analysis were extensively discussed by Gallus and Heuman (Gallus and Heumann 1996). They were illustrated by the determination of Se(IV) in water after conversion of the analyte species by piazselenol (Gallus and Heumann 1996).

In ID-GC/ICP-MS the sample is spiked with the species to be determined in which one of the isotopes of the metal or metalloid was enriched. After equilibration of the spike, the sample preparation procedure, GC/ICP-MS is run and the isotopic ratio of the metal(loid) in the species of interest is measured. The analysis principle is identical as in classical ID-ICP-MS, however, some fundamental differences occur.

The speciated isotope dilution analysis is only possible for element species well defined in their structure and composition. The species must not undergo interconversion and isotope exchange prior to separation. The equilibration of the spike and analyte, attainable in classical ID thermal ionization MS by multiple sequential dissolution and evaporation to dryness cycles, cannot be guaranteed to be achieved for speciated ID analysis in solid samples. Consequently, the prerequisite of the ID method: that the spike is added in the identical form as the analyte is extremely difficult, not to say impossible, to attain. Nevertheless, some advantages, such as the inherent corrections for the loss of analyte during sample preparation, for the incomplete derivatization yield, and for the intensity suppression/enhancement in the plasma are evident. In particular, ID quantification seems to be attractive in speciation analysis of complex matrices (e.g. gas condensates) when the different organic consituents of the sample modify continuously the conditions in the plasma and thus the sensitivity (Snell et al. 2000).

Isotopically enriched species should represent the ultimate means for specific accurate and precise instrumental calibration. Not only they are useful for routine determination by speeding analysis, but they also assist in the testing and diagnostics of new analytical methods and techniques. To date, the application examples of speciated ID-GC/ICP-MS have been relatively scarce. The determination of dibutyltin in sediment was carried out by ID analysis using a ^{118}Sn-enriched spike. No recovery corrections for aqueous ethylation or extraction into hexane were necessary and no rearrangement reactions were evident from the isotope ratios (Encinar et al. 2000). A mixed spike containing ^{119}Sn enriched mono-, di- and tributyltin was prepared by direct butylation of ^{119}Sn metal and characterized by reversed isotope dilution analysis by means of natural mono-, di- and tributyltin standards. The spike characterized in this way was used for the simultaneous determination of the three butyltin compounds in sediment CRMs (Encinar et al. 2001). Isotopically labelled Me_2Hg, MeHgCl and $HgCl_2$ species were prepared and used for the determination of the relevant species in gas condensates with detection limits in the low pg range (Snell et al. 2000).

Advances in Sample Preparation

In speciation analysis sample preparation is often troublesome and developments in this area are attracting considerable attention. The increased use of microwave-assisted extraction techniques in speciation analysis has been also reflected with regard to GC/ICP-MS (Lobinski et al. 1999, 1998; Slaets et al. 1999). The most important recent advances in sample preparation included the introduction of $NaBPr_4$ for the derivatization of organometallic species (De Smaele et al. 1998), and the use of headspace solid-phase microextraction (SPME) (Aguerre et al. 2001; De Smaele et al. 1999; Mester et al. 2001, 2000; Moens et al. 1997; Vercauteren et al. 2000), stir bar sorptive extraction (Vercauteren et al. 2001) and purge and capillary trapping for analyte recovery and preconcentration (Wasik et al. 1998a, 1998b).

Derivatization Techniques

Although some authors still use the classical DDTC extraction in the presence of EDTA followed by butylation for organolead speciation analysis (Leal-Granadillo et al. 2000), the position of tetraalkylborates allowing the derivatization in the aqueous phase, such as $NaBEt_4$ for organomercury and organotin speciation analysis and the newly introduced $NaBPr_4$ (De Smaele et al. 1998) for organolead is well established. Synthesis of $NaBPr_4$ was described in detail (De Smaele et al. 1998). The possibility of the simultaneous determination of Sn, Hg and Pb following the propylation was demonstrated (De Smaele et al. 1998). Artefact formation with hydride generation of antimony was discussed (Koch et al. 1998).

Two careful comparison studies are worth-noting. In one of them three derivatization approaches: anhydrous butylation using a Grignard reagent, aqueous etylation by means of $NaBEt_4$ and aqueous propylation with $NaBPr_4$ were compared for mercury speciation (Vercauteren et al. 2000). The absence of transmethylation during the sample preparation was checked using a 97% enriched ^{202}Hg inorganic standard (Fernandez et al. 2000). In another study two different derivatisation approaches: esterification of the carboxylic selenomethionine group using 2-propanol followed by the acylation of the amino group with trifluoroacid anhydride and the simultaneous esterification and acylation with ethyl chloroformate-ethanol were compared for the determination of selenomethionine by GC/ICP-MS (Pelaez et al. 2000). The structure of the derivatised aminoacid was confirmed in parallel by GC–MS (Pelaez et al. 2000). Derivatization techniques for GC were reviewed (Liu and Lee 1998).

Purge and Trap Using Capillary Cryofocussing

A semi-automated compact interface for time-resolved introduction of gaseous analytes from aqueous solutions into an ICP-MS without the need for a full-size GC-oven was described (Wasik et al. 1998 b). The working principle was based on purging the gaseous analytes with an inert gas, drying the gas stream using a 30-cm tubular Nafion membrane and trapping the compounds in a thick film-coated capillary tube followed by their isothermal separation on a multicapillary column. Recoveries were reported to be quantitative up to a volume of 50 ml (Wasik et al. 1998 a, 1998 b).

Solid-Phase Micro-Extraction (SPME)

Solid-phase micro-extraction (SPME) is a preconcentration technique based on the sorption of analytes present in a liquid phase or, more often, in a headspace gaseous phase, on a microfibre coated with a chromatographic sorbent and incorporated in a microsyringe. The analytes sorbed in the coating is transferred to a GC injector for thermal desorption. SPME is an emerging analytical tool for elemental speciation in environmental and biological samples (Mester et al. 2001). This solvent-free technique offers numerous advantages such as simplicity, the use of a small amount of liquid phase, low cost and the compatibility with an on-line analytical procedure.

SPME is based on an equilibrium between the analyte concentrations in the headspace and in the solid phase fibre coating. Low extraction efficiencies are hence sufficient for quantification but the amount of the analyte available may be very small. Hence, the interest for the high sensitivity of GC/ICP-MS to be combined with SPME.

The first work SPME-GC/ICP-MS concerned speciation of organomercury, -lead and -tin compounds ethylated in-situ with $NaBEt_4$ and sorbed from the headspace on a poly(dimethylsiloxane)-coated fused silica fibre. Headspace SPME using a 100 μm polydimethylsiloxane fiber at no equilibrium conditions was optimized as an extraction/preconcentration method for triphenyltin residues in tetramethylammonium hydroxide (TMAH) and KOH-EtOH extracts of potato and mussel samples (Vercauteren et al. 2000). Tricyclohexyltin was used as an internal standard. Derivatization was carried out with $NaBEt_4$ for 10-20 min (Vercauteren et al. 2000). Direct SPME (from the aqueous phase) was studied but the sensitivity was an order of magnitude lower. A detection limit of 2 pg l^{-1} was reported for an aqueous standard but a value of 125 pg l^{-1} was given for the sample extract corresponding to a DL in the low ng/g range (dry weight) (Vercauteren et al. 2000). Slightly lower detection limits (0.6 - 20 pg l^{-1}) were reported in another work (Aguerre et al. 2001).

A direct coupling of SPME with ICP-MS was described for species-specific determination of methylmercury (Mester et al. 2000). A fibre was inserted into a splitless-type GC injector which was placed directly at the base of the torch. Headspace was slightly less sensitive than immersion sampling but offered a larger linearity range. Immersion SPME was severely influenced by the matrix that leads to a 70-fold decrease in sensitivity (Mester et al. 2000).

Stir Bar Sorptive Extraction (SBSE)

SBSE applies stir bars, varying in length (1-4 cm), coated with a relatively thick (0.3-1 mm) layer of poly(dimethylsiloxane), resulting in volumes of the stationary phase varying from 55 to 220 µl (Baltussen et al. 1999 a, 1999 b, 1999 c). The stir bar is added to an aqueous sample for stirring and extraction, and after a certain stirring time, the bar is removed and thermally desorbed into a GC. Owing to the much larger volume of the stationary phase the extraction efficiency in SBSE is by far superior to that of SPME.

The SBSE technique was optimized for Bu_3Sn and Ph_3Sn determination in aqueous standard solutions with tripropyltin and tricyclohexyltin as internal standards to correct for the derivatization and extraction efficiencies. The derivatized analytes were released from the bar by thermal desorption, followed by cryofocussing of the compounds on a precolumn (- 40 °C) prior to release by flash heating onto a GC column. The instrumental detection limits reported were fabulous (10 fg l^{-1}), in practice, however, values of 0.1 pg l^{-1} could be achieved (Vercauteren et al. 2001).

Multidimensional LC with ICP-MS Detection

The Coupling HPLC/ICP-MS

The principal HPLC separation mechanisms used in natural product speciation analysis include size-exclusion, ion-exchange and reversed phase chromatography. Capillary electrophoresis is less mature but offers exciting possibilities for speciation analysis owing to the high separation efficiency, the nanolitre sample requirement, and the absence of packing susceptible to interact with metals and to affect the complexation equilibria (Kajiwara 1991; Richards and Beattie 1995). The combination of electrophoretic and electroosmotic flows provides the ability to separate a wide variety of positive, neutral and negative ions and compounds in one run. The complexity of the biological matrix may require the combination of two or more separation mechanisms in series to assure that a unique metal-species arrives at the detector at a given time.

The use of a quadrupole mass analyser in ICP-MS detection is the most widespread. The latest generation of instruments offers sub-femtogram absolute detection levels for many elements. The isobaric overlaps are generally not a problem because of the on-line separation from the potential interferents, e.g. Cl

($^{40}Ar^{35}Cl$) in the case of ^{75}As determination, but ghost peaks may appear. The application of a double focusing sector field instrument offers the higher resolution that may be required for the interference-free determination of sulfur or arsenic. An increase in resolution inevitably leads, however, to a dramatic decrease in sensitivity. It should also be noted that the sensitivity of the latest generation quadrupole instruments is only a factor of 2-3 lower than that of high resolution ICP-MS operated in the low resolution mode. A good tradeoff between sensitivity, freedom from isobaric interferences and price is offered by ICP-MS instruments equipped with a collision cell that have recently proliferated on the market (Tanner et al. 2002).

The key to a successful HPLC/ICP-MS coupling is the interface. In the simplest case the exit of an HPLC column (4.6-10 mm) is connected to a conventional pneumatic or crossflow nebulizer. The use of capillary or megabore (0.32 - 1.0 mm) HPLC systems that are becoming popular especially for reversed phase chromatography, requires the use of micronebulizers, either direct injection (DIN, DIHEN) or micronebulizers (e.g. Micromist) fitted with a small-volume nebulization chamber. The CE/ICP-MS coupling is less straightforward. The problems due to the laminar flow generated by the nebulizer suction, loss of sensitivity because of the electroosmotic flow dilution by the makeup liquid and peak broadening in the spray chamber have been resolved in the commercially available interface based on a total-consumption self-aspirating micronebulizer fitted with a small-volume spray chamber (Prange and Schaumloffel 1999; Schaumloffel et al. 2002).

Electrospray Mass Spectrometry for the Species Identification

The access to structural information for the identification of known or novel compounds is a great challenge to speciation analysis, especially that the improving sensitivity of ICP-MS instruments will inevitably increase the number of metal and metalloid species detected. Several recent reports have indicated exciting potential opportunities offered by electrospray mass spectrometry (ES/MS) for the precise determination of molecular weight and the structural characterization of molecules at trace levels in fairly complex matrices. This advantage leads to an increasing number of ES/MS reports presenting its complementary role to ICP-MS for species identification following pre-fractionation using chromatographic methods with element selective detection. The use of ES in this manner is, however, far more challenging, due in part to characteristically poorer signal-to-noise and signal-to-background ratio as well as more complicated gas-phase ion chemistry than typically encountered in an ICP. The various facets of electrospray mass spectrometry in speciation analysis have been reviewed (Barnett et al. 2000; Chassaigne et al. 2000).

Molecules containing a carbon - metal (metalloid) bond usually produce readily singly protonated ions in the electrospray source that theoretically should allow the identification of the metallocompound on the basis of the molecular mass. However, in the direct infusion mode, the attribution of a signal at a given *m/z* ratio to an elemental species is a daunting and practically impossible task for monoisotopic elements, such as arsenic. However, if an element presents a characteristic isotopic pattern such as Se (Casiot et al. 1999; Crews et al. 1996; Fan et al. 1998; Kotrebai et al. 1999, 2000) or Sn (Jones and Betowski 1993; Siu et al. 1989), recognition in the mass spectrum of a sample solution is easier, provided that the signal is not suppressed by the matrix.

A deeper insight into the species identity can be gained by the fragmentation of the protonated molecule ion (isolated at the level of the first mass filter) by collision induced dissociation (CID) followed by mass spectrometry of the product ions. The MS/MS mode allowed the identification of organoarsenic compounds in algal extracts purified by SE HPLC (McSheehy et al. 2001). For species containing an element having more than one stable isotope such as Se, valuable information can be obtained by fragmenting the two protonated molecule ions containing the adjacent most abundant isotopes (^{78}Se and ^{80}Se). Fragments that contain selenium will still be separated by the distance of two units whereas fragments that do not will remain at the same *m/z* value thus facilitating the interpretation of the mass spectra (Casiot et al. 1999).

Interpretation of mass spectra becomes easier when ES/MS is used as a chromatographic (Corr and Larsen 1996; Kotrebai et al. 1999, 2000; Le Bouil et al. 1999) or electrophoretic (Schramel et al. 1999) detector. The source collision induced dissociation mode allows the use of ES/MS as an element selective detector (Corr 1997). The elemental ES/MS mode is free from many polyatomic interferences present in ICP-MS. The concentration detection limits, are however 2-3 orders of magnitude higher than in the case of ICP-MS.

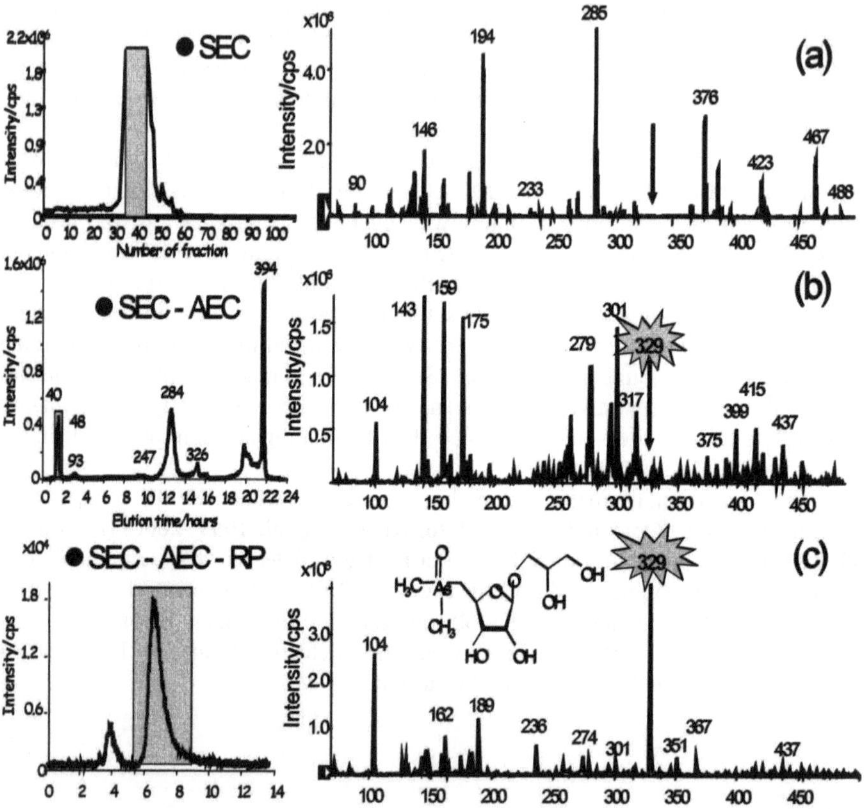

Fig. 2. Effect of the purification on the quality of a mass spectrum for arsinoyl riboside in algae. Explanation of (a) to (c) see text.

Effect of the Sample Cleanup Prior to ES-MS/MS Analysis

Fig. 2 shows the effect of the purification of an arsinoyl riboside from an algae sample on the quality (signal-to-noise ratio) of a mass spectrum acquired using a quadrupole mass analyser. No peak can be seen not only when an aqueous algal extract is directly analysed by ES/MS (data not shown) but also after a basic cleanup by size-exclusion chromatography (Fig. 2a). A mass spectrum of the fraction isolated by anion-exchange chromatography (Fig. 2b) allows the recognition of several peaks in the mass spectrum but the selection of a peak corresponding to an arsenic compound is impossible. Only a third dimension separation, by cation-exchange chromatography (Fig. 2c), allows a clear identification of a major peak, likely to correspond to an arsenocompound. Note that a quadrupole analyser does not allow an unambiguous attribution of this peak to an arsenic compound, especially if its identity is unknown. It can be achieved either by collision induced fragmentation or by a more accurate determination of the molecular mass of the protonated molecular ion, e.g. by TOF-MS or Fourier Transform Ion Cyclotron Resonance (FT-ICR)-MS.

Effect of the Mass Analyzer on the Quality of Mass Spectra

Fig. 3 shows mass spectra acquired for the arsinoyl riboside discussed in Fig. 2 using TOF (McSheehy et al. 2002 a) and FT-ICR (Pickford et al. 2002) mass spectrometers. The combination of time delayed extraction of ions with the use of an ion reflectron allows resolution above 10 000 be obtained with a mass accuracy below 10 ppm. This allows the acquisition of the molecular isotopic pattern. The molecular mass determined matches that calculated on the basis of the empiric formula: $C_{10}H_{21}O_7As$.

A finer determination of the empiric formula can be obtained by improving the resolution and mass measurement accuracy that is possible by FT-ICR-MS (Pickford et al. 2002). In this technique ions entering a chamber are trapped by a powerful magnetic field in circular orbits with a frequency independent of their velocity. The cyclotron motion of ions is excited by a RF field to generate a time dependent current. The current is converted by FT into orbital frequencies of the ions which correspond to their mass-to-charge ratios. In this way resolution exceeding 200 000 can be obtained and mass accuracy down to 0.2 ppm. In the example discussed in Fig. 3 the molecular mass of the protonated ion can be measured with the accuracy to the fifth decimal place that allows the unambiguous attribution of the molecular formula (Pickford et al. 2002).

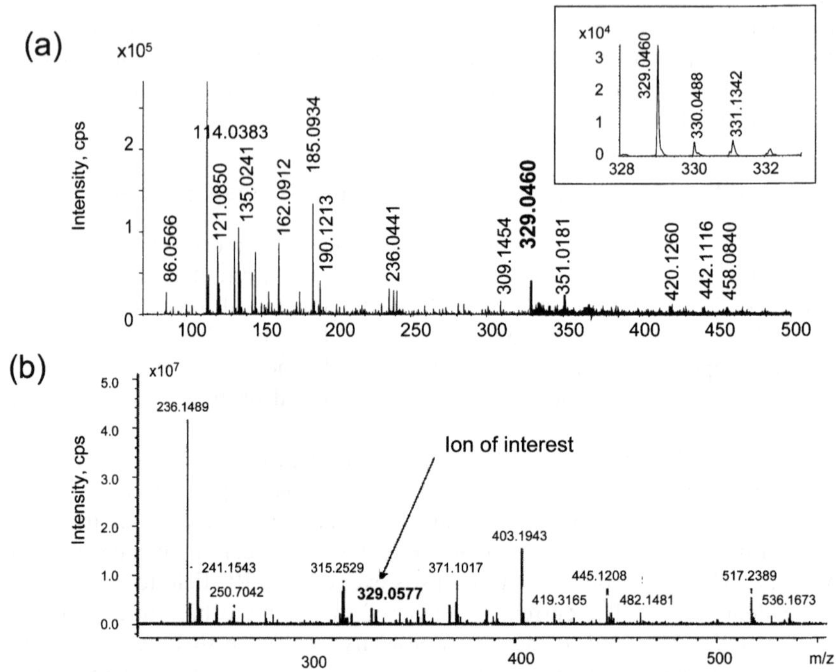

Fig. 3. Mass spectra of algae extract for arsinoyl riboside using TOF-MS a), FT-ICR-MS b).

Multidimensional Analytical Strategy for the Identification of Organoarsenic Species in Marine Biota

The complexity of elemental speciation in biological samples often makes it necessary to couple several different separation mechanisms in order to isolate the compound of interest prior to its characterization by electrospray MS. This approach has been successfully used to the characterisation of arsenic speciation in the kidney of a *Tridacna derasa* clam (McSheehy et al. 2002b). The individual arsenic compounds are isolated by tridimensional LC including size-exclusion, anion-exchange and cation-exchange mechanisms. The isolated species are analyzed by electrospray TOF-MS; the accuracy of the molecular mass measurement (especially important in the case of compounds of As that is a monoisotopic element) and the matching of the isotopic pattern of the molecular ion allow the identification of the compound. The identity confirmation or identification of an unknown compound can be achieved by CID-MS. Twenty organoarsenic compounds could be identified without the need for authentic retention time standards. Several of these compounds have not been reported in the literature yet. This example of standardless speciation analysis in an environmental biota shows that tandem MS is able to obtain virtually similar information to NMR but using about 100-1000 times lesser sample quantity.

Conclusions

Gas chromatography with ICP-MS detection has reached maturity as the analytical technique for speciation of organometallic species in a variety of matrices. It shows comparable figures of merit with that of GC-MIP-AED for standard applications including speciation of organomercury, organolead and organotin in the environment but offers a number of advantages in cases where extremely high sensitivity, multielemental screening, precise isotope ratios measurements or the analysis of complex matrices are required.

The decreasing detection limits of ICP-MS, the availability of efficient interfaces to HPLC and capillary electrophoresis, and the increasing sensitivity of electrospray MS for molecule-specific detection at trace levels open the way to the characterization of endogenous organometallic species in biological systems. The development of analytical methods for the biochemical speciation analysis is being carried out at the crossroads of interest of many disciplines and can profit from the interdisciplinarity of approach to the same degree as it can suffer from the lack of it. In order to cope with the complexity of biological matrices, multidimensional separation and mass spectrometric detection approaches are required.

Literature

Aguerre S, Lespes G, Desauziers V, Potin-Gautier M (2001) Speciation of organotins in environmental samples by SPME-GC: comparison of four specific detectors: FPD, PFPD, MIP-AES and ICP-MS. J Anal At Spectrom 16:263-269

Amouroux D, Tessier E, Pecheyran C, Donard OFX (1998) Sampling and probing volatile metal(loid) species in natural waters by in-situ purge and cryogenic trapping followed by gas chromatography and inductively coupled plasma mass spectrometry (P-CT-GC-ICP/MS). Anal Chim Acta 377:241-254

Baena JR, Gallego M, Valcarcel M, Leenaers J, Adams FC (2001) Comparison of three coupled gas chromatographic detectors (MS, MIP-AES, ICP-TOFMS) for organolead speciation analysis. Anal Chem 73:3927-3934

Baltussen E, David F, Sandra P, Janssen HG, Cramers C (1999a) Automated sorptive extraction-thermal desorption-gas chromatography-mass spectrometry analysis: Determination of phenols in water samples. J Microcolumn Sep 11:471-474

Baltussen E, Sandra P, David F, Cramers C (1999b) Stir bar sorptive extraction (SBSE), a novel extraction technique for aqueous samples: Theory and principles. J Microcolumn Sep 11:737-747

Baltussen E, Sandra P, David F, Janssen HG, Cramers C (1999c) Study into the equilibrium mechanism between water and poly(dimethylsiloxane) for very apolar solutes: Adsorption or sorption? Anal Chem 71:5213-5216

Bouyssiere B, Szpunar J, Lobinski R (2002) Gas chromatography with inductively coupled plasma mass spectrometric detection in speciation analysis. Spectrochim Acta Part B-At Spectrosc 57:805-828

Caruso JA, Ackley KL, Sutton KL (eds) (2000) Elemental Speciation. New approaches for trace element analysis. Elsevier, Amsterdam

Casiot C, Vacchina V, Chassaigne H, Szpunar J, Potin-Gautier P, Lobinski R (1999) An approach to the identification of selenium species in yeast extracts using pneumatically-assisted electrospray tandem mass spectrometry. Anal Commun 36:77-80

Chassaigne H, Vacchina V, Lobinski R (2000) Elemental speciation analysis in biochemistry by electrospray mass spectrometry. TrAC Trends Anal Chem 19:300-313

Chong NS, Houk RS (1987) Inductively coupled plasma-mass spectrometry for elemental analysis and isotope ratio determinations in individual organic-compounds separated by gas-chromatography. Appl Spectrosc 41:66-74

Cornelis R, Heumann KG, Caruso JA, Crews H (eds) (2003) Handbook of Elemental Speciation. Volume 1: Techniques and Methodology, NY, Wiley-VCH

Corr JJ (1997) Measurement of molecular species of arsenic and tin using elemental and molecular dual mode analysis by ionspray mass spectrometry. J Anal At Spectrom 12:537-546

Corr JJ, Larsen EH (1996) Arsenic speciation by liquid chromatography coupled with ionspray tandem mass spectrometry. J Anal At Spectrom 11:1215-1224

Crews HM, Clarke PA, Lewis DJ, Owen LM, Strutt PR, Izquierdo A (1996) Investigation of selenium speciation in in vitro gastrointestinal extracts of cooked cod by high-performance liquid chromatography inductively coupled plasma mass spectrometry and electrospray mass spectrometry. J Anal At Spectrom 11:1177-1182

De Smaele T, Moens L, Dams R, Sandra P, Van der Eycken J, Vandyck J (1998) Sodium tetra(n-propyl)borate: a novel aqueous in situ derivatization reagent for the simultaneous determination of organomercury, -lead and -tin compounds with capillary gas chromatography inductively coupled plasma mass spectrometry. J Chromatogr A 793:99-106

De Smaele T, Moens L, Sandra P, Dams R (1999) Determination of organometallic compounds in surface water and sediment samples with SPME-CGC-ICP-MS. Mikrochim Acta 130:241-251

Demuth N, Heumann KG (2001) Validation of methylmercury determinations in aquatic systems by alkyl derivatization methods for GC analysis using ICP-IDMS. Anal Chem 73:4020-4027

Encinar JR, Alonso JIG, Sanz-Medel A (2000) Synthesis and application of isotopically labelled dibutyltin for isotope dilution analysis using gas chromatography-ICP-MS. J Anal At Spectrom 15:1285-1285

Encinar JR, Villar MIM, Santamaria VG, Alonso JIG, Sanz-Medel A (2001) Simultaneous determination of mono-, di-, and tributyltin in sediments by isotope dilution analysis using gas chromatograpby-ICP-MS. Anal Chem 73:3174-3180

Fan TWM, Lane AN, Martens D, Higashi RM (1998) Synthesis and structure characterization of selenium metabolites. Analyst 123:875-884

Fernandez RG, Bayon MM, Alonso JIG, Sanz-Medel A (2000) Comparison of different derivatization approaches for mercury speciation in biological tissues by gas chromatography/inductively coupled plasma mass spectrometry. J Mass Spectrom 35:639-646

Gallus SM, Heumann KG (1996) Development of a gas chromatography inductively coupled plasma isotope dilution mass spectrometry system for accurate determination of volatile element species. 1. Selenium speciation. J Anal At Spectrom 11:887-892

Haas K, Feldmann J, Wennrich R, Stark HJ (2001) Species-specific isotope-ratio measurements of volatile tin and antimony compounds using capillary GC-ICP-time-of-flight MS. Fresenius J Anal Chem 370:587-596

Heisterkamp M, Adams FC (2001) Gas chromatography inductively coupled plasma time-of-flight mass spectrometry for the speciation analysis of organolead compounds in environmental water samples. Fresenius J Anal Chem 370:597-605

Heumann KG, Gallus SM, Radlinger G, Vogl J (1998) Accurate determination of element species by on-line coupling of chromatographic systems with ICP-MS using isotope dilution technique. Spectrochim Acta Part B-At Spectrosc 53:273-287

Hintelmann H, Evans RD, Villeneuve JY (1995) Measurement of mercury methylation in sediments by using enriched stable mercury isotopes combined with methylmercury determination by gas-chromatography inductively-coupled plasma-mass spectrometry. J Anal At Spectrom 10:619-624

Hintelmann H, Falter R, Ilgen G, Evans RD (1997) Determination of artifactual formation of monomethylmercury (CH_3Hg^+) in environmental samples using stable Hg^{2+} isotopes with ICP-MS detection: Calculation of contents applying species specific isotope addition. Fresenius J Anal Chem 358:363-370

Holland JF, Enke CG, Allison J, Stults JT, Pinkston JD, Newcome B, Watson JT (1983) Mass-spectrometry on the chromatographic time scale - Realistic expectations. Anal Chem 55:A997-A1010

Holz J, Kreutzmann J, Wilken RD, Falter R (1999) Methylmercury monitoring in rainwater samples using in situ ethylation in combination with GC-AFS and GC/ICP-MS techniques. Appl Organomet Chem 13:789-794

Jones TL, Betowski LD (1993) Characterization of alkyl-tins and aryl-tins by means of electrospray mass-spectrometry. Rapid Commun Mass Spectrom 7:1003-1008

Kajiwara H (1991) Application of high-performance capillary electrophoresis to the analysis of conformation and interaction of metal-binding proteins. J Chromatogr 559:345-356

Kim A, Hill S, Ebdon L, Rowland S (1992a) Determination of organometallic compounds by capillary gas-chromatography - inductively coupled plasma mass-spectrometry. HRC J High Resolut Chromatogr 15:665-668

Kim AW, Foulkes ME, Ebdon L, Hill SJ, Patience RL, Barwise AG, Rowland SJ (1992b) Construction of a capillary gas-chromatography inductively coupled plasma mass-spectrometry transfer line and application of the technique to the analysis of alkyllead species in fuel. J Anal At Spectrom 7:1147-1149

Koch I, Feldmann J, Lintschinger J, Serves SV, Cullen WR, Reimer KJ (1998) Demethylation of trimethylantimony species in aqueous solution during analysis by hydride generation gas chromatography with AAS and ICP-MS detection. Appl Organomet Chem 12:129-136

Kotrebai M, Birringer M, Tyson JF, Block E, Uden PC (1999) Identification of the principal selenium compounds in selenium-enriched natural sample extracts by ion-pair liquid chromatography with inductively coupled plasma- and electrospray ionization-mass spectrometric detection. Anal Commun 36:249-252

Kotrebai M, Tyson JF, Block E, Uden PC (2000) High-performance liquid chromatography of selenium compounds utilizing perfluorinated carboxylic acid ion-pairing agents and inductively coupled plasma and electrospray ionization mass spectrometric detection. J Chromatogr A 866:51-63

Krupp EM, Pecheyran C, Meffan-Main S, Donard OFX (2001a) Precise isotope-ratio measurements of lead species by capillary gas chromatography hyphenated to hexapole Multicollector ICP-MS. Fresenius J Anal Chem 370:573-580

Krupp EM, Pecheyran C, Pinaly H, Motelica-Heino M, Koller D, Young SMM, Brenner IB, Donard OFX (2001b) Isotopic precision for a lead species ($PbEt_4$) using capillary gas chromatography coupled to inductively coupled plasma-multicollector mass spectrometry. Spectrochim Acta Part B-At Spectrosc 56:1233-1240

Le Bouil A, Cailleux A, Turcant A, Allain P (1999) Determination of monomethylarsonic acid and dimethylarsinic acid in urine by liquid chromatography-tandem mass spectrometry. J Anal Toxicol 23:257-261

Leach AM, Heisterkamp M, Adams FC, Hieftje GM (2000) Gas chromatography-inductively coupled plasma time-of-flight mass spectrometry for the speciation analysis of organometallic compounds. J Anal At Spectrom 15:151-155

Leal-Granadillo IA, Alonso JIG, Sanz-Medel A (2000) Determination of the speciation of organolead compounds in airborne particulate matter by gas chromatography-inductively coupled plasma mass spectrometry. Anal Chim Acta 423:21-29

Liu W, Lee HK (1998) Simultaneous analysis of inorganic and organic lead, mercury and selenium by capillary electrophoresis with nitrilotriacetic acid as derivatization agent. J Chromatogr A 796:385-395

Lobinski R (1997) Elemental speciation and coupled techniques. Appl Spectrosc 51:A260-A278

Lobinski R, Adams FC (1997) Speciation analysis by gas chromatography with plasma source spectrometric detection. Spectrochim Acta Part B-At Spectrosc 52:1865-1903

Lobinski R, Sidelnikov V, Patrushev Y, Rodriguez I, Wasik A (1999) Multicapillary column gas chromatography with element-selective detection. TrAC Trends Anal Chem 18:449-460

Lobinski RS, Pereiro IR, Chassaigne H, Wasik A, Szpunar J (1998) Elemental speciation and coupled techniques - towards faster and reliable analyses - Plenary lecture. J Anal At Spectrom 13:859-867

McSheehy S, Pohl P, Lobinski R, Szpunar J (2001) Complementarity of multidimensional HPLC/ICP-MS and electrospray MS-MS for speciation analysis of arsenic in algae. Anal Chim Acta 440:3-16

McSheehy S, Szpunar J, Haldys V, Tortajada J (2002a) Identification of selenocompounds in yeast by electrospray quadrupole-time of flight mass spectrometry. J Anal At Spectrom 17:507-514

McSheehy S, Szpunar J, Lobinski R, Haldys V, Tortajada J, Edmonds JS (2002b) Characterization of arsenic species in kidney of the clam Tridacna derasa by multidimensional liquid chromatography-ICPMS and electrospray time-of-flight tandem mass spectrometry. Anal Chem 74:2370-2378

Mester Z, Sturgeon R, Pawliszyn J (2001) Solid phase microextraction as a tool for trace element speciation. Spectrochim Acta Part B-At Spectrosc 56:233-260

Mester ZN, Lam J, Sturgeon R, Pawliszyn J (2000) Determination of methylmercury by solid-phase microextraction inductively coupled plasma mass spectrometry: a new sample introduction method for volatile metal species. J Anal At Spectrom 15:837-842

Moens L, DeSmaele T, Dams R, VandenBroeck P, Sandra P (1997) Sensitive, simultaneous determination of organomercury, -lead, and -tin compounds with headspace solid

phase microextraction capillary gas chromatography combined with inductively coupled plasma mass spectrometry. Anal Chem 69:1604-1611
Pack BW, Broekaert JAC, Guzowski JP, Poehlman J, Hieftje GM (1998) Determination of halogenated hydrocarbons by helium microwave plasma torch time of flight mass spectrometry coupled to gas chromatography. Anal Chem 70:3957-3963
Pecheyran C, Quetel CR, Lecuyer FMM, Donard OFX (1998) Simultaneous determination of volatile metal (Pb, No, Sn, In, Ga) and nonmetal species (Se, P, As) in different atmospheres by cryofocusing and detection by ICPMS. Anal Chem 70:2639-2645
Pelaez MV, Bayon MM, Alonso JIGA, Sanz-Medel A (2000) A comparison of different derivatisation approaches for the determination of selenomethionine by GC/ICP-MS. J Anal At Spectrom 15:1217-1222
Pickford R, Miguens-Rodriguez M, Afzaal S, Speir P, Pergantis SA, Thomas-Oates JE (2002) Application of the high mass accuracy capabilities of FT-ICR-MS and Q-ToF-MS to the characterisation of arsenic compounds in complex biological matrices. J Anal At Spectrom 17:173-176
Prange A, Schaumloffel D (1999) Determination of element species at trace levels using capillary electrophoresis-inductively coupled plasma sector field mass spectrometry. J Anal At Spectrom 14:1329-1332
Prohaska T, Pfeffer M, Tulipan M, Stingeder G, Mentler A, Wenzel WW (1999) Speciation of arsenic of liquid and gaseous emissions from soil in a microcosmos experiment by liquid and gas chromatography with inductively coupled plasma mass spectrometer (ICP-MS) detection. Fresenius J Anal Chem 364:467-470
Quevauviller P, Adams F, Caruso JA, Coquery M, Cornelis R, Donard OFX, Ebdon L, Horvat M, Lobinski R, Morabito R, Muntau H, Valcarcel M (1999) Anal Chem, http://pubs.acs.org/journals/announcements/ letter991202.htm
Richards MP, Beattie JH (1995) Comparison of different techniques for the analysis of metallothionein isoforms by capillary electrophoresis. J Chromatogr B-Biomed Appl 669:27-37
Rodriguez I, Mounicou S, Lobinski R, Sidelnikov V, Patrushev Y, Yamanaka M (1999) Species selective analysis by microcolumn multicapillary gas chromatography with inductively coupled plasma mass spectrometric detection. Anal Chem 71:4534-4543
Rodriguez-Fernandez J, Montes-Bayon M, Pereiro R, Sanz-Medel A (2001) Gas chromatography double focusing sector-field ICP-MS as an innovative tool for bad breath research. J Anal At Spectrom 16:1051-1056
Schaumloffel D, Prange A, Marx G, Heumann KG, Bratter P (2002) Characterization and quantification of metallothionein isoforms by capillary electrophoresis-inductively coupled plasma-isotope-dilution mass spectrometry. Anal Bioanal Chem 372:155-163
Schramel O, Michalke B, Kettrup A (1999) Application of capillary electrophoresis-electrospray ionisation mass spectrometry to arsenic speciation. J Anal At Spectrom 14:1339-1342
Siu KWM, Gardner GJ, Berman SS (1989) Ionspray mass-spectrometry - quantitation of tributyltin in a sediment reference material for trace-metals. Anal Chem 61:2320-2322
Slaets S, Adams F, Pereiro IR, Lobinski R (1999) Optimization of the coupling of multicapillary GC with ICP-MS for mercury speciation analysis in biological materials. J Anal At Spectrom 14:851-857
Snell JP, Stewart II, Sturgeon RE, Frech W (2000) Species specific isotope dilution calibration for determination of mercury species by gas chromatography coupled to induc-

tively coupled plasma- or furnace atomisation plasma ionisation-mass spectrometry. J Anal At Spectrom 15:1540-1545

Szpunar J (2000) Bio-inorganic speciation analysis by hyphenated techniques. Analyst 125:963-988

Tanner SD, Baranov VI, Bandura DR (2002) Reaction cells and collision cells for ICP-MS: a tutorial review. Spectrochim Acta Part B-At Spectrosc 57:1361-1452

Tao H, Rajendran RB, Quetel CR, Nakazeto T, Tominaga M, Miyazaki A (1999) Tin speciation in the femtogram range in open ocean seawater by gas chromatography/inductively coupled plasma mass spectrometry using a shield torch at normal plasma conditions. Anal Chem 71:4208-4215

Templeton DM, Ariese F, Cornelis R, Danielsson LG, Muntau H, Van Leeuwen HP, Lobinski R (2000) Guidelines for terms related to chemical speciation and fractionation of elements. Definitions, structural aspects, and methodological approaches (IUPAC Recommendations 2000). Pure Appl Chem 72:1453-1470

Tomlinson MJ, Lin L, Caruso JA (1995) Plasma-mass spectrometry as a detector for chemical speciation studies. Analyst 120:583-589

Tu Q, Qian J, Frech W (2000) Rapid determination of methylmercury in biological materials by GC-MIP-AES or GC/ICP-MS following simultaneous ultrasonic-assisted in situ ethylation and solvent extraction. J Anal At Spectrom 15:1583-1588

Vanhaecke F, Moens L (1999) Recent trends in trace element determination and speciation using inductively coupled plasma mass spectrometry. Fresenius J Anal Chem 364:440-451

Vanloon JC, Alcock LR, Pinchin WH, French JB (1986) Inductively coupled, plasma source-mass spectrometry - a new element isotope specific mass-spectrometry detector for chromatography. Spectrosc Lett 19:1125-1135

Vercauteren J, De Meester A, De Smaele T, Vanhaecke F, Moens L, Dams R, Sandra P (2000) Headspace solid-phase microextraction-capillary gas chromatography-ICP mass spectrometry for the determination of the organotin pesticide fentin in environmental samples. J Anal At Spectrom 15:651-656

Vercauteren J, Peres C, Devos C, Sandra P, Vanhaecke F, Moens L (2001) Stir bar sorptive extraction for the determination of ppq-level traces of organotin compounds in environmental samples with thermal desorption-capillary gas chromatography - ICP mass spectrometry. Anal Chem 73:1509-1514

Wasik A, Pereiro IR, Dietz C, Szpunar J, Lobinski R (1998a) Speciation of mercury by ICP-MS after on-line capillary cryofocussing and ambient temperature multicapillary gas chromatography. Anal Commun 35:331-335

Wasik A, Pereiro IR, Lobinski R (1998b) Interface for time-resolved introduction of gaseous analytes for atomic spectrometry by purge-and-trap multicapillary gas chromatography (PTMGC). Spectrochim Acta Part B-At Spectrosc 53:867-879

Zoorob GK, McKiernan JW, Caruso JA (1998) ICP-MS for elemental speciation studies. Mikrochim Acta 128:145-168

Chapter 3

Analytical strategies for arsenic speciation in environmental and biological samples

J. Feldmann, S. Devalla, A. Raab, H. R. Hansen

Arsenic species in the environment and in biological samples

Arsenic as a metalloid has a rich inorganic and organic chemistry, because of the bond strength to sulfur and carbon. It occurs mainly in two different valencies: +III and +V. The standard redox potential for the inorganic arsenic oxo-compounds (As(III) / As(V)) is moderate ($E^o = +0.57$ V) and can therefore easily be interchanged in the natural environment. The inorganic oxo-compounds As_2O_3 in solution as $As(OH)_3$ and As_2O_5 as H_3AsO_4 are very soluble in water in contrast to arsenic-sulfur species As_2S_3, AsS, etc. This indicates that it is more important to focus on the oxides and their hydrolyzed counterparts than the sulfide species, if biogeochemical processes are studied. However, they might be important in sulfide-rich environments, in which arsenic can form sulfide-containing compounds. The sulfide replaces the oxide in their oxoanions for instance in thioarsenate (H_3AsO_3S). This is an understudied area and only few papers have ever covered this area (Schwedt and Rieckhoff 1996). In contrast to the other members of the group 15 in the Periodic Table of the Elements arsenic does not form oligoanions or polymers like phosphorous or the more metallic antimony. Halogenated arsenic compounds such as $AsCl_3$ might exist in the environment (Mester and Sturgeon 2002), but tend to hydrolyse quickly to their oxide, while arsine (AsH_3) seems to occur in small concentrations as a metabolite of microorganisms (Cullen and Reimer 1989).

On the other hand there is a rich organoarsenic chemistry which is based on the high stability and low polarity of the As-C bond due to the similar energy-level of the binding orbitals. Arsenic can bind to simple methyl groups, which can replace the arsenic-oxide bonds in trivalent and pentavalent arsenic species. The number of methyl groups can be one, two or three for both the trivalent and the pentavalent arsenic. Thus, trivalent monomethylarsonous acid MMA(III), dimethylarsinous acid DMA(III), and trimethylarsine TMA(III), and the pentavalent monomethylarsonic acid MMA(V), dimethylarsinic acid DMA(V) and trimethylarsine oxide TMAO can be formed. Only a tetramethylated species (tetramethylarsonium TMA^+) can be formed in the pentavalent state, whereas a pentamethylated species is not known, since the arsenic is too small to have a coordination of five to bind to five carbons. All of these compounds have been found in biota. According to the Challenger mechanism in which inorganic pentavelent arsenic is reduced and subsequently oxidized by an oxidative methylation most of the above mentioned compounds occur as final or as intermediate compounds of the arsenic metabolism in mammalian organisms (Challenger 1945). However, tri- and tetramethylated compounds seem to be restricted to bacteria or fungi as well as in molluscs and are absent from the mammalian organisms. Although the methyl-donor (S-adenosylmethione, SAM) is completely characterized, the catalyzing enzymes are only partially characterized yet (Zakharyan et al. 2001).

More complex compounds have been identified especially in the marine environment. The relative high abundance of arsenic in marine organisms is probably due to the complex food chains based on evolutionary primitive organisms which did not evolve to distinguish well between the nutrient phosphate and arsenate, so that arsenate, which is quite abundant (2 μg/L in open ocean) is actively taken up by similar routes and transporters as phosphate. In recent years however, it has been confirmed that the rich natural product chemistry of arsenic is not restricted to only the marine environment (e.g. Geiszinger et al. 2002). In terrestrial environment most of the same arsenic compounds have been identified, although often at much lower concentration (e.g. Quaghebeur et al. 2003).

Table 1. Selected arsenic species occurring in the abiotic environment (E) and biota (B) and their determined acute toxicity for male mice in mg/kg body mass (Tatken and Lewis 1983). R : Organic moiety bound directly to As (ribofuranoside or methyl)

Nr. of R	Trivalent Species	B	E	LD_{50}	Pentavalent Species	B	E	LD_{50}
0	$As(OH)_3$	√	√	4.5 (rat)	H_3AsO_4	√	√	
0	AsH_3	√	√	3				
1	MMA(III)	√	√	2.2[4]	MMA(V)	√	√	1,800[2]
1	$MeAsH_2$	√	√					
2	DMA(III)	√	√		DMA(V)	√	√	1,200[2]
2	Me_2AsH	√	√					
3	Me_3As (TMA(III))	√	√	8,000	TMAO	√	√	>10,000[3]
3					DMAE		√	?
3					DMAA	√	√	?
3					Me_2As-sugar-OH,-SO_4, etc.	√	(√)	>12,000[1]
3					Arsenolipids	√		?
4					TMA^+	√	√	900[3]
4					Me_3As-sugars	√		?
4					AsC	√	√	6,540[3]
4					AsB	√	√	>10,000[3]
4					AsB-2	√		?

[1] Me_2As- Sugar-OH, 0.04 % of As(III) (Kaise et al. 1996)
[2] Kaise et al. 1992a
[3] Kaise et al. 1992b
[4] Petrick et al. 2001: admin. Intraperitoneally to hamsters, (for As(III): 8.4 mg/kg).

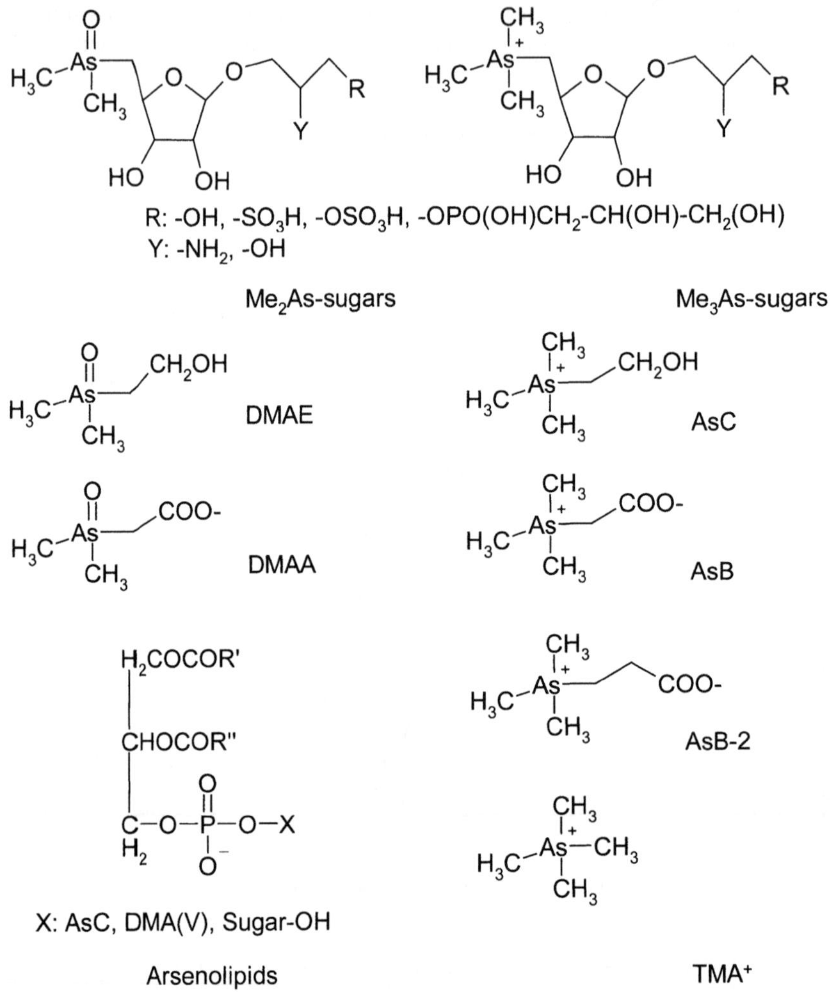

Fig. 1. Molecular structures of selected organoarsenic species most often identified in environmental and biological samples and their used abbreviations.

The most common organoarsenical is arsenobetaine (AsB), an analog of glycine betaine. It is very stable due to the full 'saturation' with four arsenic carbon bonds in which the chemical properties of arsenic are completely masked. This is probably the main reason why this compound is non-toxic. Other compounds with four arsenic-carbon bonds have been identified in biological samples such as arsenocholine, a few arsenosugars and arsenobetaine–2 (AsB-2) (Francesconi et al. 2000). In contrast to the absence of monomethylated or nonmethylated alkylarsenic compounds in biological samples, dimethylated arsenic compounds with an extra alkyl chain have been found in abundance. They are mainly arsenosugars with the dimethylarsinoyl moiety, Me_2As-sugars (Fig. 1), but also dimethylarsinoylethanol (DMAE) has been found recently in human and sheep urine as a metabolite of arsenosugars (Francesconi et al. 2001; Hansen et al. 2003). There is limited knowledge about the stability and toxicity of these compounds. Apart from the naturally occurring compounds, synthetic compounds like the growth promoters for poultry, namely, 4-hydroxy-3-nitrophenylarsonic acid, also known as Roxarsone, can occur in poultry litter (Dean et al. 1994; Jackson et al. 2003).

In addition to the variety of organoarsenic and inorganic compounds, coordinated arsenic compounds such as glutathione arsenic complexes or complexes of arsenic with metallothionein (e.g. Toyama et al. 2002) or phytochelatins (e.g. Bleeker et al. 2003) are a major focus in the field of arsenic speciation in biota.

A list of the majority of compounds found in the environment can be seen in Fig. 1 and Table 1. The large differences in acute toxicity of the different arsenic species indicate the importance of arsenic speciation work. But it is not only the toxicity which is a driving force in this field; more importantly, it is the mobility of the arsenic in biogeochemical cycles and the biochemistry of arsenic on the cellular level. For this purpose it is necessary to have strategies to identify the different arsenic species occurring simultaneously in one sample often at trace or even ultra-trace levels.

The rest of the chapter discusses a variety of analytical methods and their application scenarios. There is no single ideal method for all sample types because all methods have advantages and limitations. We believe that a good knowledge of the limitations of different methods is absolutely necessary to interpret the analysis correctly.

Analytical strategies

Sample preparation

Most biological materials to be analyzed are found as solids and have to be transferred to solutions, because most analytical and clean-up methods need the samples in solution for the analysis of different compounds. The solid samples have to be transferred into solution by an extraction process. Most arsenic species are in the cytoplasm and therefore water-soluble. The samples are extracted with methanol/water mixtures in which the methanol is used to open the phospolipid membranes to have access to the cytoplasm. Only recently, more emphasis has been made to determine the non-water soluble arsenic species by using methanol/chloroform mixtures in order to dissolve the so-called arsenolipids (Hanaoka et al. 1999). Sample preparation steps, which alter the solution chemistry of the cytoplasm, might have an influence on the stability of the arsenic species. Thus, the extraction should be as gentle as possible and under physiologically identical conditions. But at the same time, it should be ensured that the extraction is vigorous enough so that quantitative extraction is achieved. This by itself seems to be a contradiction. The procedure has to be optimized so that no species transformations occur and quantitative extraction is achieved. The best strategy would be to do speciation directly on the intact organism, however most methods suitable for this suffer from high detection limits and interference problems. A few speciation studies have been done directly on the solid unchanged biological material using X-ray absorption spectroscopy and especially EXAFS (extended X-ray absorption fine structure) and XANES (X-ray absorption near edge spectroscopy) to unravel the electronic environment around the arsenic atom, which can give evidence whether arsenic is trivalent or pentavalent and bound to sulfur rather than oxygen (Langdon et al. 2002). However, the identification of the entire species is not possible.

Further clean-up methods of the extracts are often not necessary if very sophisticated separation techniques and detection methods are used. However, it should be emphasized here that the extract will contain all water-soluble compounds from the biological samples including many organic compounds of different sizes along with the few arsenic-containing compounds. If the nature of the arsenical at the given pH is known, for instance cationic, then certain clean up strategies such as solid-phase extractions using anion exchange cartridges can be employed to remove salts and other anionic organic compounds which can interfere with the analysis of the arsenic species. Other methods utilize size exclusion columns, which make it possible to separate macromolecules from small water-soluble compounds, like the majority of arsenicals in biological samples.

For the analysis of volatile arsenic species such as arsine or the methylated arsines in water samples chemofocussing methods have been used (Cullen and Reimer 1989). For example the headspace of a microorganism culture was swept through a solution of $HgCl_2$ in hydrochloric acid in order to trap $(CH_3)_3As$ as an $(CH_3)_3As*2HgCl_2$ adduct. The volatile arsine can be characterized as trimethylhydroxylarsonium nitrate or picrate. More recently, purge and trap methods coupled to GC/MS, GC/AAS or ICP-MS have been employed for the quantification and identification of the volatile arsine compounds in water samples (e.g. Hirner et al. 1998) or directly for gaseous samples (Drugov 1998). The separation of the volatile arsenicals is therefore mainly based on the different volatility of the arsenic species compared to the matrix. However, also solid-phase micro-extraction can be used for the separation of volatile arsenicals from the matrix (Wooten et al. 2002).

Separation methods

Having ensured that the species in the solution are stable and not transformed during the sample preparation steps, the solution can be separated by conventional chromatographic and electrophoretic methods. As initially mentioned, in most cases only water-soluble species have been analyzed and only aqueous-based mobile phases have been used. Most separation techniques can directly be coupled to sensitive detectors, which are either species-specific or arsenic-specific. If a separation of crude extracts takes place the detection technique has to be sensitive enough to identify the small concentrations of individual arsenic species in the extract of biological samples or water and gas samples. Although capillary electrophoresis (CE) has excellent separation capacities, which have been demonstrated for standards in a series of publications (e.g. Koellensprenger et al. 2002; Chen et al. 2003), it lacks the sufficient low detection limits for natural samples and furthermore the separation is strongly dependent on the matrix of the sample. This makes CE not the first choice for the separation of arsenic species in environmental and biological samples. However, recently Prange and co-workers (2001) have shown that metallothionein metal interaction can be determined in real samples.

Gas chromatography is used mostly for the separation of arsines, either directly as volatile arsenic compounds in the sample (for example in landfill gas (Feldmann et al 1998) or as chemical vapor deposition precursors (Bartram 2001), or generated by hydride generation methodology using $NaBH_4$ or by derivatisation with toluol-3,4-dithiol (Chen et al. 2001). Although some reports (e.g. Clamagirand et al. 1993) suggest that ethylation of organoarsenical using $NaBEt_4$ is possible, the reaction seem to be not quantitative and therefore of limited use for analytical purposes. Since the arsenic species diversity is enormous, the use of gas chromatography is mostly restricted to the determination of purely volatile arsenicals.

Liquid chromatography is widely used for the separation of water-soluble arsenic species. The pK_a values of all arsenic species vary greatly. At neutral pH some species occur as anions whereas others occur as cations or as neutral species. It would be highly desirable to be able to separate all occurring arsenic species with one method, therefore reverse phase liquid chromatography has been employed for a variety of compounds with different ion-pairing reagents. This technique involves the co-ordination of the ion-pair reagent such as tetraethylammonium hydroxide (TEAH) or tetrabutylammonium hydroxide (TBAH) with the arsenic species. However, the separation is much better for anions than cations. This is the reason why most researchers use two different methods for water-soluble arsenic speciation studies; one anion and one cation exchange method. Both techniques are complementary. A two-dimensional matrix of the retention times of the different species give much better confidence in the correct assignment of arsenic species, as when only one retention time comparison with standards is used for species-identification in combination with arsenic-specific detection (Fig. 2). If these separation techniques are directly coupled to a detector, the mobile phase and flow rate has to be optimized not only for the best separation of the arsenic species, but also to the requirements of the detectors. For instance it is not a routine method to use high volatile organic solvents such as acetonitrile when ICP-MS (inductively-coupled plasma mass spectrometry) is coupled, or it is not advisable to use a non-volatile mobile phase such as a phosphate buffer if ESI-MS (electrospray ionization mass spectrometry) is employed.

In addition to all these constrains it is necessary to test if the target compounds are stable enough in the mobile phase and do elute unchanged from the chromatographic columns. It is not only necessary to have an inside knowledge of the separation power of these techniques, more importantly, it is necessary to know which arsenic species will be hidden by using the separation technique chosen.

Arsenic species cannot be detected due to:

- Co-elution with other more abundant arsenic species
- Decomposition of labile arsenic compounds during extraction or separation
- Formation of non-volatile arsenic species by using hydride generation methodology
- Irreversible reaction with the stationary phase or
- Formation of macro-molecules which stick to the columns

The loss of arsenic species during chromatography can however be tested. The measurement of spiked extracts and the determination of the chromatographic recovery (ratio between the sum of the eluted individual arsenic species to the total arsenic concentration in the extract) will help to assist in the identification whether arsenic compounds are hidden by the used analytical method.

Species integrity is also an important point during the separation process. Not only can labile species disintegrate in contact with the mobile phase, they can also fall apart during the separation due to strong interaction with the stationary phase. This should always be tested, although it is difficult to identify, if the arsenic species are not known or not available as standards. An indication of transformation of arsenic species during chromatography is the low chromatographic recovery and an increase of the baseline if arsenic-specific detection is used.

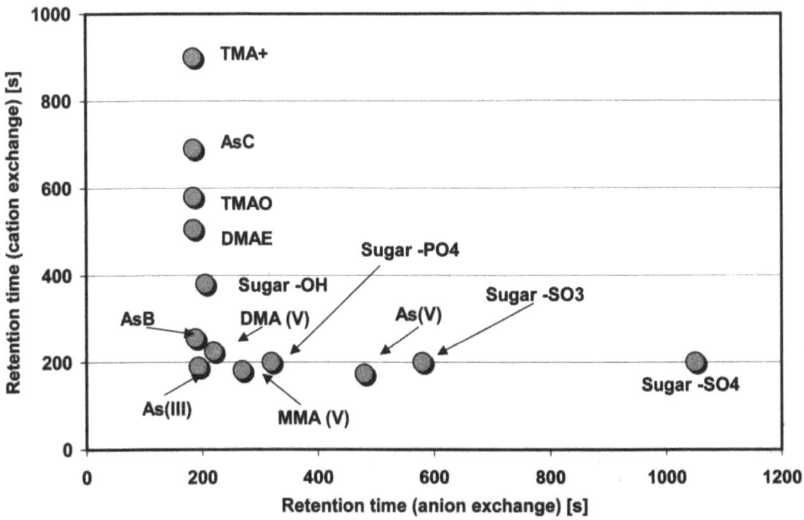

Fig. 2. Two dimensional retention time plot for two complimentary LC methods (strong cation exchange and strong anion exchange, see Fig. 6).

The availability of standards is a major problem in the field of arsenic speciation, which hampers not only the correct identification of arsenic species but more importantly the study of environmental and molecular biological processes. Only a few organoarsenicals are available off the shelves. Therefore compounds have to be synthesized or extracted from mostly marine samples, so that they can be used for spiking experiments in order to assign chromatographic peaks to the right arsenic species. For the characterization of those standards soft ionization mass spectrometry is used in addition to ^{1}H-NMR and ^{13}C-NMR if the extract is concentrated and clean enough to identify the arsenic-containing compound within a matrix of many organic compounds. If the purity of compounds are not known these compounds cannot be used for any type of toxicity tests.

Identification

Fully interpreted ^{1}H-NMR and ^{13}C-NMR spectra, as well as mass spectra with an uncertainty better than 5 ppm on the molecular mass for compounds < 500 Da and 10 ppm for compounds > 500 Da are normally the minimal criteria of publishing a newly characterized compound isolated from biological samples if no crystal structure is available. These criteria are often not fullfilled for the identification of new arsenic species in biological extracts. In many cases the identification of an arsenic species is only based on retention time comparison with standards of a single method in addition to an arsenic-specific detection using mostly atomic fluorescence spectrometry (AFS) or inductively-coupled plasma mass spectrometry (ICP-MS).

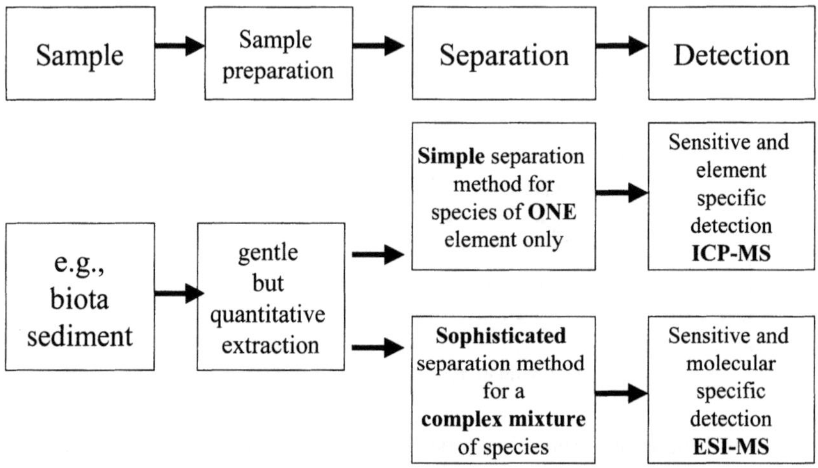

Fig. 3. General concept for metal speciation in biological samples.

When arsenic-specific detectors are coupled with chromatography and no available arsenic standard has the same retention time in a spiked sample then the eluting peaks are interpreted as unknown arsenic compounds. In the past five to ten years numerous unknown arsenic-containing species (occurring as unassigned peaks in LC chromatograms) were reported in a variety of sample extracts. Newly synthesized arsenic compounds with identical chromatographic behavior enabled a retrospective identification of these peaks as arsenic compounds. However, it is insufficient to identify a compound just by its chromatographic behavior. Thus another analytical strategy has to be followed: the use of an species-specific detector for chromatography electrospray ionization mass spectrometry (ESI-MS) and if possible either with accurate mass determination or NMR data.

Electrospray ionization is a soft ionization with makes it possible to obtain indirectly the molecular mass of the species (e.g. $M+H^+$ occurs often in the positive mode). Inorganic arsenic species cannot easily be detected with this method. Most organoarsenicals form molecular peaks besides all the other organic compounds in the separated crude extract of biological samples. Since arsenic is monoisotopic no particular isotopic pattern reveals the occurrence of an organoarsenical. It is not unequivocal to identify an arsenic-containing compound in a matrix of many non-arsenic compounds. Although ESI-MS is species-specific, it demands high separation power to separate the different compounds in the extract in order to identify possible arsenic species, by monitoring the molecular mass or a major common fragment in the SIM (Selected Ion Monitoring) mode of the MS. However, it is often possible to fragment species further and gain molecular information about the compound using the fragmentation information, in particular, if a certain fragment can be analyzed and further fragmented. The latter technique can only be realized if a triple quadrupole mass spectrometer or an ion trap MS is employed (Miguens-Rodriguez et al. 2002). Users of these relatively expensive instruments have the opportunity to identify arsenic-species without having access to arsenic standards. This is seen today as a major breakthrough in the arsenic speciation field. Other developments are the QTOF-MS (quadrupole-time of flight mass spectrometer) and lately FT-ICR-MS (Fourier transformation ion cyclonic resonance mass spectrometer), which are able to detect the molecular mass very accurately. Due to the mass defect of arsenic (m/z 74.92) it is often possible to identify an arsenic-containing compound in a mixture of other organic compounds (Pickford et al. 2002).

The risk that the integrity of the samples is not guaranteed, increases with the degree of sophistication of the separation methods employed. Since the necessity of the use of highly advanced separation methods (gel electrophoretic techniques combined with multi-dimentional chromatographic techniques) in order to separate all organic compounds in an extract when soft ionization mass spectrometric techniques increase the risk of species transformation. This is a good reason to generate only crude extracts and one or two separation techniques with an arsenic-specific detection. One advantage of element specific detection is that arsenic-containing compounds are easy to recognize, thus the arsenicals do not have to be separated from all the other organic compounds, which do not contain any arsenic. Furthermore, it should be mentioned that when ICP-MS is used as an arsenic-specific detector it is necessary to monitor not only arsenic (m/z 75) but also m/z 77 in order to identify the occurrence of an argonchloride cluster which can be formed if a chlorine-containing compound elutes from a column and forms an interference on m/z 75 ($^{40}Ar^{35}Cl^{+}$) and on m/z 77 ($^{40}Ar^{37}Cl^{+}$). The latter one has also to be checked for selenium compounds (m/z 78, or 82). In addition to these precautions, the plasma stability of the ICP-MS has to be monitored during the entire run, since plasma instabilities can result in the generation of peaks on m/z 75, which can easily be interpreted as arsenic containing compounds. These can be identified by the addition of a continuous internal standard which can be added post-column before the eluent enters the nebulizer.

Quantification

The emphasis of this essay has so far been on identification, although quantification is none the less important. In general, the first step is the identification of an arsenic species before it can be quantified. This is however, not always necessary when ICP-MS is used as an arsenic-specific detector for chromatography, since most compounds contain only one arsenic atom per molecule and the hard ionization of an argon plasma decomposes all arsenic species to As^+. This results in very similar responses for the different arsenic species eluting at different times from the LC columns. Some differences in the responses however, have been recorded, which might be explainable by the co-elution of major matrix compounds, which results in a significant change in the ionization conditions at that time of the chromatogram. Furthermore, it might be the result of a different nebulization efficiency, the transport of the sample into the plasma, which is also strongly dependent on the matrix of the eluting solution. Under these considerations with all the uncertainties, it is however still possible to quantify a compound, which has not been identified. This is in particular helpful when the significance of an unknown compound in the sample is described, whether it is a minor or a major arsenic species.

The quantification of known compounds is however not easier. There are very few certified reference materials available for which arsenic species are certified or published data available in order to show whether the analytical method used performs similar to those of researchers who published data on the CRM before. For the different arsenic species there are however huge differences in the availability of data for CRM in published work. Arsenobetaine data are available of a series of CRMs (e.g. DOLT and DORM (both NRC) or NIES No.18), while arsenosugars are not identified in CRMs. The lack of arsenic standards and the lack of CRMs for arsenic species are today the major obstacles in the evaluation of the quality of arsenic speciation analysis in environmental and biological samples.

Although the positive identification of arsenic species in an environmental sample needs the confirmation using ESI-MS or similar techniques, the quantification of the arsenic species is however very difficult using ESI-MS solely.

Analytical aspects of selected arsenic speciation studies

Gas chromatography coupled to MS or ICP-MS

Cryotrapping for volatile arsenic species

For the identification of ultra-trace levels of volatile arsenic compounds evolving from biological active environments such as landfill sites or algal mats on hot spring pools, it is necessary to have a series of clean-up steps if gas chromatography is coupled to electron impact mass spectrometry (GC/MS), since the gaseous samples contain dozens of different organic compounds in high concentrations. The resulting MS spectra are difficult to interpret. Due to the anticipated low concentration of volatile arsenic or other metal compounds in the gas samples a pre-concentration method is necessary. Cryotrapping (CT) on a packed column filled with non-polar chromatographic material (SP-2100 on Chromosorb) with subsequent separation of the different species according to their boiling point can be used to preconcentrate the sample and select only the gas fraction in which the volatile arsenic species occur. (This can be checked by coupling online to an ICP-MS as an arsenic-specific detector). The fraction can be collected and injected again onto a second column. This time the column has a higher separation power and a different polarity, so that the gas fraction will be separated further and an unequivocal identification of the volatile metal species can be made by fragmentation pattern of the eluting molecule. This procedure has been demonstrated previously for the determination of volatile tin, antimony and bismuth compounds in landfill and sewage gas (Feldmann et al. 1998). Once the structures are fully established using this method, capillary-GC can be coupled directly to ICP-MS. This detection is much more sensitive and specific for the different metal species, which makes the preconcentration and clean up steps redundant. In addition CT/GC/ICP-MS can be used as a multielement technique since the sample preparation is always the same for the different metal and metalloid species (separation of less volatile species from the more volatile major constituents of the gas sample like N_2, CH_4 or CO_2). This technique was routinely used for the stability test of volatile arsenic, antimony and tin species in moisturized air at room temperature and at 50°C in the dark and under intense UV radiation (Haas and Feldmann 2000). Fig. 4 illustrates the sensitivity of the CT/GC/ICP-MS with detection limits at the sub-picogram level. Using NaOH cartridge, CO_2 can be absorbed before the gas sample is cryotrapped. The loss of volatile arsines is below 15 % (Feldmann et al. 2001). Concentrations of about 10 µg element per m^3 are a realistic concentration in order to identify if the volatile species have a high enough atmospheric lifetime to diffuse in the vicinity of point sources such as landfill sites. The arsenic species seem to be very stable, although increased methylation grade, temperature and more significantly UV radiation reduces the atmospheric lifetime significantly (see table 2).

Table 2. Estimates atmospheric lifetime of volatile arsenic compounds in moisturized air at a starting concentration of 10 µg/m^3 (Haas and Feldmann 2002)

	Dark, 20°C	Dark, 50°C	UV, 30°C (5000 Lux)
AsH_3	>2000 h	> 600 h	150 h
$MeAsH_2$	1440 h	600 h	1.5 h
Me_2AsH	480 h	84 h	0.1 h
Me_3As	---------	30 h	0.1 h

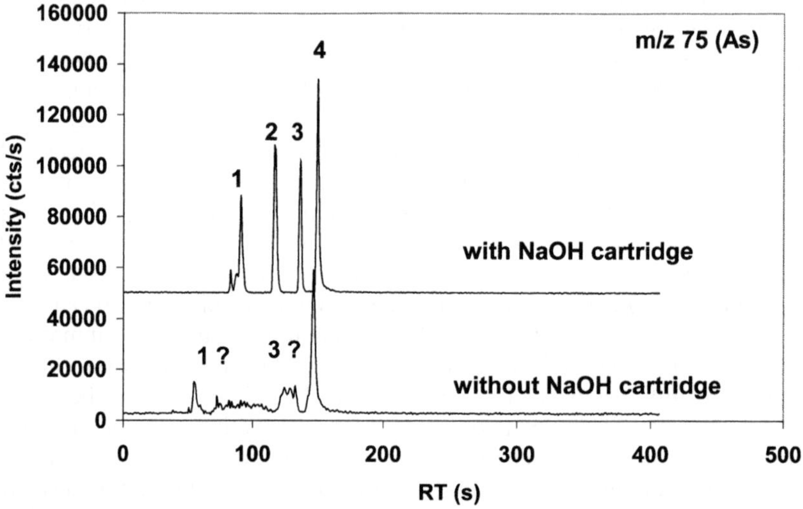

Fig. 4. CT/GC/ICP-MS chromatogram of an air sample containing large amounts of CO_2 with and without NaOH cartridges before cryotrapping of 1: AsH_3, 2: $MeAsH_2$, 3: Me_2AsH and 4: Me_3As of a concentration of 10 µg arsenic /m^3 each. (Feldmann et al. 2001).

Hydride generation methodology

Many arsenic species can be easily and quantitatively derivatized to volatile arsines by using sodium borohydride ($NaBH_4$) and gas chromatography can be utilized for their separation. Other commonly used derivatization reagents such as sodium tetraethylborate ($NaBEt_4$) or Grignard reagents (e.g. $C_5H_{11}MgBr$) do not work successfully with arsenic, since the alkyl groups are transferred as carbanions which is not the best route to approach the nucleophilic arsenic. Using derivatisation, it is necessary that the chemical reaction is quantitative and has no side reactions. This is indeed the case when $NaBH_4$ is used in hydride generation (HG). The strong As-C bond is stable during HG in contrast to the As-O bond; O is replaced by H. Compare Table 3 and Fig. 5 for the differences in the derivatization reaction. The species information about the alkyl groups bound to arsenic remains. By changing the pH of the solution from neutral to acidic, $NaBH_4$ can also be used for the determination of most pentavalent arsenic species. This makes it possible to discriminate between trivalent and pentavalent arsenic species which form the same volatile arsine species. Pentavalent arsenic species which have more than three As-C bonds like TMA^+, AsB, AsC or trimethylated arsenosugars cannot be reduced to volatile neutral arsines. In addition it is not possible to volatilise arsenosugars with a dimethylated arsinoyl moiety, although the arsenic can be reduced to trivalent arsines. The ribofuranoside group is simply too large and increases the boiling point of the generated arsine enormously. In summary, only those arsenic species that form volatile arsines are purged from the solution and cryo-trapped on a non-polar GC column. With subsequent heating the trapped arsines separate according to their boiling points and lead to the identification of their trivalent and pentavalent water-soluble non-volatile arsenic species.

This technique is very sensitive and has the advantage that preconcentration and clean up steps are incorporated. For example small concentration of methylated arsenic species can be determined in seawater with more than 3.5% salt (see Fig. 5). Although, the matrix (e.g. high concentration of redox-active elements such as iron and manganese in porewater samples) has an influence on the arsine generation, this technique can be used for quantitative purposes in the standard-addition mode. The major limitation is that most of the more complex organoarsenicals and tetra-alkyl arsenicals cannot be determined by this method and are therefore described as hidden arsenic species.

Although this technique can be used for a variety of different elements and their methylated counterparts (e.g. antimony, mercury, germanium and tin) when it is coupled directly to ICP-MS as a multi-element detector, it cannot be used for lead speciation and only to a certain degree for bismuth, tellurium and selenium speciation. The reasons are the low stability of, for instance, the lead hydrides and the limited knowledge of the occurrence of partly methylated species of bismuth, tellurium and selenium. The stability of certain metal species during hydride generation is unknown, so that the deduction of which species had been in the solution before $NaBH_4$ was added is very speculative. This is highlighted in the controversially discussed demethylation process of antimony (Jenkins et al.1998; Koch et al. 1998), whether or not the Sb-C bond is cleaved during the derivatization conditions or not. However, attempts have even been made to use it for screening analysis directly on sediment suspensions (Grueter et al. 2000), which can only give an overview whether methylated metal species do occur in the sample or not.

Examples of the use of hydride generation methodology:

- Identification of arsenic metabolites in body fluids of mammals which were exposed to inorganic arsenic [As(III), As(V), MMA(III), MMA(V), DMA(III), and DMA(V)].
- Arsenic speciation in seawater and porewater of marine and freshwater sediments from non-contaminated sites [As(III), As(V), MMA(V), DMA(V) and TMAO].
- Screening method for soil porewater.
- Identification of methylated intermediates of arsenic by microorganism cultures.

Table 3. Hydride generation methodology of methylated trivalent and pentavalent arsenic species in solutions.

Species	Solution pH 1	Solution pH 5 (buffered)
$As(OH)_3$	AsH_3	AsH_3
H_3AsO_3	AsH_3	-------
MMA(III)	$MeAsH_2$	$MeAsH_2$
MMA(V)	$MeAsH_2$	-------
DMA(III)	Me_2AsH	Me_2AsH
DMA(V)	Me_2AsH	-------
TMAO	Me_3As	Me_3As

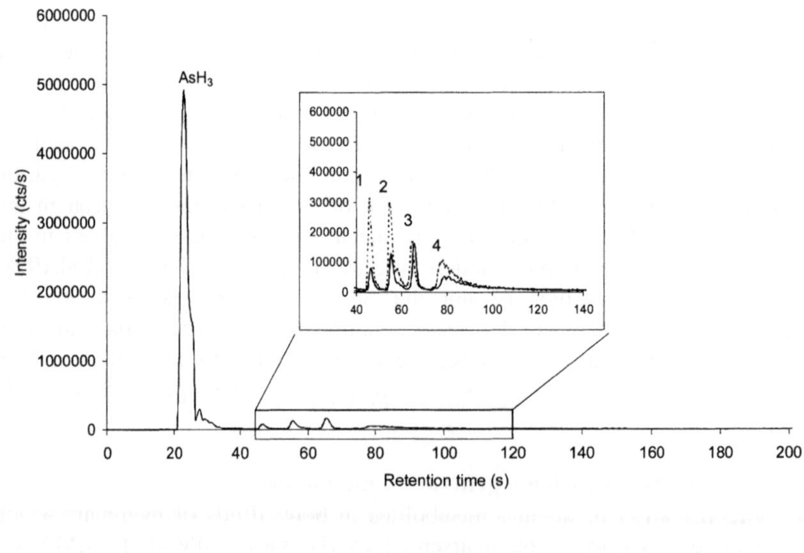

Fig. 5. Hydride generation CT/GC/ICP-MS of an anaerobic pore water sample containing less than 0.05 µg As/l as 1: MMA(V), 2: DMA(V), 3: TMAO, 4: unknown; besides large amounts of inorganic arsenic 2 µg/l of an harbor sediment in Aberdeen (Scotland) (Pengprecha 2002).

Liquid chromatography coupled to ESI-MS and ICP-MS

In order to identify arsenic species not accessible for HG/GC analysis, liquid chromatography has often been employed. In most cases liquid chromatographic separation based on ion exchange enabled researchers in the past ten years to separate cationic and anionic species in two separate runs. In our lab routinely two different methods are used; a strong anion exchange method (Hamilton PRP X-100 column) with either a 30 mM ammonium phosphate buffer (pH 5.3 or 6.0) or a 20 mM ammonium carbonate buffer (see Fig. 6) and a strong cation exchange method (Supelcosil) with a 20 mM pyridine buffer at pH 3.0. Both methods enable us to separate at least 13 different arsenic species and assign peaks according to their retention times as allocated on the two dimensional map shown in Fig. 2. Using this method in combination with ICP-MS or ESI-MS it is possible to study samples that contain more complex organoarsenicals than the methylated species. It should be noted however, that only carbonate buffers can be used for anion exchange chromatography when ESI-MS is used as a detector.

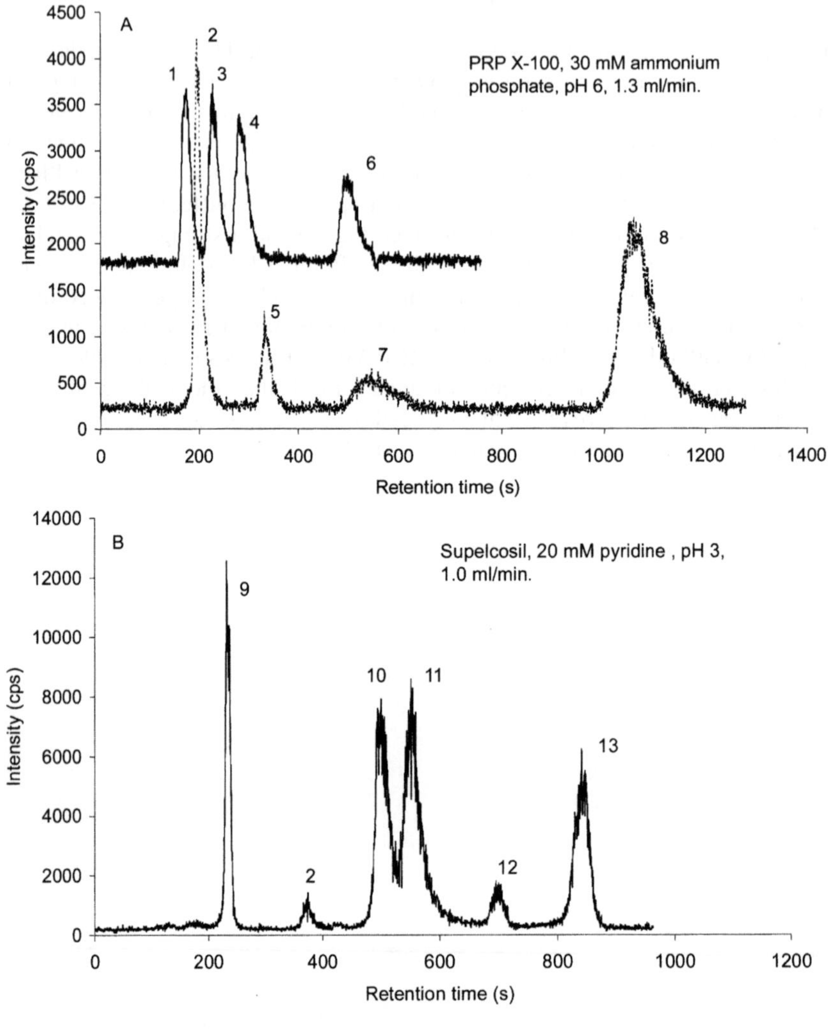

Fig. 6. LC/ICP-MS chromatograms for 13 different arsenic species: A: anion exchange and B: cation exchange. 1: As(III), 2: Sugar-OH, 3: DMA(V), 4:MMA(V), 5: Sugar-PO_4, 6: As(V), 7: Sugar-SO_3, 8: Sugar-SO_4, 9: AsB, 10: DMAE, 11: TMAO, 12: AsC, 13: TMA^+.

Examples of projects utilizing liquid chromatography

- Degradation of arsenosugars under anaerobic marine conditions
- Degradation of arsenosugars in soil
- Metabolism of arsenosugars by sheep and humans
- Arsenic uptake and transformation in rice

In the following paragraphs interesting analytical aspects of arsenic speciation studies using LC methods are discussed.

Identification of arsenic compounds by R_t comparison

Arsenosugars are the major arsenic species in marine macroalgae (seaweed). When they decompose in the marine environment, it has often been described that anaerobic degradation of the arsenosugars will generate DMAE and eventually AsB (Edmonds and Francesconi 1982). This has never been proven, and is increasingly doubtful. Although DMAE has been identified in microcosm experiments as the major degradation product in the sediment besides large amounts of DMA(V), no AsB has ever been determined; neither in microcosms nor in the natural environment. The arsenosugars rather breakdown to give inorganic arsenic and interestingly in the deep marine sediment in which a high level of free sulfide is generated, thioarsenate was identified by comparison of the retention times of a synthetic thioarsenate and the signal of the sample. This can be seen in Fig. 7. This arsenic species has however, only been found in sediments of a microcosm and not in natural environment, although many studies exist which determined the arsenic species in sediments. One reason might be that mostly hydride generation methods have been employed for arsenic speciation in the marine environment and this species is one of the hidden arsenic species using this technique. This illustrates the importance of using liquid chromatography when the study is not only focused on a few target species. This technique also allows the determination of unexpected arsenic species as long as they are soluble, elutable and do not undergo degradation during the chromatographic run. If standard species are available, arsenic species in samples can be identified by comparing the retention time of the species in the sample with that of a standard, if there are no other co-eluting arsenic species.

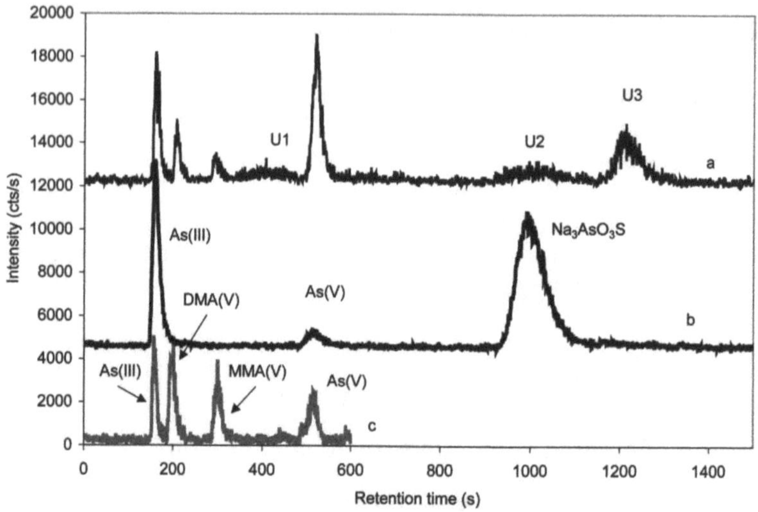

Fig. 7. Anion exchange separation (HPLC/ICP-MS, m/z 75 trace is shown) of a pore water sample from 42 cm depth of a sediment microcosm in which seaweed was degraded (a): anaerobically, (b): thioarsenate, arsenite and arsenate standards and (c): inorganic and methylated arsenic standards (Pengprecha et al. 2002).

Identification with ICP-MS and ESI-MS

The assignment of a peak to a specific compound based on a single retention time is not unequivocal. Therefore, good research practice is to use not only ICP-MS but also ESI-MS for the identification of "new" arsenicals. This approach is even more appropriate when no standard is available. The detection of the unknown arsenic-containing metabolites is only possible when an element-specific detector such as ICP-MS is employed for online coupling to the LC system. The sole use of ESI-MS for arsenic metabolism studies of unknown samples is practically impossible since it is difficult to pin-point arsenic-containing compounds from a mixture of many organic species. Although the use of different fragmentor voltages and low oxygen content in the curtain gas (Kuehnelt et al. 2002) results in the molecules getting stripped down to naked As^+, it cannot be excluded that other organic fragments with m/z 75 would not interfere with the As^+, and this in turn could lead to wrong assignments of peaks. Only a combination of ESI-MS and ICP-MS can reveal the occurrence of an arsenic-containing species and its molecular mass. Fig. 8 shows the use of both, ICP-MS and ESI-MS for the identification of arsenosugars in seaweed. ICP-MS identifies the metal and metalloid, whereas ESI-MS gives molecular information ($M+H^+$, and molecular fragments).

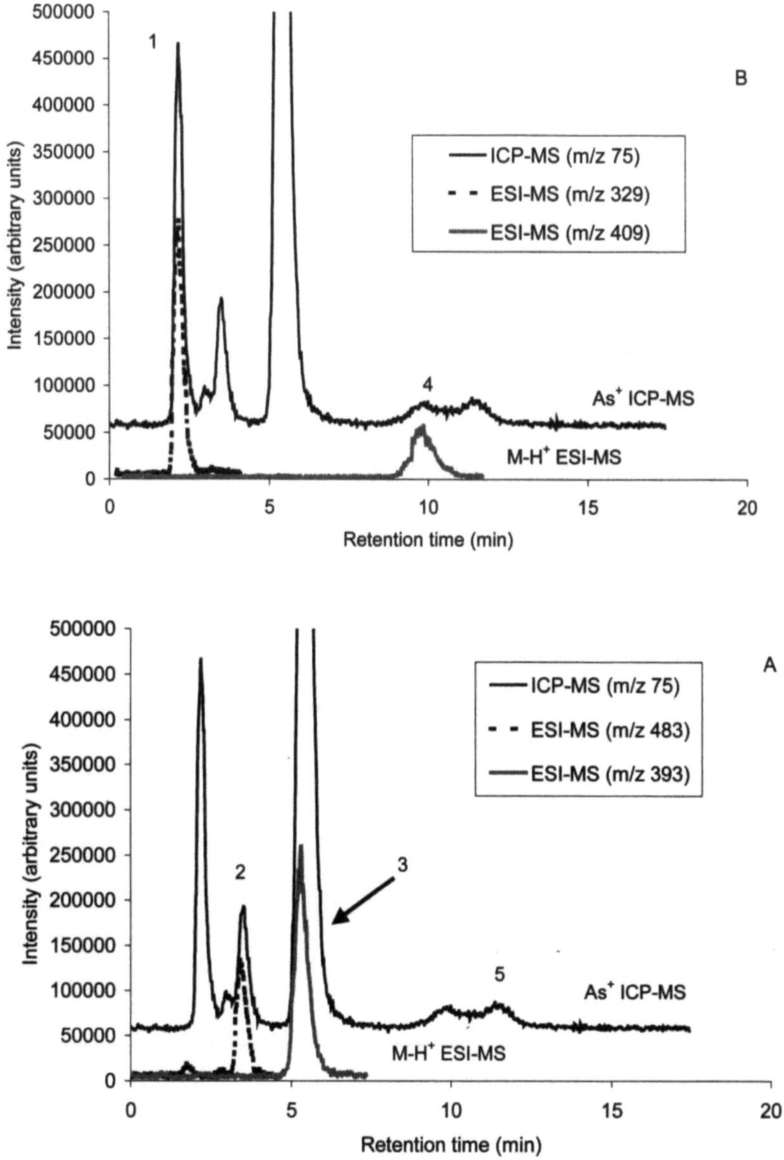

Fig. 8. Combination of soft and hard ionization by using anion exchange chromatography with carbonate buffer as mobile phase parallel ESI-MS and ICP-MS for the identification of arsenic species in *Laminaria digitata*. 1: Me_2As- Sugar-OH (m/z 329), 2: Me_2As-Sugar-PO_4 (m/z 483), 3: Me_2As-Sugar-SO_3 (m/z 393), 4: Me_2As-Sugar-SO_4 (m/z 409), 5: arsenate. The monitored m/z are the M+H masses of the molecules.

Identification of chromatographic loss

Although it has been acknowledged that LC methods can detect most of the species listed in Fig. 1 and Table 1, it should be noted that there is increasing evidence that the low chromatographic recovery rate of arsenic species in biological extracts or body fluids may be the result of protein-arsenic coordination compounds, which are larger in size and tend to "stick" to the column and can therefore be described as hidden species when LC is employed. It can be seen from Table 4, that similar samples may show different recoveries; while the human hair sample (GBW 9101) shows a quantitative recovery, the sheep wool from Orkney shows a poor chromatographic recovery of about 30 % (Raab et al. 2002). These compounds are probably the most difficult hidden arsenic-species to identify, but the difference between the total arsenic in the extract and the sum of the species eluted give a quantitative measure of the chromatographic loss. The addition of CuCl increased the chromatographic recovery in particular of the methylated species (Raab et al. 2003). This treatment has often been used to cleave the labile arsenic-protein coordination.

Table 4. Arsenic species concentration and chromatographic recovery of arsenic species in hair and wool extract using anion exchange chromatography (Raab et al. 2002).

n=3	CRM GBW 9101 human hair extract (certified arsenic, 0.59±0.07 μg g^{-1})	North Ronaldsay sheep wool extract (total arsenic: 14.7±2.4 μg g^{-1})
As(III)	0.14 ± 0.05 μg g^{-1}	0.23 ± 0.06 μg g^{-1}
As(V)	0.27 ± 0.05 μg g^{-1}	0.35 ± 0.04 μg g^{-1}
DMA(III)	< 0.07 μg g^{-1}	0.89 ± 0.10 μg g^{-1}
DMA(V)	0.17 ± 0.04 μg g^{-1}	2.89 ± 0.10 μg g^{-1}
MMA(V)	< 0.07 μg g^{-1}	0.12 ± 0.02 μg g^{-1}
SUM	0.58 ± 0.12 μg g^{-1}	4.49 ± 0.19 μg g^{-1}
Chromatographic Recovery	98.3[1] %	29.8 %

[1] total recovery

Extraction efficiency and species integrity

When the arsenic species in foodstuff is of interest, it is necessary to extract arsenic quantitatively, in particular for risk assessment purposes. Conventional extraction methods like methanol/water with variable ratios from 10:1 to pure water, fail often to give quantitative extractions. Only 10 % of the total arsenic has been extracted from rice shoots using this method (Abedin et al. 2002). More vigorous methods like the use of trifluoro-acetic acid (TFA) were successfully used for the extraction of arsenic from the rice shoots. If such strong methods are necessary to be employed, it is inevitable to test the species stability during the extraction process. For instance, the TFA extraction method shows a reduction of 25% arsenate to arsenite, where as the methylated pentavalent arsenic species [DMA(V) and MMA(V)] are stable (see Fig. 9). However, it should be tested how stable other arsenic species are during this treatment

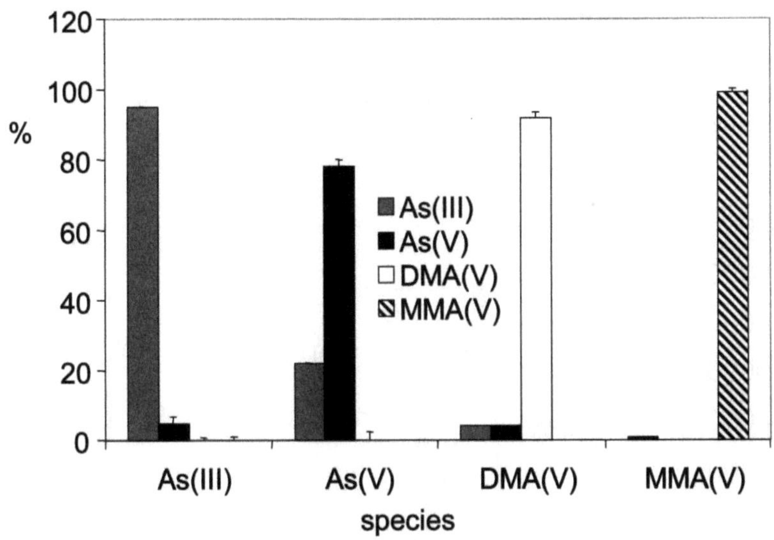

Fig. 9. Recovery of spiked arsenic species and the species transformation during the extraction of rice shoots using TFA (Abedin et al. 2002).

Determination of non-water soluble arsenic species

When seaweed is extracted, often only 70-90% can be extracted by the gentle methanol/water extraction. But some of the arsenic species are not water-soluble. These can be extracted by using chloroform/methanol solvent mixture, which is a common method of extraction of phospholipids and sphingomyelins. Once in the organic phase ESI-MS can be used, but due to the difficulties in the identification of arsenic-containing compounds unequivocally, it is necessary to use an element-specific detector as well. However, the use of organic mobile phases is not possible on a routine basis with ICP-MS since the high load of organic solvent leads to the formation of elemental carbon on the sampler cone of the MS interface. To overcome this obstacle one strategy is to cleave the lipid part from the water-soluble part in order to transfer the water-soluble part into an aqueous media. Phospolipase D enzyme has been used to cleave complex lipids at a certain bond (P - X bond to yield phosphatidic acid and the arsenic species; see Fig. 10). The arsenic-containing water-soluble X group can be determined by conventional LC methods for arsenic speciation (see Fig. 6). For instance it has been established that *L. digitata* contains arsenic in the form of arsenolipids with X group as DMA(V) and arsenosugars-OH, according to the chromatogram shown in Fig. 10 (Devalla et al. 2002). The lipid part has to be determined by LC/ESI-MS or after derivatisation using GC/MS.

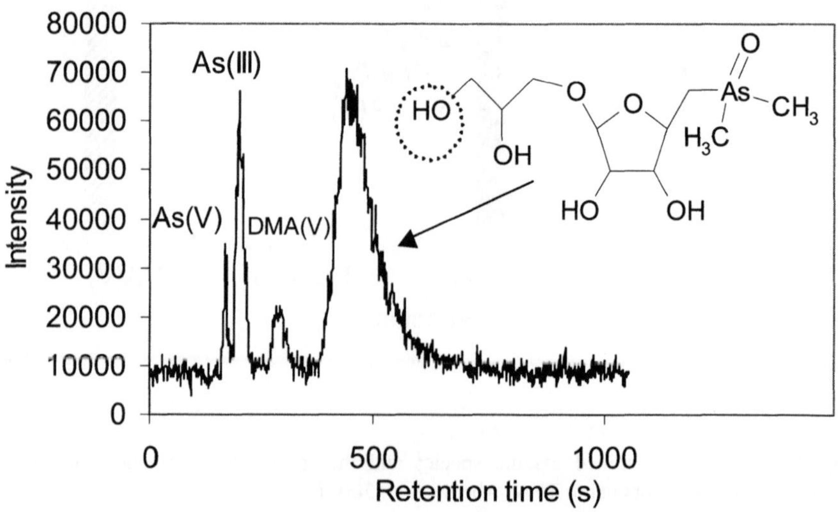

Fig. 10 Analysis of arsenolipids using LC/ICP-MS after methanol/chloroform extraction and hydrolisation with Phospholipase D of seaweed (*L. digitata*). The shown species are bound to a phospholipids as indicated in Fig 1

Determination of labile arsenic compounds in biota

The stability of labile arsenic compounds is emphasized in the following paragraph. Arsenic-coordination complexes such as arsenic glutathione species ($AsGS_3$, $MeAsGS_2$, Me_2AsGS) can easily be separated on a reverse phase column using acetonitrile/formic acid (Kala et al. 2000) but not under the conventional conditions. Fig. 11 shows a chromatogram on a conventional strong anion exchange column (PRP X-100) in which all three arsenic-glutathione complexes fall apart and elute as their degradation products (As(III), DMA(V) and MMA(V)).

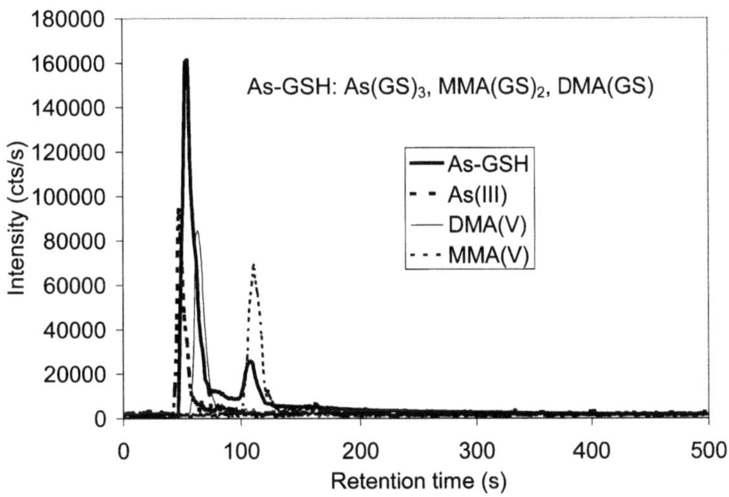

Fig. 11. The decomposition of an aqueous solution containing three arsenic-glutathione complexes ($AsGS_3$, $MeAs\text{-}GS_2$, $Me_2As\text{-}GS$) on a strong anion exchange column (PRP X-100, with 30 mM ammonium phosphate at pH 6).

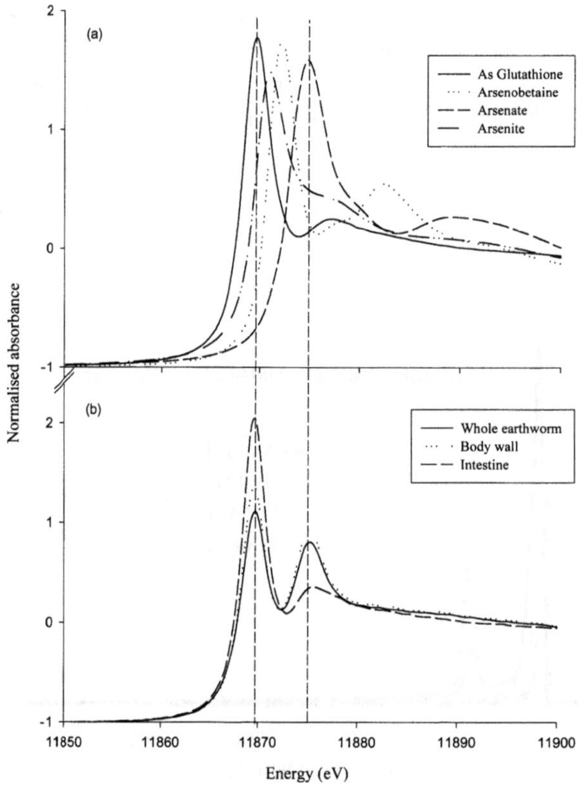

Fig. 12. XANES spectra of arsenic standards (a) and arsenic-resistant earthworms (*Lumbricus rubellus*) (taken from Langdon et al. 2002)

Arsenic-sulphur complexes are expected to occur in those organisms which accumulate large quantities of arsenic, where phytochelatin and/or metallothionein might bind to the metalloid. Earthworms (*L. rubidus)* collected from contaminated soils contained large amounts of arsenic in the form of As(III), and As(V) and some AsB (identified by conventional methanol/water extraction followed by strong anion exchange and cation exchange chromatography-ICP-MS and ESI-MS). The question arises as to why would the earthworm store the arsenic in one of the more toxic forms as As(III) at such a high concentrations. One possibility is that the As(III) is in fact a co-ordination complex in the cellular environment which decomposed during extraction and separation similar to those of $AsGS_3$ complexes. How can that be proved? The only way is to perform arsenic speciation directly on the intake organism. This is possible by using X-ray absorption spectroscopy. Figure 12 shows the XANES-spectra of a variety of arsenic standards (a) and that of earthworm compartments (b). It becomes obvious that the energy of the absorption edge is slightly shifted to higher energies if the oxidation number is increased. Trivalent species do absorb at lower energy. The absorption edge however is further shifted to lower energy when the arsenic environment is electron richer, which is the case when it is coordinated with sulfur instead of oxygen. The earthworm shows in addition to the arsenate absorption peak also an absorption peak at the same energy as the synthesized $AsGS_3$ compound. This difference between chromatographic identified species and XANES identified species is clear evidence that the arsenic species are altered when a conventional extraction/separation approach is used.

Conclusions

Arsenic speciation in biological and environmental samples is anything but routine analysis. This is indicated in the small number of reference material available with certified concentration of arsenic species. Any attempt so far to produce a standard reference material for arsenosugars failed and these are compounds, which are considered as relatively stable during extraction and separation procedures. The real challenge today is to identify unknown arsenic-containing compounds detected by LC/ICP-MS. For this purpose other techniques such as ESI-MS or MALDI-MS are used to identify the molecular mass. QTOF-MS and FT-ICR-MS may bring a new dimension in the field of arsenic speciation, due to their accurate mass measurements. It is expected that traditional biochemical methods such as gel electrophoresis will find more use in this field, especially for the determination of arsenic-protein complexes. In addition to this new fleet of conventional techniques, X-ray absorption methods such as XANES or EXAFS can be utilized for the detection of very labile arsenic species directly in the unchanged organism. It is just emerging that besides the wealth of organoarsenic species, many labile coordination species may occur, which might play a key role in the complex biochemistry of arsenic.

Acknowledgements

The authors thank the Department of Chemistry of the University of Aberdeen for a studentship (HRH), and the Faculty of Science for the RA support (AR), the Leverhulme Trust (F/00/152/A), the Royal Society (21685), EPSRC (GR/M 10755, GR/M91853) and the Sheep Court of North Ronaldsay (Scotland) for funding and permission.

Literature

Abedin MJ, Cresser M, Meharg AA, Feldmann J, Cotter-Howells J (2002) Arsenic accumulation and metabolism in rice (Oryza sativa L.). Environ Sci Technol 36:962-968

Bleeker PM, Schat H, Vooijs R, Verkleij JAC, Ernst WHO (2003) Mechanisms of arsenate tolerance in Cytisus striatus. New Phytol 157:33-38

Bartram ME (2001) GC/MS analyses of chemical vapor deposition precursors. Anal Chem 73:534A-539A

Challenger FE (1945) Biological methylation. Chem Rev 36:315-361

Chen J, Zhou LM, Qu GL (2001) Derivatization by toluol-3,4-dithiol for arsenic speciation and gas chromatographic analysis. Chin J Anal Chem 29:1276-1279

Chen ZL, Lin JM, Naidu R (2003) Separation of arsenic species by capillary electrophoresis with sample-stacking techniques. Anal Bioanal Chem 375:679-684

Clamagirand V, Marr IL, Wardell JL (1993) Ethylation of methylarsenic(III) compounds by sodium tetraethylborate. Appl Organmet Chem 7:577-581

Cullen WR, Reimer KJ (1989) Arsenic speciation in the environment. Chem Rev 89:713-764

Dean JR, Ebdon L, Foulkes, ME, Crews, HM, Massey RC (1994) Determination of the growth promoter, 4-hydroxy-3-nitrophenyl-arsonic acid in chicken tissue by coupled high-performance liquid-chromatography inductively-coupled plasma-mass spectrometry. J Anal At Spectrom 9:615-618

Devalla S, Raab A, Feldmann J (2002) unpublished data

Drugov YS (1998) Gas chromatography of inorganic substances. J Anal At Spectrom 53:606-620

Edmonds JS, Francesconi KA, Hansen JA (1982) Dimethyloxarsylethanol from anaerobic decomposition of brown kelp Ecklonia radiata: a likely precursor of arsenobetaine in marine fauna. Experientia 38:643-644

Feldmann J, Koch I, Cullen WR (1998) Complementary use of capillary gas chromatography-mass spectrometry (ion trap) and gas chromatography-inductively coupled plasma mass spectrometry for the speciation of volatile antimony, tin and bismuth compounds in landfill and fermentation gases. Analyst 123:815-820

Feldmann J, Naëls L, Haas K (2001) Cryotrapping of CO_2-rich atmospheres for the analysis of volatile metal compounds using capillary GC-ICP-MS. J Anal At Spectrom 16:1040-1043

Francesconi K, Khokiattiwong S, Goessler W, Pedersen SN, Pavkov, M (2000) A new arsenobetaine from marine organisms identified by liquid chromatography-mass spectrometry. Chem Comm 12:1083-1084

Francesconi KA, Tanggaard R, McKenzie CJ, Goessler W (2001) Arsenic metabolites in human urine after ingestion of an arsenosugar. Clin Chem 48:92-101

Geiszinger A, Goessler W, Kosmus (2002) Organoarsenic compounds in plants and soil on top of an ore vein. Appl Organmet Chem 16:245-249

Grüter UM, Kresimon J Hirner AV (2000) A new HG/LT-GC/ICP-MS multi-element speciation technique for real samples in different matrices. Fresenius J Anal Chem 368:67-72

Haas K, Feldmann J (2000) Sampling of trace volatile metal(loid) compounds in ambient air using polymer bags: A convenient method. Anal Chem 72:4205-4211

Haas K, Feldmann J (2002) unpublished data

Hanaoka K, Goessler W, Yoshida K, Fujitaka Y, Kaise T, Irgolic KJ (1999) Arsenocholine- and dimethylated arsenic-containing lipids in starspotted shark Mustelus manazo. Appl Organomet Chem 13:765-770

Hirner AV, Feldmann J, Krupp E, Grumping R, Goguel R, Cullen WR (1998) Metal(loid)organic compounds in geothermal gases and waters. Org Geochem 29:1765-1778

Jackson BP, Bertsch PM, Cabrera ML, Camberato JJ, Seaman JC, Wood CW (2003) Trace element speciation in poultry litter. J Environ Qual 32:535-540

Jenkins RO, Craig PJ, Miller DP, Stoop LCAM, Ostah N, Morris TA (1998) Antimony biomethylation by mixed cultures of micro-organisms under anaerobic conditions. Appl Organomet Chem 12:449-455

Kala SV, Neely MW, Kala G, Prater CI, Atwood DW, Rice JS, Lieberman MW (2000) The MRP2/cMOAT transporter and arsenic-glutathione complex formation are required for biliary excretion of arsenic. J Biol Chem 275:33404-33408

Kaise T, Horiguchi Y, Fukui S (1992a) Acute toxicity and metabolism of arsenocholine in mice. Appl Organomet Chem 6:369-373

Kaise T, Fukui S (1992b) The chemical form and acute toxicity of arsenic compounds in marine organisms. Appl Organomet Chem 6:155-160

Kaise T, Oya-Ohta Y, Ochi T, Okubo T, Hanaoka K, Irgolic KJ, Sakurai T, Matsubara C (1996) Toxicological study of organic arsenic compound in marine algae using mammalian cell culture technique. J Food Hyg Soc Jpn 37:135-141

Hansen HR, Raab A, Francesconi KA, Feldmann J (2003) Metabolism of arsenic by sheep chronically exposed to arsenosugars as a normal part of their diet. 1. Quantitative intake, uptake, and excretion. Environ Sci Technol 37:845-851

Koch I, Feldmann J, Lintschinger J, Serves SV, Cullen WR, Reimer KJ (1998) Demethylation of trimethylantimony species in aqueous solution during analysis by hydride generation gas chromatography with AAS and ICP MS detection. Appl Organmet Chem 12:129-136

Koellensperger G, Nurmi J, Hann S, Stingeder G, Fitz WJ, Wenzel WW (2002) CE-ICP-SFMS and HPIC-ICP-SFMS for arsenic speciation in soil solution and soil water extracts. J Anal At Spectrom 17:1042-1047

Kuehnelt D, Goessler W, Francesconi KA (2002) Workshop Arsenic Speciation September 2002, Gent, Belgium

Langdon C, Meharg AA, Feldmann J, Balger T, Charnock J, Farquhar M, Piearce T, Semple K, Cotter-Howells J (2002) Arsenic-speciation in arsenate-resistant and nonresistant populations of the earthworm, Lumbricus rubellus. J Environ Monit 4:603-608

Mester Z, Sturgeon RE (2002) Detection of volatile organometal chloride species in model atmosphere above seawater and sediment. Environ Sci Technol 36:1198-1201

Miguens-Rodriguez M, Pickford R, Thomas-Oates JE, Pergantis SA (2002) Arsenosugar identification in seaweed extracts using high-performance liquid chromatography/electrospray ion trap mass spectrometry. Rapid Comm Mass Spectrom 16:323-331

Pickford R, Miguens-Rodriguez M, Afzaal S, Speir P, Pergantis SA, Thomas-Oates JE (2002) Application of the high mass accuracy capabilities of FT-ICR-MS and Q-ToF-MS to the characterisation of arsenic compounds in complex biological matrices. J Anal At Spectrom 17:173-176

Prange A, Schaumloffel D, Bratter P, Richarz AN (2001) Species analysis of metallothionein isoforms in human brain cytosols by use of capillary electrophoresis hyphenated to inductively coupled plasma-sector field mass spectrometry. Fresenius J Anal Chem 371:764-774

Pengprecha P (2002) PhD thesis, University of Aberdeen, Aberdeen, UK

Pengprecha P, Raab A, Wilson M, Feldmann J (2003) Biodegradation of arsenosugars in marine sediment. Appl Organmet Chem (submitted)

Petrick JS, Jagadish B, Mash EA, Aposhian, HV (2001) Monomethylarsonous acid (MMA(III)) and arsenite: LD50 in hamsters and in vitro inhibition of pyruvate dehydrogenase. Chem Res Toxicol 14:651-656

Quaghebeur M, Rengel Z, Smirk M (2003) Arsenic speciation in terrestrial plant material using microwave-assisted extraction, ion chromatography and inductively coupled plasma mass spectrometry. J Anal At Spectrom 18:124-234

Raab A, Hansen HR, Zhuang L, Feldmann J (2002) Arsenic accumulation and speciation analysis in wool from sheep exposed to arsenosugars. Talanta 58:67-76

Raab A, Genney, DR, Meharg AA, Feldmann J (2003) Identification of arsenic species in sheep-wool extracts by different chromatographic methods. Appl Organomet Chem 17:684-692

Schwedt G, Rieckhoff M (1996) Separation of thio- and oxothioarsenates by capillary zone electrophoresis and ion chromatography. J Chromatogr A 736:341-350

Tatken RL, Lewis RJ (eds) (1983) Registry of toxic effects chemical substances. US Department of Health and Human Services, Cincinnati OH.

Toyama M, Yamashita M, Hirayama N, Murooka Y (2002) Interactions of arsenic with human metallothionein-2. J Biochem 132:217-221

Wooten JV, Ashley DL, Calafat AM (2002) Quantitation of 2-chlorovinylarsonous acid in human urine by automated solid-phase microextraction-gas chromatography-mass spectrometry. J Chromatogr B 772:147-153

Zakharyan RA, Sampayo-Reyes A, Healy SM, Tsaprailis G, Board PG, Liebler DC, Aposhian HV (2001) Human monomethylarsonic acid (MMA(V)) reductase is a member of the glutathione-S-transferase superfamily. Chem Res Toxicol 14:1051-1057

Chapter 4

Occurrence and speciation of arsenic, antimony and tin in specimens used for environmental biomonitoring of limnic ecosystems

H. Emons, Z. Sebesvari, K. Falk, M. Krachler

Environmental biomonitoring

At present most of the regular environmental monitoring is focused on the observation of local emission sources and the operation of systems for air and water monitoring in several countries. But modern environmental observation has to provide more effect-related information about the state of our environment and its changes with time. Therefore, it can not only be based on the analysis of abiotic environmental samples such as air, water, sediment, or soil. Rather environmental studies and control have to take into account much more the situation of the biosphere. This includes the transfer of contaminants, mainly of anthropogenic origin, into plants, animals, and finally also into human beings.

Therefore, biomonitoring plays an increasing role in modern environmental observation programs (Phillips and Rainbow 1993; Emons et al. 1997). Here, selected biological organisms, called bioindicators, are used for the monitoring of pollutants either by observation of effects (e.g. phenomenological effects such as loss of needles or discoloring of leaves) or by measurement of taken-up chemical compounds by the specimens. The latter approach is based on the chemical analysis of appropriate bioindicators which accumulate the pollutant during exposure and offer the possibility of integrating the pollutant burden over time (as well as local integration with animals). In the following we will only consider aspects of such compound-oriented biomonitoring.

Obviously, effect- and assessment-related biomonitoring can not be based only on the determination of total element concentrations in environmental samples. Speciation is without any doubt indispensable because important properties of chemical compounds such as transport and distribution behavior, bioavailability, chemical reactivity as well as toxicity (in all of its modes) depend on their oxidation state, binding form, and binding partners.

Specimens which have been used for environmental biomonitoring until now represent a very broad variety with respect to matrix composition. For active monitoring different plant or animal species, mainly grass cultures or freshwater mussels, are exposed at selected locations for a limited time. Passive monitoring is performed with the help of environmentally representative plants as well as animals which are naturally present in the ecosystem of interest. Typical examples for specimens are algae, mussel tissue, fish muscles, needles or leaves of trees, bird eggs, lichen, etc.

For a scientifically sound description of the environmental situation well composed sets of biomonitoring specimens have to be selected in dependence on the ecosystem of interest and available biological populations. From an analytical point of view such samples are very complex and different regarding concentration levels of trace elements, matrix influence on sample preparation and analyte determination as well as chemical stability. In the past the criterion of accumulation of pollutants by the respective organism played an important role for its application. But the tremendous progress in analytical chemistry with respect to the determination of many compounds at the trace and even ultratrace level has decreased the importance of this aspect. Nowadays, the amount of knowledge about biological properties and ecological functions of the available specimens appears to be the limiting factor and decides about its utility.

In previous years most of the studies and applications of environmental biomonitoring have been performed in coastal marine and terrestrial ecosystems. Obviously there is a lack of corresponding knowledge for river and lake areas. But the further understanding and control of freshwater ecosystems is not only crucial for the supply of drinking water for human consumption. The proper functioning of limnic ecosystems is of utmost importance for the whole biosphere. Therefore, this contribution gives a brief overview about current aspects and future challenges regarding environmental biomonitoring of three selected metalloids and their chemical species in limnic ecosystems with specific examples studied in the authors' laboratory.

Occurrence of As, Sb and Sn in freshwater ecosystems

Metals and metalloids occur naturally in the freshwater environment as a result of rock weathering, soil runoff and atmospheric deposition. Elevated levels are mostly caused by anthropogenic activities or as a result of specific geological conditions. Transport of metals can occur relatively fast in aquatic ecosystems. Aquatic sediments can act as a sink or a source for contaminants. Long-term input of contaminants can lead to very high sediment concentrations. Therefore, organisms living on (epifaunal) or living in (infaunal) the sediment are in close contact and interaction to a medium which can be highly polluted.

Metal exposure can cause molecular responses (e.g. changes in stress protein level), individual responses (e.g. reduction in growth or abnormal development), and responses at the community level such as decrease in the number of taxa and total abundance, decrease in the abundance of metal-sensitive taxa and increase in the relative abundance of metal-tolerant taxa. Exposure to metals results often also in their accumulation in the organisms. Freshwater organisms in contaminated environment can be exposed to metals in surface and interstitial water, sediment and food. Therefore, metal accumulation occurs *via* uptake across the body surface and *via* food. The degree of accumulation depends on several variables including concentration and bioavailability of the metal species, surface size and feeding group of the organisms and their ability to metabolise the metal compounds.

One has to take into account that metals occur also in limnic ecosystems in a large variety of chemical species which are differerent in their bioavailability and toxicity. Therefore, measuring the distribution and concentration of such species in several compartments of the freshwater ecosystem including biota, as well as investigating the ecological community structure, provides information on the effect of metal exposure upon freshwater organisms.

Arsenic

The arsenic level in fresh water is often below 2 µg/l, which is comparable to that in seawater (Cullen and Reimer 1989). Elevated levels of arsenic content in freshwater result mostly from anthropogenic activities like ore-milling practices, coal mining or the use of pesticides in agriculture. But geogenic sources can also contribute to high As concentrations in groundwater as seen in parts of Asia (Matschullat 2000). Arsenic undergoes redox reactions and methylation in natural waters and can be taken up by aquatic organisms. Arsenate is the most stable form in oxic aquatic environments, while arsenite and methylated arsenic species are the major compounds in eutrophic regions (Cullen and Reimer 1989). The distribution of methylated arsenic species seems to be correlated with the biological activity in the water column (Hasegawa et al. 2001). However, there are significant differences in the concentrations and chemical forms of arsenic between limnic and marine organisms. Limnic organisms contain usually less than a few µg As/g dry mass (d.m.), whereas the arsenic content of marine organisms ranges from several micrograms to more than 100 µg/g. For example, Lai et al. (1997) reported arsenic species in marine algae in the 8-49 µg/g (d.m.) range whereas only 3 µg As/g (d.m.) was found in the freshwater alga *Nostoc. sp.*.

Probably due to the generally smaller arsenic concentrations in freshwater organisms only few information is available about corresponding arsenic species. Even total arsenic contents are rarely reported compared with the large amount of data in the literature for marine ecosystems.

Maeda et al. (1992a, 1992b) have reported that arsenic is combined with proteins in the living cells of the fresh water alga *Chlorella vulgaris*. They also showed that inorganic arsenic was transformed in the alga to monomethylarsonous acid, dimethylarsinic acid and trimethylated compounds. Lai et al. (1997) have detected the presence of an arsenosugar in the fresh water alga *Nostoc sp.*. Shiomi et al. (1995) investigated two fish species from limnic ecosystems and have identified arsenobetaine (AB) as the major As species (60-80% of total arsenic content). Arsenobetaine found in cultured rainbow trout (*Salmo gairdneri*) seems to be caused from the commercial feed containing arsenobetaine as the major arsenical. Studies of the Japanese smelt (*Hypomesus nipponensis*) suggested that AB is a naturally occurring compound in the freshwater environment. Gomez-Ariza et al. (2000) have also found arsenobetaine in the fresh water alga *Zyngogomium sp.*.

Falk (1999) has detected 75-120 ng As/g in the muscles of bream (*Abramis brama*) from the rivers Elbe, Rhine and Saar (Germany). This concentration range is about a factor of 300 lower than the arsenic content of the seawater fish eelpout (*Zoarces viviparus*) (3 µg As/g).

Burger et al. (2002) reported concentrations of arsenic in the muscles of 11 fish species from the Savannah River (USA). They found that fishes from higher trophic levels showed generally higher levels of arsenic but bottom-dwelling fishes could sometimes have higher levels than carnivores. From 1984 to 1985 the U.S. Fish and Wildlife Service collected 315 composite samples of whole fish from 109 stations in major rivers throughout the USA and in the Great Lakes. The geometric mean concentration was 0.14 µg As/g fresh mass (f.m.) with maximum values of 1.5 µg/g. This study reported also a decline of the mean As concentration between 1976, the year of the first analysis, and 1984 (Schmitt and Brumbaugh 1990).

Antimony

Antimony is ubiquitously present in the environment from natural and anthropogenic sources. The metalloid is used in various industrial products such as component of lead alloys, additive in glassware, ceramics and plastics and finds increasing application as fire retardant. Some organic antimony compounds are used in medicine as antiparasitic drugs. But there is presently only little information available on the distribution, speciation, and bioaccumulation of antimony in freshwater ecosystems.

Antimony occurs naturally in the freshwater environment as a result of rock weathering and soil runoff. Typical concentrations of total dissolved antimony in unpolluted waters range from a few ng/l to a few µg/l. Antimony was found nearly exclusively in the dissolved fraction of the water and not in the particulate fraction. Usual concentrations in the sediment are in the order of a few µg/g, higher concentrations are related to anthropogenic sources like mining and smelting (Filella et al. 2002).

Whereas nowadays total concentrations of antimony can be readily determined in environmental matrices, the speciation of this metalloid in biological samples represents still an analytical problem (Krachler et al. 2001). Also organic Sb species in freshwater have been rarely analyzed and a recent review of Filella et al. (2002) cited only one corresponding reference (Andreae et al. 1981). It was found in some rivers that the methylated Sb species usually account for less than 10% of the total dissolved antimony. Moreover, the further evaluation of the analytical procedure suggested that some chemical species were, in fact, methodical artefacts (Dodd et al. 1992). Krupp et al. (1996) have detected mono-, di-, and trimethyl- as well as triethyl antimony in river sediments.

Mann et al. (1988) found no evidence for the bioconcentration of antimony in aquatic algae. Dodd et al. (1996) reported for the first time the presence of organoantimony compounds in an extract from the freshwater plant pondweed (*Potamogetan pectinatus*) growing in Canadian lakes which were influenced by mine effluents. Maeda et al. (1997) showed that the accumulation of antimony in living cells of the freshwater alga *Chlorella vulgaris* was associated to proteins. They have found evidence that higher toxic Sb(III) was converted to less toxic Sb(V) by the living alga.

Tin

At present organotin compounds are among the most intensively studied organometallic species because of their widespread use and hazardous environmental effects even at very low concentration. Organotins are mainly used as catalysts to stabilize polymers like PVC, as wood preservatives against fungal damage and as biocides for agricultural applications and in antifouling paints. Tri-n-butyltin (TBT), di-n-butyltin (DBT), triphenyltin (TPT), and diphenyltin (DPT) are the most frequently applied species. These compounds can be converted in the aquatic environment to other forms *via* several chemical and biological reactions. The half-life of TBT is up to 4 months in freshwater and up to 4-5 months in freshwater sediments (Maguire and Tkacz 1985). While inorganic forms of tin have relatively low toxicity for aquatic organisms, the more lipophilic organotins can be very bioavailable and highly hazardous. Generally, trisubstituted organotins are more toxic to aquatic organisms than di- and monosubstituted compounds (Gadd 2000). At large, organotins with short alkyl chains degrade slowly in the aquatic environment (Crompton 1998). Due to the widespread use of large amounts of these compounds have entered aquatic ecosystems. Up to now most attention has been directed to TBT with an emphasis on pollution in marine ecosystems because this species is the principal biocidal ingredient of many antifouling paints. But TBT pollution in freshwater harbors or in rivers with active shipping activity have been also reported (Kalbfus et al. 1991, Becker et al. 1992, Ansari et al. 1998). In addition, agrarian and industrial emissions play an important role for limnic ecosystems. Therefore, it is possible to detect hot spots of organotins in rivers or lakes with local pollutant sources like an organotin producing company.

The input of tributyltin compounds in the aquatic environment has resulted in a wide range of adverse effects to several groups of organisms such as molluscs, crustaceans, micro-algae, bacteria, fish, marine mammals, etc. Low-level TBT exposure to non-target aquatic organisms may cause sublethal responses like reduction in growth or abnormal development like imposex, the development of male characteristics in females (Huggett 1992). But there is a taxonomically correlated sensitivity to TBT. Molluscs react partially sensitive to very low levels of TBT (10-50 ng/l). Because of their high sensitivity to these chemicals molluscs are often used as indicators of TBT pollution. Generally, the larvae of any tested species were more sensitive to tributyltin exposure than the adults (Laughlin et al. 1996).

Although TBT is often the most toxic species of the butyltins, monobutyltin (MBT) shows for some aquatic microorganisms a comparable or sometimes even higher toxicity (Cooney 1995). Thus, debutylation in the living organism *via* TBT $\rightarrow$ DBT $\rightarrow$ MBT $\rightarrow$ Sn does not reduce toxic effects for all organisms. Some microorganisms can methylate inorganic or organic tin under aerobic or anaerobic conditions (Cooney 1995). Huang et al. (1993) reported that TBT was metabolised by the limnic alga *Scenedesmus obliquus* to DBT. Methylation and debutylation alter the adsorptivity and solubility of tin compounds. Thus, microorganisms can influence the environmental mobility of tin.

Yang et al. (2001) showed that concentrations of TBT in caged mussels (*Elliptio complanata*) were related to the mussel size and the concentrations of TBT in contaminated aquatic areas, and that concentrations in mussel tissue increased dramatically over winter. Organotin compounds were also determined in muscles and liver from bream. The concentration of tributyltin (TBT) was 202 ± 17 ng/g (fresh mass = f.m.) in muscles and 128 ± 9 ng/g (f.m.) in liver. The concentration of total organotin compounds in water samples along the River Elbe up to the Elbe estuary was in the range of 30-96 ng/l (Shawky and Emons 1998). Kannan et al. (1997) described a total butyltin concentration of up to 2 μg/g (f.m.) in Ganges river dolphin. The biomagnification factor for butyltins in the river dolphin from its food ranged between 0.2-7.5. One should note that this paper describes one of the rare investigations which include several levels of the food chain. The majority of TBT toxicity studies has been directed to the effect of dissolved-phase TBT on epifaunal organisms. There are only a few investigations of the hazardous effects of sediment-bound TBT on infaunal organisms (Austen and McEvoy 1997).

Speciation analysis of As, Sb and Sn in limnic samples

An increasing number of approaches for the trace analysis of metalloid species has been reported in the literature (Caruso et al. 2000 ; Ebdon et al. 2001; Cornelis et al. 2003). Therefore, the following description serves as an example and is based on our own developments.

Analytical chemistry is regarded as a problem-solving science and one has to consider the specific aspects of the questions of interest from the beginning of the design for a total analytical process (Figure 1). The main purpose of environmental biomonitoring is to obtain information about the original species pattern in bioindicators. Therefore, it is not sufficient to take into account only the operations in the analytical laboratory. Rather one has to design the complete process starting with sampling of such delicate material up to an assessment of results.

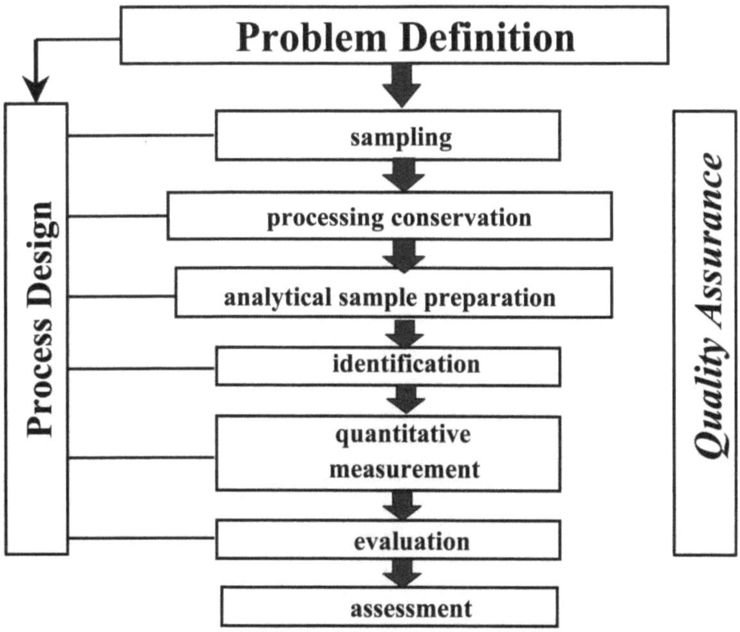

Fig. 1. General steps of the total analytical process

Species transformation and loss can already occur during sampling at the sampling site. Degradation depends on the chemical nature of the species, may be influenced by enzyme activity and is usually more critical in animal organs than for plant samples. A key parameter for reaction rates is the temperature. Therefore, we have diminished species transformations by decreasing the temperature as low and early as possible. The samples were collected, dissected and immediately shock-frozen above liquid nitrogen at the place of collection as start of a cryochain which has been continued until extraction. After grinding of the deep-frozen material in a special vibrating cryomill (cooled with liquid nitrogen) where the samples had only contact to titanium or Teflon™ surfaces the resulting fine powders were stored in cryocontainers at temperatures below -150 °C above liquid nitrogen until analysis.

The majority of samples collected for biomonitoring has to be treated by mechanical operations such as dissection of organs, grinding, mixing, and aliquotation. Most of this sample processing can be performed at very low temperatures. Working near liquid nitrogen temperature offers also the opportunity to grind and homogenate fresh biological material with very different mechanical properties resulting for instance from varying water and fat contents. We have generally processed our samples to a fine powder with particle sizes below 200 μm. One should avoid freeze-drying because of possible species transformations (e.g. demethylation as reported by Shawky et al. (1996)) and loss of volatile species.

Presently, one of the most critical steps from the aspect of retaining the species information is the analytical sample preparation. The necessary transfer of the species of interest into liquid and/or gas phases introduces a major uncertainty about unwanted changes of the speciation pattern. Most of the current preparation protocols include a step for the extraction of the metal(loid) species from the biomatrices. Our As speciation procedure is based on the extraction of the finely ground sample powder with methanol/water mixtures under conditions optimized from an approach reported by Kuehnelt et al. (1997). Depending on the As concentration of the matrix 1-3 g of the fresh sample was weighed into polyethylene-centrifuge tube and 10–25 ml of a methanol-water mixture (9:1 v/v) was added. The tubes were shaken for approximately 17 h. The extracts were centrifuged at 17000 rpm for 15 min and the clear supernatants were transfered into round bottomed flasks. The extraction residues were washed three times with methanol-water (9:1 v/v). The resulting liquid extracts were combined and the methanol was evaporated on a rotary evaporator at temperatures below 35 °C. Each sample was filled up to 10 ml with water. Before injection the sample was filtered through a 0.22 μm cellulose filter and diluted with water to an appropriate concentration. Each matrix was extracted three times and the arsenic species in each extract were also determined three times.

The extraction of organotin species from biological matrices was also performed with fresh (i.e. non-dried) finely ground sample powder from the cryostorage. A sample in-weight of 0.3-1.0 g was treated with either 10 ml 0.5 M HCl or 0.5 M acetic acid in methanol (pH 4) at 50 °C for one hour in an ultrasonic bath. The extract was centrifuged at 3350 rpm for 15 min.

The preparation of bioindicator samples for speciation analysis of antimony represents still a major challenge. Various extraction solutions and procedures have been tested, including 0.2 M acetic acid, EDTA solutions, methanol/water as well as acetonitrile/water mixtures, and 0.66 M NaOH (Krachler and Emons 2000a). But all of the approaches have provided only Sb extraction efficiencies of less than 10 % for most of the biological materials and therefore, further studies are underway.

This points to thc fact that onc has always to control the extraction yield on the basis of the amount of extracted metal atoms from the biological sample. The scheme for such mass balance studies is shown in Figure 2.

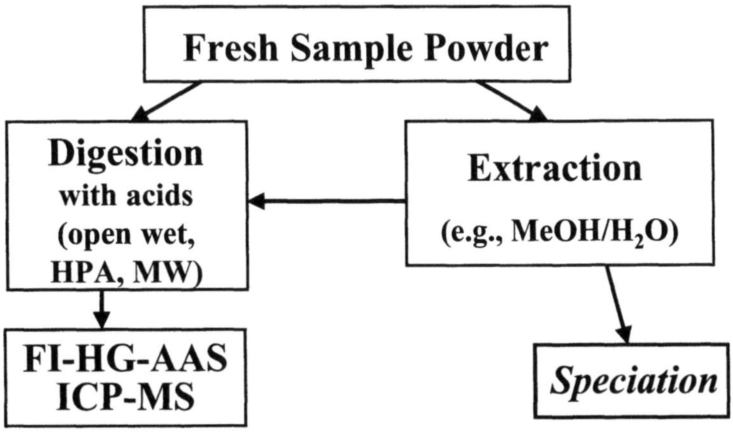

Fig. 2. Scheme of mass balance studies for extraction

It is based on powerful and reliable analytical procedures for the determination of the total As, Sn, and Sb content, respectively, in the bioindicator samples. Details of corresponding methods which detect the metalloids with the help of hydride generation-atomic absorption spectrometry (HG-AAS) or inductively coupled plasma-mass spectrometry (ICP-MS) and in case of Sn also with radiochemical neutron activation analysis (RNAA) have been published elsewhere (Shawky and Emons 1998; Krachler et al. 1999a and 1999b; Krachler and Emons 2000a). We were able to obtain extraction yields for As species from fresh biological matrices above 95 % and excellent recoveries (always within the range of uncertainty of the certificates, i.e. about 100 %) of organotin species from fish muscles reference material. But one has to take into account that the preservation of the original molecular structure or the stoichiometry and completeness of chemical reactions for derivatisation (e.g. alkylation) can not be controlled at the moment.

Figure 3 shows the flow scheme used here for the speciation analysis of the three metalloids of interest. Species separation is achieved by chromatographic techniques based on dissolved or gaseous compounds. The on-line coupling with powerful detection techniques such as ICP-MS or AAS offers sufficient quantification capability also for biomonitoring purposes if the dilution factor during the whole analytical procedure is minimized.

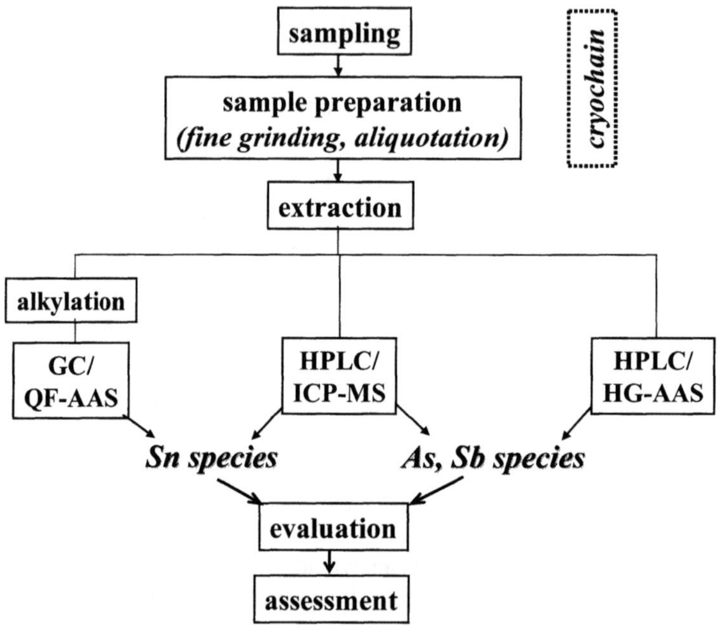

Fig. 3. Applied procedures for speciation analysis of As, Sb, and Sn

Arsenic species

The arsenic species As(III), As(V), monomethylarsonic acid (MMA), dimethylarsinic acid (DMA), arsenobetaine (AB), arsenocholine (AC), tetramethylarsonium ion (Tetra) and trimethylarsine oxide (TMAO) occur as neutral, positively or negatively charged compounds depending on the pH. Therefore, ion exchange chromatography appears to be a suitable method for the determination of these compounds. Own developments of chromatographic methods by using anion-exchange and cation-exchange chromatography for the separation of these eight natural occurring arsenic species were presented in detail elsewhere (Falk and Emons 2000). In addition the analysis of another group of arsenic species the so-called arsenosugars (methylated As-ribosides) was included because of their occurence in aqueous ecosystems. Reversed-phase (RP) chromatography on a column Inertsil ODS was used to avoid co-elution of arsenosugars and other As species in ion-exchange chromatography. The original method described by Shibata and Morita (1992) had to be modified and extracts of fresh biomonitoring samples were exposed separately to two mobile phases containing tetraethylammonium hydroxide (TEAH) and the more sterically discriminating ion-pair reagent tetrabutylammonium hydroxide (TBAH), respectively. Naturally occuring arsenosugars from algae could be well separated by these methods (Falk 1999).

Finally the four different HPLC conditions summarized in Table 1 were employed in the HPLC/ICP-MS approach. All mobile phases were degassed with helium before use. The separation was performed at 25 °C and at a flow rate of 1 ml min^{-1}. The injection volume was 20 µl.

Table 1. Chromatographic conditions for the analysis of As species by HPLC/ICP-MS (indices: A... anion-exchange, C... cation-exchange, E... tetraethylammonium, B... tetrabutylammonium)

	column	mobile phase
E_A	Ion-120 (anion-exchange)	30 mmol l^{-1} NH_4CO_3, 2 % (v/v) MeOH, pH 10.0 (adjusted with NH_3.aq)
E_C	Nucleosil 5SA (cation-exchange)	30 mmol l^{-1} pyridine, 2 % (v/v) MeOH, pH 3.0 (adjusted with HCOOH)
RP_E	Inertsil ODS (reversed-phase)	10 mM tetraethylammonium hydroxide (TEAH), 4.5 mM malonic acid, 0.1 % (v/v) MeOH, pH 6.0 (adjusted with HNO_3)
RP_B	Inertsil ODS (reversed-phase)	10 mM tetrabutylammonium hydroxide (TBAH), 4.5 mM malonic acid, 0.1 % (v/v) MeOH, pH 4.0 (adjusted with HNO_3)

The detection of arsenic with HG-AAS was mainly used in connection with anion-exchange chromatography. In that case 100 µl of extracted sample were injected into a mobile phase of 8 mM disodiumhydrogenphosphate and 8 mM sodiumdihydrogenphosphate (pH 6; 1.4 ml/min) and separated on a column PRP-X100 (Hamilton) according to Begerow et al. (2001). The limits of detection (LOD) which could be achieved by HPLC/HG-AAS and HPLC/ICP-MS, respectively, are summarized in Table 2.

Table 2. Limits of detection (in ng(As) l^{-1}) for As speciation with HPLC/HG-AAS and HPLC-ICP-MS

Species	HPLC/ICP-MS		HPLC/HG-AAS
	Cross-flow nebulizer	HHPN	
As(III)	1600	470	100
As(V)	550	34	90
MMA(V)	800	330	500
DMA(V)	550	300	600
AB	650	36	
AC	650	53	
TMAO	350	79	
Tetra	900	100	

It has been shown for the HPLC/ICP-MS coupling that the use of efficient sample introduction devices such as a hydraulic high pressure nebulizer (HHPN) can decrease the limits of detection. The linear ranges were 2-500 ng(As) g^{-1} and 0.5-250 ng(As) g^{-1} for HPLC/ICP-MS with cross-flow nebulizer and HHPN, respectively. The analytical reproducibility was better than 3 % (for 20 µg l^{-1} or 2 µg l^{-1} in case of the HHPN) with the exception of As(III) with reproducibilities between 3.5-8.0 %. But in real sample solutions memory effects and species instability are often influencing the applicability of these nebulizers and one has to design the optimum set-up in dependence on the specific sample and concentration ratios.

Antimony species

Various hyphenated methods have been developed and optimized for the separation and quantification of Sb species from different matrices (Krachler et al. 2001). It turned out that HG-AAS as well as ICP-MS offer powerful detection after HPLC separation. Systematic investigations (Krachler and Emons 2000b) have revealed that in case of HPLC/HG-AAS the anion exchange column Dionex AS14 provided best results for the separation of Sb(V) and Sb(III) with 1.25 mM EDTA at pH 4.7. The ION-120 column was the best to separate TMSb and Sb(V) with a mobile phase of 2 mM NH_4HCO_3 and 1 mM tartaric acid at pH 8.5 (flow rate 1.5 ml/min, injection volume 100 µl). Also the hydride generation parameters had to be optimized for this coupling of techniques. Finally a $NaBH_4$ solution concentration of 0.6% (w/v), stabilised with 0.04% (w/v) NaOH, a $NaBH_4$ solution flow rate of 4 ml min^{-1}, a HCl solution concentration of 0.5% (w/v), a HCl solution flow rate of 6 ml min^{-1} and a carrier gas flow rate of 50 ml min^{-1} argon were selected.

The coupling of HPLC with ICP-MS was optimized for two different sample introduction systems (Krachler and Emons 2001a, 2001b). Samples containing Sb species in dominantly aqueous solutions (either original or after extraction) can be directly aspirated after their chromatographic separation (as above) into the plasma of an ICP-MS using ultrasonic nebulization (USN) with membrane desolvation. For the nebulizer U-6000AT$^+$ (Cetac Technologies) the following optimum conditions were found: sweep gas (Ar) with 2.35 l min^{-1}, heating temperature 80 °C, desolvation temperature 80 °C, and cooling temperature 2 °C. Typical operating conditions for the ICP-MS (Elan 5000, PerkinElmer, Shelton, CT) are listed in Table 3.

Table 3. Experimental parameters for Sb measurements with the ICP-MS Elan 5000

Forward Power	1200 W
Cones	Nickel
Plasma gas	15.0 l min^{-1}
Nebulizer gas	~ 1.0 l min^{-1}, daily optimised to obtain maximum $^{121}Sb(V)$ signal intensity
Auxiliary gas	~ 1.0 l min^{-1} daily optimized to obtain maximum $^{121}Sb(V)$ signal intensity
Data acquisition	^{121}Sb, peak hoping mode, replicate time 1 s, dwell time 20 ms, 50 sweeps/reading, 1 reading/replicate

For samples with more complex composition, especially with higher content of various ions such as sodium or chloride, the USN was replaced by a HG-system (Krachler and Emons 2001b). Identical parameters ($NaBH_4$ and HCl concentrations, flow rates, gas pressure), as optimized for HPLC/HG-AAS, proved to give best results also for HPLC/HG-ICP-MS. Sb(V) and Sb(III) were separated on a PRP-X100 column (Hamilton) using 20 mM EDTA at pH 4.7 as mobile phase. $TMSbCl_2$ cannot be eluted under the aforementioned chromatographic conditions, as alkaline pH values are required for that purpose. But the ION-120 anion exchange column provided a good separation between $TMSbCl_2$ and Sb(V) using 2 mM NH_4HCO_3 and 1 mM tartaric acid at pH 8.5 as mobile phase within less than 3 min. Corresponding limits of detection for the various methods of Sb speciation are summarized in Table 4. The linear dynamic ranges are extended to at least 100 µg l^{-1} for HG-AAS and 1000 ng l^{-1} for ICP-MS. Reproducibilities of 3 % or better have been achieved at the 100 ng l^{-1} level with ICP-MS detection.

Table 4. Limits of detection for speciation analysis of antimony achieved with different detection methods coupled on-line to HPLC

Species	LOD HG-AAS ng (Sb) l^{-1}	LOD USN-ICP-MS ng (Sb) l^{-1}	LOD HG-ICP-MS ng (Sb) l^{-1}
Sb(III)	700	14	8
Sb(V)	1000	12	20
TMSb	400	9	12

Tin species

The dissolved Sn species obtained by extraction of bioindicator or sediment samples were analyzed with the help of a home-made on-line coupling of gas chromatography and quartz furnace-atomic absorption spectrometry (GC/AAS). The system has been already described in detail elsewhere (Shawky et al. 1996) and is based on the volatization of organotin compounds by ethylation in an aqueous medium containing 70 mM sodium tetraethylborate. The resulting volatile species were purged from the solution with helium into a chromatographic glass column (3 mm inner diameter, 95 cm length) packed with 3 % SP 2100 on 60-80 mesh chromosorb G AW DMCS (Supelco, Bellefonte, PA). The column was immersed in liquid nitrogen as cold trap. After collection of the ethylated Sn species the column was heated to 200 °C in 4 min to release the compounds into the quartz furnace. The detection was performed with a PerkinElmer 4000 AAS equipped with an electrodeless discharge lamp at 224.6 nm. It was possible to analyse monomethyl-, dimethyl- and trimethyltin (MMT, DMT, TMT) as well as monobutyl-, dibutyl- and tributyltin (MBT, DBT, TBT) species with this procedure. Limits of detection for the whole analytical procedures were found to be 2.7 ng(Sn) g^{-1} (fresh mass) for MMT, 1.5 ng g^{-1} for DMT, 3.5 ng g^{-1} for TMT, 2.6 ng g^{-1} for MBT, 2.5 ng g^{-1} for DBT, and 4.2 ng g^{-1} for TBT with reproducibilities between 2 % (20 ng l^{-1} standard solutions) and 7.5 % (TBT in fish muscles).

Quality assurance

The determination of metalloid species in environmental samples is still a relatively young area of analytical chemistry. Its further acceptance both in the analytical and in the environmental science communities depends strongly on the reliability of the applied methods and the relevance of the resulting environmental information. Therefore, appropriate measures of analytical quality assurance (AQA) are of utmost importance. Specific challenges for AQA within the speciation analysis of metalloids result in particular from two facts: the often limited stability of the analytes of interest outside their natural microenvironment and the necessity to use so-called hyphenated instrumental techniques, i.e. delicate on-line couplings of various separation and detection methods for ultratrace analysis (Emons 2002). At present method validation in trace element speciation is mostly limited to characterize analytical figures of merit for the method by measuring solutions of species standards and to study the repeatability and reproducibility of the procedures with real-world samples. Unfortunately, the problem of trueness control is seldomly tackled because really independent methods are mostly not available for the majority of the present speciation analyses. This points to the fact that certified reference materials (CRMs) have to play here an even more important role than in trace element analysis for analytical quality assurance during method development and the following permanent quality control in the application laboratories. Unfortunately, their limited availability for environmental speciation limits partially the further progress of the field (Emons 2001). We have used both suitable CRMs and independent analytical approaches as much as possible for AQA in our laboratories as described briefly with the following examples.

The accuracy of the determination of the total As concentration was proved by measurements of three marine reference materials with HG-AAS. All results show a good agreement to the certified value (CRM DORM-2 Dogfish Muscle: 17.5 ± 0.4 mg kg^{-1} / *18.0 ± 1.1 mg* kg^{-1} (measured / *certified*); CRM BCR 278 Mussel Tissue: 6.01 ± 0.11 mg kg^{-1} / *5.9 ± 0.2 mg* kg^{-1}, CRM BCR 279 Sea Lettuce: 3.02 ± 0.11 mg kg^{-1} / *3.09 ± 0.2 mg* kg^{-1}). To verify the results for the total As concentration in the real samples two different analytical methods were applied. The results obtained by HG-AAS and ICP-MS were statistically identical.

To check the procedure for the determination of arsenic species by HPLC/ICP-MS the reference material CRM BCR 627 Tuna Fish with certified values for AB and DMA was investigated. Furthermore the determined concentrations of arsenic compounds in the CRM DORM-2 were compared to values in the literature (Goessler et al. 1998; Corr 1997). The obtained results for both materials were in good agreement with the certified and published values, respectively, and have been presented in detail elsewhere (Falk and Emons 2000).

The analytical procedures developed for the ultratrace analysis of antimony in plant and animal samples (Krachler et al. 1999a, 1999b) allow limits of detection of 20 pg g^{-1} and 7 pg g^{-1}, respectively, for the two groups of biological materials. Analytical quality assurance has been supported by the use of the CRMs BCR 281 Rye Grass, GBW 07602 Bush Branches and Leaves, GBW 07604 Poplar Leaves and GBW 07605 Tea Leaves. The found Sb concentrations were always within the certified ranges for Sb in these CRMs. At present no reference material exists for the speciation of antimony. Therefore, the developed methods could only be checked by intermethod comparison between the HPLC/HG-AAS and the approaches. The analysis of standard mixtures with both methods provided statistically identical results for the species Sb(III), Sb(V) and TMSb (Krachler et al. 2002).

The total analytical procedure for organotin species has been validated with the help of the CRMs NIES-11 Fish and PACS-1 Harbor Sediment. The determined concentrations of TBT in the fish material (found: 1.26 ± 0.05 µg g^{-1}; certified: 1.3 ± 0.1 µg g^{-1}) and of butyltins in the sediment (MBT: 243 ± 81 ng g^{-1} (found) and 280 ± 170 ng g^{-1} (certified); DBT: 1170 ± 27 ng g^{-1} (found) and 1160 ± 180 ng g^{-1} (certified); TBT: 1182 ± 43 ng g^{-1} (found) and 1270 ± 220 ng g^{-1} (certified)) reveal the applicability of the Sn speciation method described above (Shawky et al. 1996).

One should always take into account that the integration of appropriate methods for quality assurance is an important prerequisite for the useful application of analytical procedures in environmental speciation and should not be underestimated.

Speciation of As, Sb, Sn in limnic bioindicators – first results

As mentioned above the majority of speciation studies with bioindicators has been performed for marine ecosystems. But we have also investigated some of the samples obtained in the frame of systematic biomonitoring programs (Emons et al. 1997) of the German rivers Elbe, Rhine and Saar with respect to As and Sn species. Corresponding findings are summarized in the following.

Arsenic

Bream (*Abramis brama*) are fishes which live close to the sediment phase within a very restricted area of many European freshwater ecosystems. After collection and processing of breams the resulting finely ground powder which represented always homogenates from organs of 20-30 individual specimens, was analyzed according to Fig. 3. Fresh material of bream muscles was extracted with methanol/water and investigated afterwards by HPLC/ICP-MS. The hydraulic high-pressure nebulizer (HHPN) had to be used for introduction of the eluted As species into the ICP because of the very low As content of all samples. Depending on the species pattern of the sample either all four or just some of the chromatographic methods summarized in Table 1 were employed. Very small HPLC peaks, which contributed just insignificant percentages to the distribution pattern, were not quantified. Significant but unknown peaks whose retention times did not match with any of the As species discussed in Section 3 were determined by external calibration with known species and labelled as P2 or P?. It is worth to note that the arsenic species P2 was also found in common mussel tissue from the German Wadden Sea (Falk 1999). The results shown in Fig. 4 demonstrate that both arsenobetaine (AB) and P2 were detected in all bream samples.

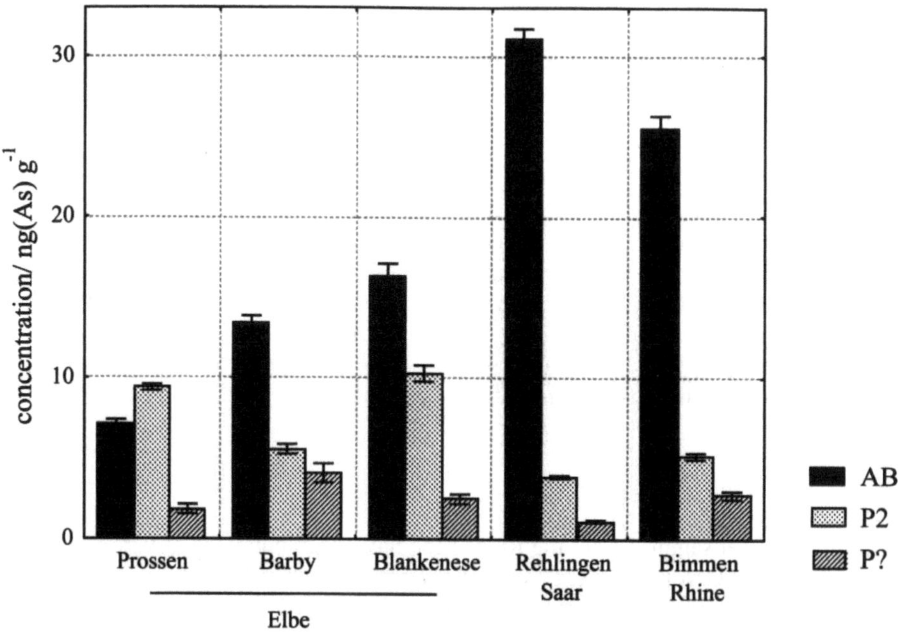

Fig. 4. Arsenic species in bream muscles from the rivers Elbe, Saar and Rhine; total As concentrations: 75.6 ng g^{-1} f.m. (Prossen), 93.7 ng g^{-1} f.m. (Barby), 119.0 ng g^{-1} f.m. (Blankenese), 87.6 ng g^{-1} f.m. (Rehlingen), 85.1 ng g^{-1} f.m. (Bimmen)

But different concentration ratios occur in dependence on the sampling site. AB was the main As compound in breams from the rivers Saar and Rhine, whereas the concentration ratio AB:P2 was much smaller for samples from the River Elbe. Breams from Prossen located at the Czech/German border had even more P2 than AB in their muscles. This is remarkable because this arsenic species P2 was not found in marine fishes, even not in eelpout (*Zoarces viviparus*) living in the Wadden Sea near the Elbe estuary. On the contrary AB was clearly the dominating As species (about 80 %) in eelpout muscles (Falk 1999).

For the evaluation of mass balances the sum of the determined arsenic species in each sample was compared to its total As concentration in the fresh material (75-120 ng g^{-1} f.m.). While for other fish samples such as eelpout muscles and the CRM DORM-2 most of the arsenic (> 90 %) could be quantified in the extracts by HPLC/ICP-MS, for bream samples only recoveries of about 25 % (Elbe) and about 40 % (Saar, Rhine) could be achieved. Therefore, it can be concluded that the bream muscles contain unknown arsenic species which are strongly retained on the HPLC columns either directly or *via* other molecules. For instance, hydrophobic interactions of arsenic lipids or proteins with the stationary phase of the column could be responsible for the lower recoveries. Both the much lower total

As content in bream (300 times smaller than in eelpout) and the very different As species patterns support the assumption that the limnic bioindicator reflects not only the smaller As concentration in its aquatic environment and the corresponding food chain but points also to differences of species accumulation and transformation in limnic ecosystems in comparison to the marine environment.

Antimony

Because of the present problems concerning the extractability of Sb species from biological material (see above) a further differentiation of the antimony compounds in bioindicators is not feasible. Moreover first biomonitoring studies in various German ecosystems have revealed that total Sb concentrations in limnic bioindicators are extremely low (e.g. 4 ng g^{-1} in bream liver from the River Saar) in contrast to terrestrial bioindicators from the same region (Krachler et al. 1999c).

Tin

Unfiltered water samples from the River Elbe have been collected at various sites from the Czech/German border (Prossen) downstream to Cuxhaven where actually a mixture of river water and North Sea water (German Wadden Sea) was taken (Shawky and Emons 1998). Speciation analysis with respect to organotin compounds resulted in organotin patterns along the river as shown in Fig. 5.

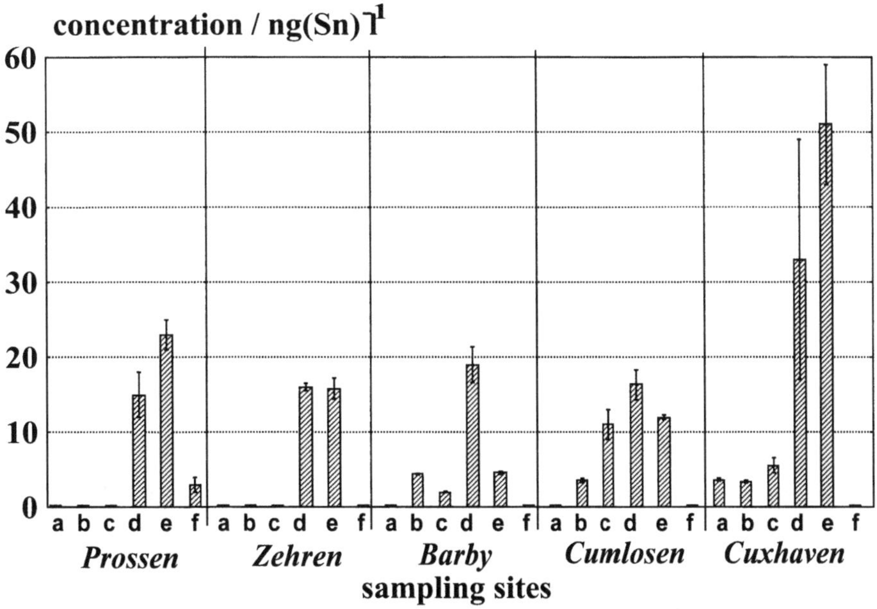

Fig. 5. Organotin species in water/suspended matter from the River Elbe at different sampling sites; Sn species: (a) MMT, (b) DMT, (c) TMT, (d) MBT, (e) DBT, (f) TBT

The most critical species TBT could only be determined at the Czech/German border which indicated immission sources in the Czech Republic. The detectable butyltin species along the River Elbe were attributed to industrial discharges from organotin production (mainly tetrabutyl tin) and their degradation products as well as from antifouling paints on river vessels. The methylation of Sn(II) and Sn(IV) ions by microorganism activity may explain the presence of methyltin species between Barby and Cuxhaven.

The accumulation of organotin species in the aquatic food chain is of major concern for environmental and human health. It turned out that bream organs can serve as a very sensitive bioindicator for butyltin compounds (Shawky and Emons 1998). No methyltin species were detectable in the bream samples in contrast to corresponding eelpout samples from the German Wadden Sea (Shawky et al. 1996). The organotin content in bream muscles from the River Elbe at Hamburg-Blankenese was higher than in bream liver (290 ng(Sn) g^{-1} and 223 ng(Sn) g^{-1}, respectively, all related to fresh mass). The corresponding species distributions are shown in Fig. 6.

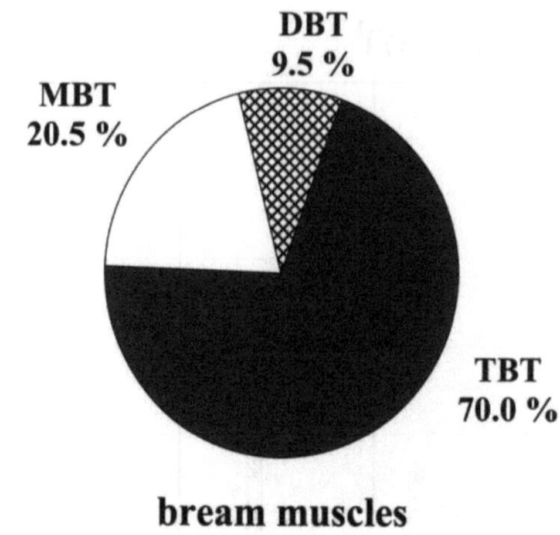

Fig. 6a. Distribution of organotin species in bream muscles from the River Elbe at Hamburg-Blankenese; total Sn concentration: 290 ng(Sn)/g (f.m.)

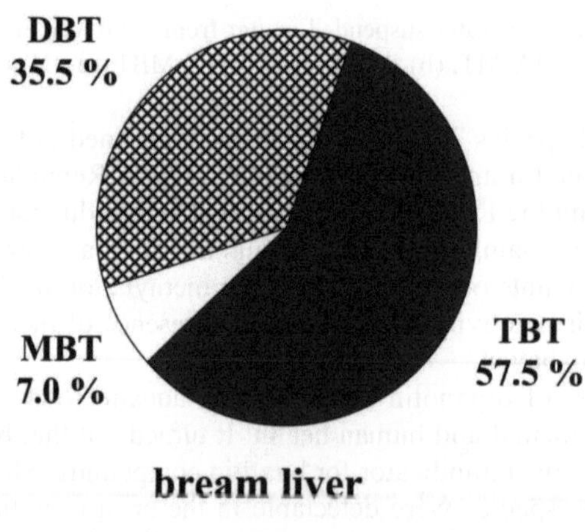

Fig. 6b. Distribution of organotin species in bream liver from the River Elbe at Hamburg-Blankenese; total Sn concentration: 223 ng(Sn)/g (f.m.)

An estimation of bioconcentration factors (BCF) on the basis of organotin species concentrations related to fresh mass of the bioindicator and the species concentrations in water/suspended matter revealed that bream muscles possessed the highest BCF for TBT (up to 47000) among all investigated biological matrices such as marine algae, common mussel tissue, eelpout muscles and liver. Bream liver had a somewhat lower BCF (about 30000) for TBT, but a higher BCF value for DBT in comparison to bream muscles (4950 and 1750, respectively). Overall bream organs have a great potential as bioindicators for butyltin compounds.

Outlook and future challenges

As shown in this brief overview the metalloids arsenic, antimony and tin are barely investigated with respect to their elemental concentration, speciation and bioaccumulation in freshwater organisms in comparison to marine specimens. Moreover only a few components of the freshwater food chain have been considered until now. Some groups of biological organisms such as higher plants or the highly diverse group of insects, from which representatives are always present, are even not mentioned in the existing literature. But there are several reasons which recommend insect larvae as attractive biomonitors. For instance, many taxa are sedentary and benthic which means that they are closely associated with the local conditions and sediments and that they are near to the base of food chains (Goodyear et al. 1999; Hare 1991, 1992).

Ecosystem-oriented speciation studies are also complicated by the fact that rarely more than one author has studied the same biological species or genus of animal. Therefore, a systematic comparison of data available in the literature is almost impossible. Furthermore, quantitative relationships between concentrations of trace metal species in freshwater organisms and their immediate environment are not well understood because of the lack of available data. Further investigations of a larger range of biological specimens at various trophic levels and metal(loid) species are needed to facilitate the understanding of bioaccumulation in freshwater organisms.

Concerning the further development of speciation analysis one has to confess that the needs of environmental biomonitoring are demanding (Emons 2002). Corresponding challenges include, for example, the separation of the analyte from the matrix, the check of species stability during the whole analytical procedure, the chemical identification of naturally occuring metal(loid) species and the availability of independent analytical speciation procedures for quality assurance. Moreover, microanalytical techniques for metalloid speciation will be necessary to investigate small biological samples as insects, specific animal organs or even plant cells.

Acknowledgment

The co-operation of the ESB Team of the Research Center Jülich and in particular the work of S. Shawky concerning the Sn speciation and the ICP-MS measurements of C. Mohl are gratefully acknowledged. Parts of these studies have been supported by the Deutsche Forschungsgemeinschaft (projects EM 51/2 and EM 51/3).

Literature

Andreae MO, Asmode JF, Foster P, Vantdack L (1981) Determination of antimony(III), antimony(V), and methylantimony species in natural waters by atomic absorption spectrometry with hydride generation. Anal Chem 53:1766-1771

Ansari AA, Singh IB, Tobschall HJ (1998) Organotin compounds in surface and pore waters of Ganga Plain in the Kanpur-Unnao industrial region, India. Sci Total Environ 223:157-166

Austen MC, McEvoy AJ (1997) Experimental effects of tributyltin (TBT) contaminated sediment on a range of meiobenthic communities. Environ Poll 96:435-444

Becker K, Merlini L, Debertrand N, Dealencastro LF, Tarradellas J (1992) Elevated levels of organotins in Lake Geneva: Bivalves as sentinel organisms. Bull Environ Contam Toxicol 48:37-44

Begerow J, Dunemann L, Sur R (2001) in Angerer J, Schaller KH (eds) Analyses of hazardous substances in biological materials. Vol. 7, Deutsche Forschungsgemeinschaft, Wiley-VCH, Weinheim, 97-117 pp

Burger J, Gaines KF, Boring CS, Stephens WL, Snodgrass J, Dixon C, McMahon M, Shukla S, Shukla T, Gochfeld M (2002) Metal levels in fish from the Savannah River: potential hazards to Fish and Other Receptors. Environ Res Section A 89:85-97

Caruso JA, Sutton KL, Ackley KL (eds) (2000) Elemental speciation. Elsevier, Amsterdam

Cooney JJ (1995) Organotin compounds and aquatic bacteria - a review. Helgolander Meeresuntersuchungen 49:663-677

Cornelis R, Caruso JA, Crews H, Heumann KG (eds) (2003) Handbook of elemental speciation. Wiley, Chichester

Corr JJ (1997) Measurement of molecular species of arsenic and tin using elemental and molecular dual mode analysis by ionspray mass spectrometry. J Anal At Spectrom 12:537-546

Crompton TR (1998) Occurrence and analysis of organometallic compounds in the environment. John Wiley & Sons, Chichester

Cullen WR, Reimer KJ (1989) Arsenic speciation in the environment. Chem Rev 89:713-764

Dodd M, Grundy SL, Reimer KJ, Cullen WR (1992) Methylated antimony (V) compounds - synthesis, hydride generation properties and implications for aquatic speciation. Appl Organomet Chem 6:207-211

Dodd M, Pergantis SA, Cullen WR, Li H, Eigendorf GK, Reimer KJ (1996) Antimony speciation in freshwater plant extracts by using hydride generation-gas chromatography-mass spectrometry. Analyst 121:223-228

Ebdon L, Pitts L, Cornelis R, Crews H, Donard OFX, Quevauviller Ph (eds) (2001) Trace element speciation for environment and health. Royal Society of Chemistry, Cambridge

Emons H, Schladot JD, Schwuger MJ (1997) Environmental specimen banking in Germany - Present state and further challenges. Chemosphere 34:1875-1888

Emons H (2001) Challenges from speciation analysis for the development of biological reference materials. Fresenius J Anal Chem 370:115-119

Emons H (2002) Artefacts and facts about metal(loid)s and their species from analytical procedures in environmental biomonitoring. Trends Anal Chem 21:401-411

Falk K (1999) Speciation von Arsen in biologischen Umweltproben aus aquatischen Ökosystemen mittels HPLC-ICP-MS. Ph D Thesis, Berichte des Forschungszentrums Jülich 3695, ISSN 0944-2952

Falk K, Emons H (2000) Speciation of arsenic compounds by ion-exchange HPLC/ICP-MS with different nebulizers. J Anal At Spectrom 15:643-649

Filella M, Belzile N, Chen YW (2002) Antimony in the environment: a review focused on natural waters I. Occurence. Earth-Science Rev 57:125-176

Gadd GM (2000) Microbial interactions with tributyltin compounds: detoxification, accumulation, and environmental fate. Sci Total Environ 258:119-127

Goessler W, Kuehnelt D, Schlagenhaufen C, Slejkovec Z, Irgolic KJ (1998) Arsenobetaine and other arsenic compounds in the National Research Council of Canada Certified Reference Materials DORM 1 and DORM 2. J Anal At Spectrom 13:183-187

Gomez-Ariza JL, Sanchez-Rodas D, Giraldez I, Morales E (2000) Comparison of biota sample pretreatments for arsenic speciation with coupled HPLC-HG-ICP-MS. Analyst 125:401-407

Goodyear KL, McNeill S (1999) Bioaccumulation of heavy metals by aquatic macroinvertebrates of different feeding guilds: a review. Sci Total Environ 229:1-19

Hare L, Tessier A, Campbell PGC (1991) Trace element distributions in aquatic insects: variations among genera, elements and lakes. Can J Fish Aquat Sci 48:1481-1491

Hare L (1992) Aquatic insects and trace metals: bioavailability, bioaccumulation and toxicology. Crit Rev Toxicol 22:327-369

Hasegawa H, Sohrin Y, Seki K, Sato M, Norisuye K, Naito K, Matsui M (2001) Biosynthesis and release of methylarsenic compounds during the growth of freshwater algae. Chemosphere 43:265-272

Huang GL, Bai ZP, Dai SG, Xie QL (1993) Accumulation and toxic effect of organometallic compounds on algae. Appl Organomet Chem 7:373-380

Huggett RJ, Unger MA, Seligman PF, Valkirs AO (1992) The marine biocide tributyltin: Assessing and managing the environmental risks. Environ Sci Technol 26:232-237

Kalbfus W, Zellner A, Frey S, Stanner E (1991) Gewässergefährdung durch organozinnhaltige Antifouling-Anstriche. UBA-Forschungsbericht 126 05 010, Berlin (Texte 44/91)

Kannan K, Senthilkumar K, Sinha RK (1997) Sources and accumulation of butyltin compounds in ganges river dolphin, *Platanista gangetica*. Appl Organomet Chem 11:223-230

Krachler M, Burow M, Emons H (1999a) Development and evaluation of an analytical procedure for the determination of antimony in plant materials by hydride generation – atomic absorption spectrometry. Analyst 124:777-782

Krachler M, Burow M, Emons H (1999b) Optimized procedure for the determination of antimony in lipid-rich environmental matrices by flow injection hydride generation atomic absorption spectrometry. Analyst 124:923-926

Krachler M, Burow M, Emons H (1999c) Biomonitoring of antimony in environmental matrices from terrestrial and limnic ecosystems. J Environ Monit 1:477-481

Krachler M, Emons H (2000a) Extraction of antimony and arsenic from fresh and freeze-dried plant samples as determined by HG-AAS. Fresenius J Anal Chem 368:702-707

Krachler M, Emons H (2000b) Potential of high performance liquid chromatography coupled to flow injection hydride generation atomic absorption spectrometry for the speciation of inorganic and organic antimony compounds. J Anal At Spectrom 15:281-285

Krachler M, Emons H, Zheng J (2001) Speciation of antimony for the 21st century: promises and pitfalls. Trends Anal Chem 20:79-90

Krachler M, Emons H (2001a) Speciation analysis of antimony by high-performance liquid chromatography – inductively coupled plasma – mass spectrometry using ultrasonic nebulization. Anal Chim Acta 429:125-133

Krachler M, Emons H (2001b) Urinary antimony speciation by HPLC-ICP-MS. J Anal At Spectrom 16:20-25

Krachler M, Falk K, Emons H (2002) HPLC/HG-AAS and HPLC/ICP-MS for speciation of arsenic and antimony in the frame of biomonitoring. Amer Lab 34:10-14

Krupp EM, Gruemping R, Furchtbar URR, Hirner AV (1996) Speciation of metals and metalloids in sediments with LTGC/ICP-MS. Fresenius J Anal Chem 354:546-549

Kuehnelt D, Goessler W, Irgolic KJ (1997) Arsenic compounds in terrestrial organisms .1. Collybia maculata, Collybia butyracea and Amanita muscaria from arsenic smelter sites in Austria. Appl Organomet Chem 11:289-296

Lai VWM, Cullen WR, Harrington CF, Reimer KJ (1997) The characterization of arsenosugars in commercially available algal products including *Nostoc* species of terrestrial origin. Appl Organomet Chem 11:797-803

Laughlin RB, Thain J, Davidson B, Valkirs AO, Newton FC (1996) Experimental study of chronic toxicity of tributyltin compounds. In: Champ MA and Seligman PF (eds) Organotin. environmental fate and effects. Chapman & Hall, London

Maeda S, Arima H, Ohki A, Naka K (1992a) The association mode of arsenic accumulated in the fresh-water alga *Chlorella vulgaris.* Appl Organomet Chem 6:393-397

Maeda S, Kusadome K, Arima H, Ohki A, Naka K (1992b) Biomethylation of arsenic and its excretion by the alga *Chlorella vulgaris.* Appl Organomet Chem 6:407-413

Maeda S, Fukuyama H, Yokoyama E, Kuroiwa T, Ohki A, Naka K (1997) Bioaccumulation of antimony by *Chlorella vulgaris* and the association mode of antimony in the cell. Appl Organomet Chem 11:393-396

Maguire RJ, Tkacz RJ (1985) Degradation of tri-n-butyltin species in water and sediment from Toronto Harbour. J Agric Food Chem 33:947-953

Mann H, Fyfe WS, Kerrich R (1988) The chemical content of algae and waters: bioconcentration. Toxic Assess 3:1-16

Matschullat J (2000) Arsenic in the geosphere – a review. Sci Total Environ 249:297-312

Phillips JH, Rainbow PS (1993) Biomonitoring of trace aquatic contaminants. Elsevier, London

Schmitt CJ, Brumbaugh WG (1990) National contaminant biomonitoring program: concentrations of arsenic, cadmium, copper, lead, selenium, and zinc in U.S. freshwater fish, 1976-1984. Arch Environ Contam Toxicol 19:731-747

Shawky S, Emons H, Dürbeck HW (1996) Speciation of organotin compounds in fish samples. Anal Commun 33:107-110

Shawky S, Emons H (1998) Distribution pattern of organotin compounds at different trophic levels of aquatic ecosystems. Chemosphere 36:523-535

Shiomi K et al. (1995) Arsenobetaine as the major arsenic compound in the muscle of two species of fresh-water fish. Appl Organomet Chem 9:105-109

Shibata Y, Morita M (1992) Characterization of organic arsenic compounds in bivalves. Appl Organomet Chem 6:343-349

Yang F, Maguire RJ, Chau YK (2001) Occurrence of butyltin compounds in freshwater mussels (Elliptio complanata) from contaminated aquatic areas in Ontario, Canada. Water Qual Res J Canada 36:805-814

Chapter 5

Methylated metal(loid) species in biological waste treatment

R.A. Diaz-Bone, B. Menzel, A. Barrenstein, A.V. Hirner

Introduction

Biomethylation of metal(loid)s by microorganisms is a process, that fundamentally changes the physico-chemical properties of elements by the addition of methyl groups. In particular, the toxicity, mobility and bioavailability are often increased. Due to bioaccumulation even small environmental concentrations can have harmful effects on the upper levels of the food chain.

Methylation in the environment has been proven for S, Cl, Ge, As, Se, Br, Sn, Sb, I, Te, Cd, Hg, Tl, Pb and Bi (Craig 1986, Hirner et al. 2000). The main prerequisites for the biomethylation of metal(loid)s are microbiological activity and bioavailable metal(loid)s. Maximal rates of biomethylation are usually found under anaerobic conditions with a redox potential between -100 to +150 mV, but biomethylation is also possible under aerobic conditions (Craig 1986, Jenkins et al. 2002).

Biomethylation can be found in both natural environments like sediments, wetlands (Wickenheiser et al. 1998) or geothermal exhalations (Hirner et al. 1998), but also in anthropogenic environments like sewage treatment plants and waste deposits (Feldmann et al. 1994, Feldmann und Hirner 1995). Grüter et al. (2000) found up to 10 µg/kg methylated arsenic and up to 20 µg/kg methylated tin species in deposited waste after hydride generation.

Although biological waste treatment offers optimal conditions for biomethylation due to the high biological activity, little investigation has been done. Krupp (1999) found volatile methylated and hydride species of Sb, Bi, As, Te, Sn and Hg (in decreasing order) in a sewage sludge fermenter. Maillefer et al. (2003) investigated the air in compost heaps from compost facilities, but besides methylated iodine, no volatile species were detected.

The scope of this study was to examine, whether methylated metal(oid)s are formed during composting processes. First, the content of methylated metal(loid) species was surveyed in compost from 34 composting facilities. In order to investigate the time dependence of organometal(loid) concentrations, compost heaps in different stages of composting were sampled from a compost facility. In addition to compost material, gases were sampled from the compost heaps and the enclosed plant by direct cryotrapping in order to check for volatile metal(oid) species, as these species represent a direct health threat to workers in composting facilities. In order to investigate whether biomethylation is restricted to industrial composting, garden compost experiments were set up and the concentrations of organometal(loid)s in different stages of composting were studied.

Biological waste treatment

Development of source separation and composting in Europe

The biological waste treatment in Europe has been continuously growing since the mid-1980´s. The advantages of the separate collection and processing of organic waste are the recycling of organic waste and the minimization of the organic fraction in waste deposits. Composting is already established in the countries of central Europe like Austria, Germany and Sweden, and fast growing in other countries of the EU and eastern Europe. Barth (2001) assumed that at least 32% of urban waste and approximately 40% of the total waste production in Europe could be biologically treated *via* composting and anaerobic digestion resulting in a theoretical recovery potential of 49 Mio tons per year.

Composting systems in Germany

In Germany 80% of the potential organic waste is processed by biological waste treatment (Barth 2001). German composting facilities distinguish between two types of source material for composting: The first type is source-separated organic household waste (3.3 Mio tons) which consist of a mixture of kitchen and garden waste. It is separately collected by communes. The second type is green waste (3.1 Mio tons). It consists typically of grasses, leaves, branches or other plant tissues collected in parks, gardens and cemeteries. Green waste often contains appreciable amounts of soil, which is collected along with the plant waste and can act as a major source for metal(loid)s.

Most German biological waste treatment facilities process waste aerobically *via* composting; less than 10% use anaerobic fermentation (ANS 2002).

A number of composting systems with different technical complexity are in use. In 1998 415 of 550 facilities used the most simple windrow system (Kern et al. 1998). The facilities can also further be distinguished by aeration of the compost piles (Fig. 1). Most facilities (314) do not support the aeration by technical means (Kern et al. 1998), but also aeration by suction and pressure or combinations of both are in use.

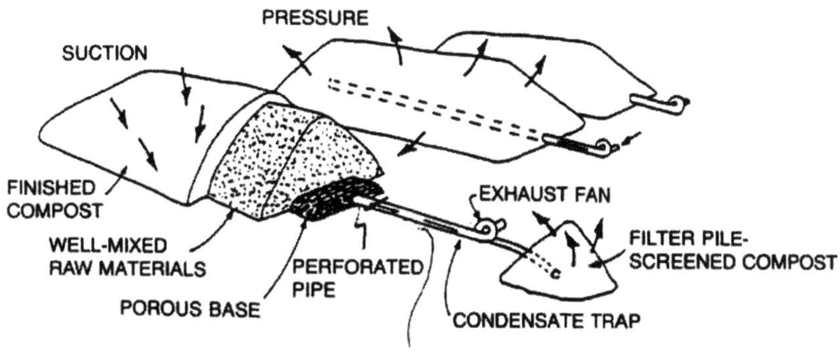

Fig. 1: Aeration types of static pile composting (Diaz et al. 1993)

Organic waste is typically composted in watered compost piles for six weeks. After this period the compost is labeled "fresh compost" and can be sold. Usually fresh compost is stored for several weeks in order to reach a higher degree of maturity.

Heavy metals quality standards for compost

The content of heavy metals in compost is regulated in most countries of the EU for the elements cadmium, chromium, copper, nickel, mercury, and zinc (Table 1); only in the USA is a limit for arsenic proposed.

Due to consequent separate collection of biowaste in Germany, compost from most composting facilities matches biowaste ordinance and contains less heavy metals than compost from home composting (Kern et al. 1998).

Table 1: Heavy metal limits in the EU and the USA (mg/kg dry mass) (EU: Barth 2001, USA: Diaz et al. 1993)

Country	Quality Standard of	Cd	Cr	Cu	Hg	Ni	Pb	Zn	As
Austria	Biowaste Ordinance Class A	1	70	150	0.7	60	120	500	-
Belgium (Fland.)	Agricultural Ministry	1.5	70	90	1	20	120	300	-
Denmark	Agricultural Ministry	0.4	-	1000	0.8	30	120	4000	-
Germany	Biowaste Ordinance Type II	1.5	100	100	1	50	150	400	-
Ireland	Draft	1.5	100	100	1	50	150	350	-
Luxembourg	Environmental Ministry	1.5	100	100	1	50	150	400	-
Netherlands	Second Class "Compost"	1	50	60	0.3	20	100	200	-
Spain (Catalonia)	Class A (draft)	2	100	100	1	60	150	400	-
Sweden	Quality assurance organization	1	100	100	1	50	100	300	-
United Kingdom	TCA Quality Label	1.5	100	200	1	50	150	400	-
USA	Proposed Public health Related Standard Class 1	5	100	300	3	50	250	500	10

Experimental

Derivatisation of ionic species

The analytical method used for the analysis of ionic species was hydride-generation purge&trap gas chromatography followed by inductively-coupled plasma mass spectrometry (HG-PT-GC/ICP-MS). Hydride generation allows the direct analysis of a large number of organic species derived from different elements with low detection limits. It can therefore be regarded as a good screening method.

Because many of the compound standards are not available, quantification is carried out by inter-element calibration based on a method developed by Feldmann (1997). Though the method is semi-quantitative with an uncertainty of +/-30%, it allows the determination of a wide range of compounds.

pH 1 hydride generation

Two different derivatisation techniques were used in this study. For the survey of 34 compost facilities the hydride generation procedure published by Grüter et al. (2000) was applied. A 1 g sample was suspended in 20 ml water. The sample was adjusted to ca. pH 1 by addition of 2 ml 3% HCl. To start the derivatisation, 2 ml of 5% $NaBH_4$ were manually added by syringe *via* the septum. The analytes were purged by a helium flow of 300 ml/min and cryotrapped on U-shaped glass tubes half-filled with 10% SP-2100 on Supelcoport (80/100 mesh), which were immersed in liquid nitrogen. Separation was carried out by heating both the trap and a GC column filled with the same material; the analytes were detected by ICP-MS (VG Plasmaquad PQ2).

pH-gradient hydride generation

Both for the detailed investigation of a compost facility and the garden compost experiments, a recently developed pH-gradient hydride generation procedure was applied (Diaz-Bone et al. 2003). Compounds such as Me_3AsO and Me_3SbCl show a derivatisation optimum near to neutral pH and high dismutation at low pH (Grüter et al. 2001, Koch et al. 1998). $MeAsO(OH)_2$ and $Me_2AsO(OH)$ on the other hand, require acidic conditions for derivatisation. In order to measure all these species in a single run, hydride generation is commenced at pH 7 (adjusted by a citrate buffer system) and gradually decreased to pH 1, whilst the derivatisation agent is added continuously.

Matrix components can lower the derivatisation efficiency by reacting with the derivatisation reagent. Therefore we increased the amount of $NaBH_4$ by a factor of four in order to lower the matrix-dependence of the derivatisation yields in real samples.

The main modifications of the apparatus in comparison to the pH 1 hydride generation described by Grüter et al. (2000) are the automated addition of both the derivatisation reagent and the acid to obtain a higher reproducibility, and a double cryotrapping system, which enables parallel cryotrapping and measurement (Fig. 2). Detection was carried out by ICP-MS (ICP-MS 7500a, Agilent Technologies).

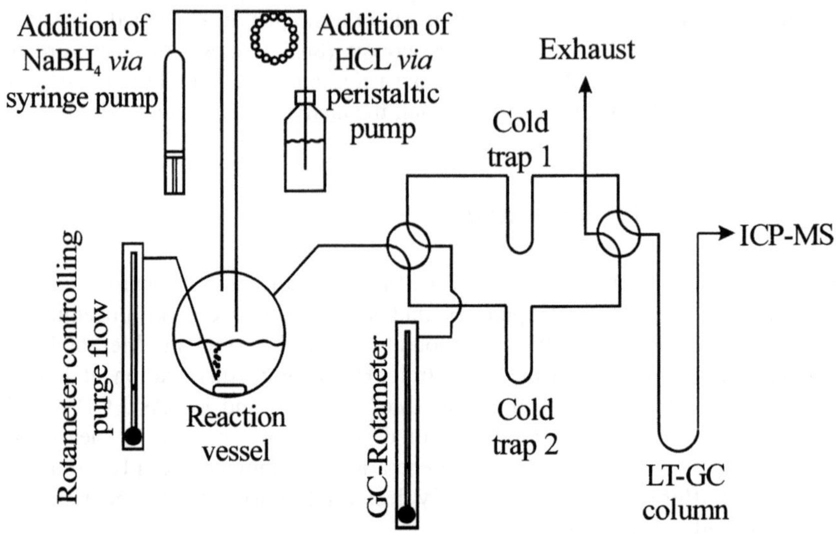

Fig. 2: Schematic of pH-gradient HG-PT-GC/ICP-MS

In comparison to the hydride generation at pH 1 the pH-gradient improves the derivatisation yields and reduces demethylation especially for antimony but also for arsenic species. A more detailed description of the pH-gradient method and a comparison to the pH 1 hydride generation procedure will be reported soon (Diaz-Bone et al. in preparation).

Sampling of gaseous species

Gaseous species were measured by cryotrapping followed by GC/ICP-MS. For sampling of gases within a compost pile a PVC-tube was inserted approximately 30 cm into the compost piles of the compost facilities or approximately 15 cm in the case of the smaller garden compost heaps. 5 l gaseous sample was cryotrapped at a flow rate of approximately 1 l/min on a trap filled with chromatographic material (Supelcoport 2100) immersed in ethanol cooled by liquid nitrogen to –80°C. The gas was sucked through the trap by an electric pump and both the suction velocity and the total volume were measured by a gas meter (Ritter Corporation). Detection was carried out by ICP-MS (ICP-MS 7500a, Agilent Technologies).

Metal analysis

For the analysis of compost aqua regia extractable metal content approximately 0.5 g sample material was digested with 6 ml HNO_3 and 2 ml HCl at a temperature of 180°C for 20 minutes in a Mars 5 microwave digester (CEM Corporation). The extracts were diluted with deionised water to 100 ml and measured by ICP-MS (ICP-MS 7500a, Agilent Technologies).

Set-up of garden composting experiments

A commercially available garden compost container with a volume of 230 l was filled with a mixture of approx. 50 kg of fresh organic urban household waste from the compost facility, which was sampled in more detail, and 5 kg self-collected green waste, mainly garden trimming. The heaps were sampled weekly at a depth of 15 cm in the hot core of the heap. After sampling the compost was watered but not turned.

Ionic species were analyzed by pH-gradient hydride generation followed by GC/ICP-MS. The measurements were carried out in triplicate. Gaseous samples were taken after one and two weeks by cryotrapping using the procedure described above.

Results and discussion

Ionic organometal(loid) species in compost from waste treatment facilities

Samples from 34 different composts of industrial composting facilities were investigated in cooperation with the State Environmental Protection Agency of North Rhine-Westphalia. At least 16 sub-samples were pooled and mixed, approximately 500 g of sample were transported in PVC bottles into the laboratory and stored at –20°C until analysis.

Fig. 3 shows the average and the range of measured concentrations of organometal(loid) compounds obtained by the hydride generation procedure according to Grüter et al. (2000).

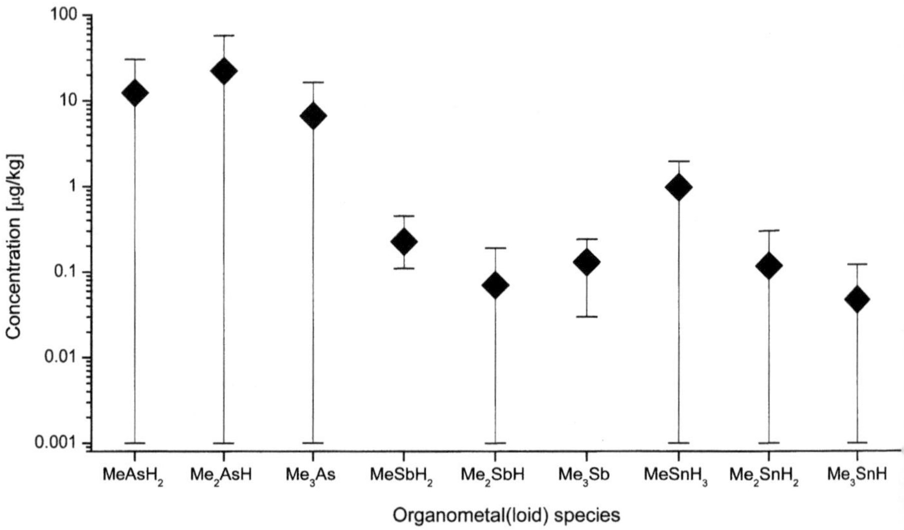

Fig. 3: Minimum, arithmetic mean and maximum concentration of metal(loid) compounds in compost from composting facilities measured by pH 1-HG-GC/ICP-MS. Concentrations below the detection limit are indicated as 0.001 µg/kg.

Arsenic was found in highest amounts of up to 100 µg/kg total methylated arsenic species. Dimethylarsenic acid was the dominant species with average concentrations of more than 10 µg/kg. Methylated tin, mercury and antimony species were found up to the µg/kg range. In comparison to samples from various contaminated soils and waste materials measured with the same methodology (Grüter et al. 2000) the concentrations of methylated arsenic and antimony found in the compost samples were significantly higher.

The compost samples from different composting facilities showed large differences in the concentrations of methylated arsenic, tin and mercury species; only antimony species were found in all samples analyzed.

Although compost from some composting facilities showed extreme concentrations above 10 µg/kg, in compost from other facilities concentrations were below detection limit (shown as 0.001 µg/kg), the reason for these differences remaining unclear. Several explanations are possible, in particular the stage of composting, the type of raw material processed, the metal content or the applied composting technique. Because the samples had to be treated anonymously, further information on the facilities was very limited.

The only information available was the type of waste processed, but according to the concentration ranges in Fig. 4 the source materials does not seem to have a major influence. Compost from a mixture of both source materials showed the highest variability.

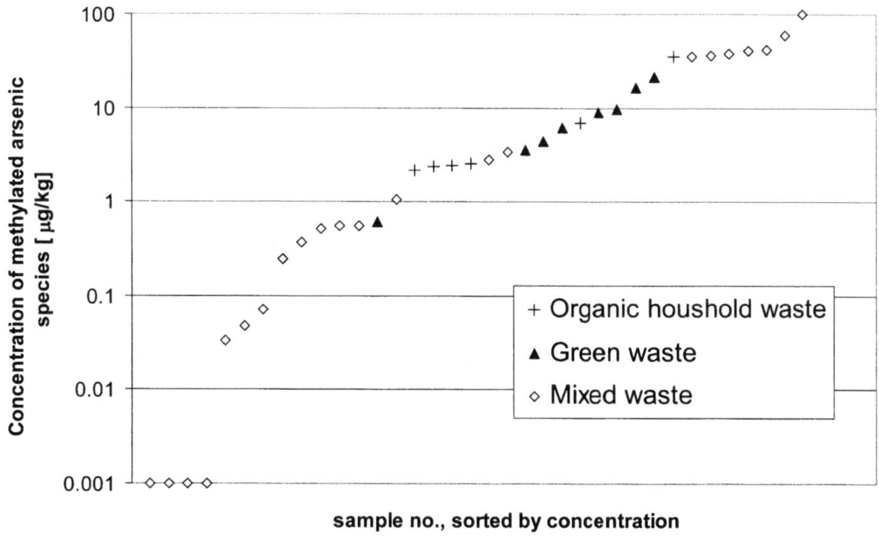

Fig. 4: Concentrations of metal(loid) arsenic species in compost from different types of composting facilities measured by pH 1-HG-GC/ICP-MS. The symbol indicates whether the facility processed only green waste, only organic household waste or a mixture of both. Concentrations below the detection limit are shown as 0.001 µg/kg.

Another possible explanation for the different amounts of methylated species is the difference in total metal content, but no correlation was found between the concentration of total and methylated metal(loid) contents.

Metal(loid) species in different aged compost

In order to investigate the time dependence of organometal(loid) concentrations, we studied a compost facility in detail. In addition to compost volatile metal(oid) species were sampled by direct cryotrapping. In contrast to other biological waste treatment such as sewage treatment or fermentation plants workers in composting facilities are directly exposed to potential gaseous emissions of volatile organometal(loid) compounds.

The facility is located in an urban area, processes 10% green and 90% biodegradable domestic waste, and has a capacity of 24000 tons per year. The organic waste is processed in large compost heaps of approximately 40 m length, 3 m width and 1.5 m height. The compost piles are aerated by suction and the factory floor is additionally ventilated. The facility was easy to access and no special safety was necessary.

The source material was analyzed for aqua regia extractable metal content (Table 2). The concentration of arsenic in the source material was significantly below the concentration of the proposed public health related standard class 1 of the USA (Table 1).

	Concentration [µg/kg]	Standard deviation
Ge	713	3.71%
As	1852	4.22%
Sn	1480	1.20%
Sb	1495	1.36%
Te	7	27.47%

Table 2: Aqua regia extractable metal content of source material from a compost facility

We sampled a profile through compost piles in different stages of composting that were set-up at 2 hours 10, 20, 30 and 40 days respectively before sampling. Each compost heap was sampled at a depth of approximately 30 cm. Approximately 500 g of sample were transported in PVC bottles into the laboratory and stored at –20°C until analysis.

After 40 days the compost piles were removed and the material labeled "fresh compost" was deposited for further maturing. We sampled two fresh composts of different maturity from the same facility. The stated compost ages are estimations based on information from the operators of the facility. The samples were analyzed by the pH-gradient hydride generation method (Fig. 5).

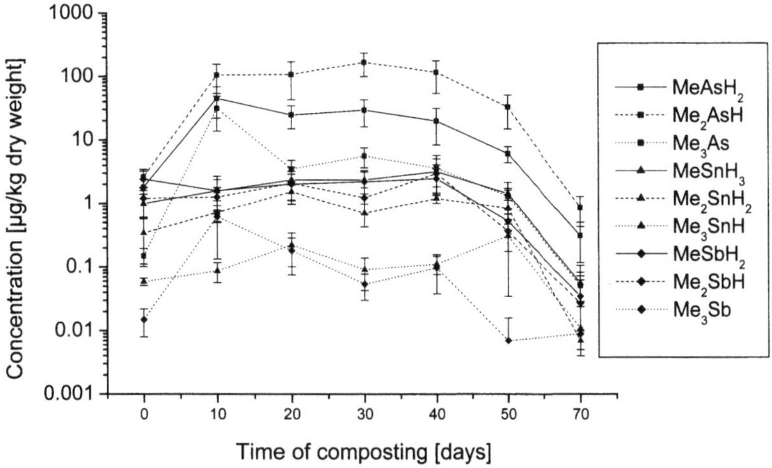

Fig. 5: Methylated metal(loid) species in compost piles from a compost facility of different ages measured by pH gradient-HG-GC/ICP-MS. The composts aging 50 and 70 days are fresh composts. The compost ages are estimations based on information of the operators of the facility.

The profile shows that the concentrations of most organometal(loid) species in the compost increase rapidly within the first week and then remain on a high level. Dimethylarsenic ranges up to more than 160 µg/kg, but also mono- and trimethylarsenic species are found in concentration of more than 30 µg/kg. Antimony and tin species are in the µg/kg range. The concentrations of most species remain on the same level during six weeks of composting. Trimethylarsenic shows a peak concentration after one week.

In contrast to the compost from the first six weeks, the concentration of organometal(loid)s in the fresh compost decreased by approximately two orders of magnitude. Fresh compost is not watered and therefore the temperature and the biological activity decrease, and the redox potential increases. The decrease of organometal(loid) concentration may be due to decay of the species *via* decay or biological demethylation. Because of the low concentration measured in the gas phase (see below), volatilization is regarded as less important.

We also sampled the air within the enclosed plant, the gases within the compost piles and the exhaust gases of the facility by cryotrapping. Due to the aeration of the compost piles by suction it was not surprisingly that almost no organometal(loid) species were found in the air in the plant. Within the compost piles up to 400 ng/m^3 trimethylarsine was detected. Methyliodide and dimethylselenide were found in concentrations of more than 100 ng/m^3, but traces of tetramethyltin, trimethylbismuth, trimethylstibine and dimethyltelluride were also detected. The concentrations of volatile organometal(loid) compounds in the compost heaps increased with the composting time. The compost facility has an exhaust of approximately 60000 m^3 per hour, due to the high dilution, only 44 ng/m^3 trimethylarsine (corresponding to an emission of 2.6 mg trimethylarsine per hour) was detected in the exhaust gas.

The absence of detectable volatile species in compost air reported by Maillefer et al. (2003) may be explained by two factors. On the one hand, it is possible that a facility with a low metal content or biomethylation potential was investigated. On the other hand, compost air was sampled and transported into the laboratory in a glass chamber, which may lead to a decay of organometal(loid) species, as compost air contains high amounts of oxygen.

Biomethylation in garden compost

In order to investigate whether the very high biomethylation potential of industrial composting can also be found in normal garden composts, we set up a commercially available garden compost container and filled it with source-separated urban organic household waste from the compost facility described above (Table 2). The compost was sampled and measured before the set-up and after one, two and three weeks in order to investigate the production of organometal(loid)s and after 12 weeks in order to investigate whether the components are stable in the compost. The samples were analyzed using the pH-gradient hydride generation method (Fig. 6).

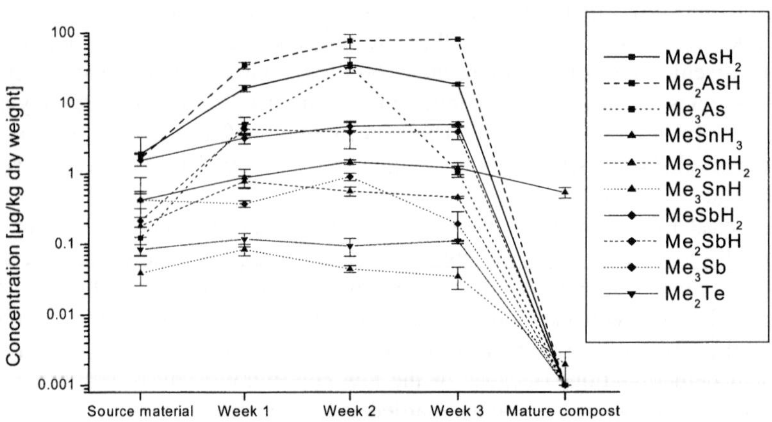

Fig. 6: Methylated metal(loid) species in garden compost experiment measured by pH gradient-HG-GC/ICP-MS. Concentrations below the detection limit are shown as 0.001 µg/kg.

As in the compost facility a rapid increase of organometal(loid) concentration was observed within one week. The most abundant species was again dimethylarsenic with concentrations of 80 µg/kg after three weeks. Both for monomethylarsenic and trimethylarsenic the highest concentrations of more than 30 µg/kg were found after two weeks. Antimony and tin species were in the µg/kg range. Also up to 100 pg/kg dimethyltellurium were found.

The highest concentrations of dimethylarsenic and methyltin species were about 50% lower in comparison to the compost facility, for antimony species twice as high (Table 3).

Table 3: Comparison of maximal concentrations of organometal(loid) species in compost from a compost facility and garden compost measured by pH gradient-HG-GC/ICP-MS.

Species [µg/kg]	Compost facility	Garden compost
$MeAsH_2$	44.9	35.8
Me_2AsH	167.7	81.2
Me_3As	31.2	33.8
$MeSnH_3$	3.2	1.5
Me_2SnH_2	1.5	0.8
Me_3SnH	0.2	0.1
$MeSbH_3$	2.4	5.0
Me_2SbH	2.1	4.3
Me_3Sb	0.6	0.9

The lower concentrations of the mature compost in comparison to fresh composts from the compost facility may be simply due to the increased age. Surprisingly, the concentration of monomethyltin decreased only by half in mature garden compost.

A number of volatile permethylated species were detected within the compost heap (Table 4). The concentration ranges are again in good agreement with the compost facility, larger differences were only found for methyliodide and trimethylstibine.

Table 4: Comparison of maximal concentrations of volatile organometal(loid) species in compost gas from a compost facility and garden compost measured by GC/ICP-MS.

Species [ng/m^3]	Compost facility	Garden compost
Me_3As	389.1	657.5
Me_2Se	25.1	20.8
Me_4Sn	0.3	0.9
Me_3Sb	1.0	37.4
Me_2Te	6.9	16.4
MeI	130.3	5.0
Me_3Bi	1.4	1.1

Conclusions

In this study, it was shown for the first time that high concentrations of methylated metal(loid) compounds can be produced by composting of biomaterials. More than 100 µg/kg methylated arsenic species, several µg/kg methylated tin and antimony species and 100 pg/kg Me_2Te were detected in the compost by hydride generation. Volatile compounds of these elements were detected in the gas phase within the compost heaps. In addition, methylated species of selenium, iodine and bismuth were found in the compost gas.

Organometal(loid) compounds were formed rapidly within the compost and remained stable over a period of several weeks. In fresh compost the concentration dropped rapidly and almost disappeared after twelve weeks in the garden compost. The mechanisms of this decrease are so far unknown; demethylation, decomposition and volatilization of the organometal(loid) species are all possible.

The results of samples from a composting facility using compost piles for composting were very similar to normal garden composts.

The high biomethylation potential was surprising, as composting is a predominantly aerobic process. Methylation may be restricted to the micro-anaerobic compartments within the compost, but it is unlikely that such a high biomethylation is caused by only a fraction of the compost.

Various industrial biological waste treatment systems utilizing a range of different environments for the decomposition of organic waste are in use. These systems should be investigated in regard to their biomethylation potential and organometal(loid) emissions. By appropriate technical modifications concerning safety requirements the potential environmental hazard may be minimized.

Further laboratory experiments are necessary to investigate the microbial fauna involved and the parameters influencing biomethylation in organic waste treatment processes.

The conversion of inorganic metal(loid)s to highly mobile or even volatile compounds may have general implications for the use of compost. In respect to minimization of organometal(loid) emissions mature compost should be preferred to fresh compost.

Literature

ANS - Arbeitskreis zur Nutzbarmachung von Siedlungsabfällen e.V. ANS (2002) Kompostatlas Online. Internet address: http://www.ans-ev.de/

Barth J (2001) Biological waste treatment in Europe – technical and market developments. Internet address: http://www.bionet.net

Craig PJ (ed) (1986) Organometallic compounds in the environment. Longman, Harlow.

Craig PJ, Jenkins RO, Dewick R, Miller DP (1999) Trimethylantimony generation by *Scopulariopsis brevicaulis* during aerobic growth. Sci Tot Environ 229: 83-88

Diaz LF, Savage GM, Eggerth LL, Golueke CG (1993) Composting and recycling – Municipal solid waste. Lewis Publ., Boca Raton

Diaz-Bone RA, Felix J, Hirner AV (2003) Multi-element organometal(loid) analysis by pH-gradient hydride generation. Book of Abstracts of the European Winter Plasma Conference on Plasma Spectrochemistry 2003: 296

Feldmann J (1997) Summary of a calibration method for the determination of volatile metal(oid) compounds in environmental gas samples by using gaschromatography-inductively coupled plasma mass spectroscopy. J. Anal. At Spectrom 12: 1069-1076

Feldmann J, Grümping R, Hirner AV (1994) Determination of volatile metal and metalloid compounds in gases from domestic waste deposits with GC/ICP-MS. Fresenius J Anal Chem 350: 228-234

Jenkins RO, Forster SN, Craig PJ (2002) Formation of methylantimony species by an aerobic prockaryote: *Flavobacterium* sp.. Arch Microbiol 178: 274-278

Grüter UM, Kresimon J, Hirner AV (2000) A new HG/LT-GC/ICP-MS multi-element speciation technique for real samples in different matrices. Fresenius J Anal Chem 368: 67-72

Grüter UM, Hitzke M, Kresimon J, Hirner AV (2001) Derivatisation of organometal(loid) species by sodium borohydride – Problems and solutions. J Chromatogr A 938: 225-236.

Hirner AV, Rehage H, Sulkowski M (2000) Umweltgeochemie. Steinkopff, Darmstadt

Hirner AV, Feldmann J, Krupp E, Grümping R, Goguel R, Cullen WR (1998) Metal(loid)organic compounds in geothermal gases and waters. Org Geochem 29:1765-1778

Kern M, Fund K, Mayer M (1998): Stand der biologischen Abfallbehandlung in Deutschland. Müll und Abfall 11

Koch I, Feldmann J, Lintschinger J, Serves SV, Cullen WR (1998) Demethylation of trimethylantimony species in aqueous solution during analysis by hydride generation/gas chromatography with AAS and ICP-detection. Appl Organomet Chem, 12: 129-136

Krupp E (1999) Analytik umweltrelevanter Metall(oid)spezies mittels gaschromatographischer Trennmethoden. Dissertation, Universität-GH Essen

Maillefer S, Lehr CR, Cullen WR (2003) The analysis of volatile trace compounds in landfill gases, compost heaps and forest air. Apll Organomet Chem 17: 154-160

Michalke K, Wickenheiser EB, Mehring M, Hirner AV, Hensel R (2000) Production of volatile derivates of metal(loid)s by microflora involved in anaerobic digestion of sewage sludge. Appl Environ Microbiol 66: 2791-2796

Wickenheiser EB, Michalke KT, Hensel R, Drescher C, Hirner AV, Brutishauer B, Bachofen R (1998) Volatile compounds in gases emitted from the wetland bogs near Lake

Cadagno. In: Peduzzi R, Bachofen R, Tonella M (ed). Lake Cadagno: A Meromictic Lake. Documenta del Ístituto Italiano di Idrobiologia, No 63

Chapter 6

Volatile mercury species in environmental gases and biological samples

J. Hippler, J. Kresimon, A. V. Hirner

Introduction

Mercury is emitted into the atmosphere from a number of natural as well as anthropogenic sources. Anthropogenic emissions of mercury are thought to be of the same order of magnitude as those from natural sources (Ebinghaus et al. 1998). In addition to elementary mercury in the atmosphere, dimethyl and diethylmercury, as well as particle-bound mercury have been detected in the very low pg/l concentration range (Kaiser and Tölg 1980).

Organic mercury compounds are one to two orders of magnitude more toxic than inorganic mercury (Clarkson 1992), and have caused much loss of life (Hirner et al. 2000). However, while there are numerous publications concerning monomethylmercury generated by biomethylation, only a few studies have concerned themselves with dimethylmercury. Problems associated with the sampling of volatile compounds may account for this (Puk and Weber 1994a) and this also holds true for more recent reports (e.g. Guidotti and Vitali 1998; Barat and Poulos 1998; Dunemann et al. 1999; Cai et al. 2000). The number of analytical techniques for dimethylmercury is however steadily increasing (e.g. Puk and Weber 1994b; Mothes and Wennrich 1999; Dietz et al. 2000, 2001; Välimäki and Perämäi 2001; and papers cited in the discussion section).

In this study, using two contrasting analytical methods, scenarios describing the occurrence of volatile elemental mercury and volatile dimethylmercury in environmental and biological samples will be described and broadly discussed with regard to relevant literature.

Analytical methods

Instrumental techniques

Mercury species were determined by two different methods: low temperature gas chromatography coupled with plasma mass spectrometry (GC/ICP-MS, method A) as described elsewhere (Grüter et al. 2000), and atomic fluorescence spectrometry (GC/AFS, method B).

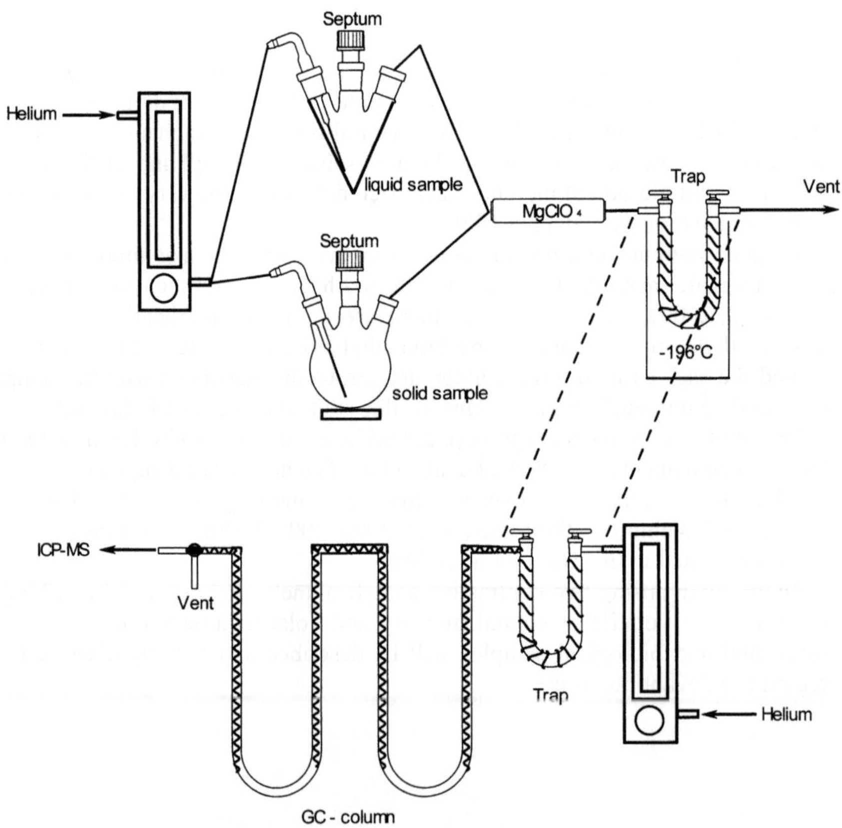

Fig. 1. Analytical scheme of method A (GC/ICP-MS)

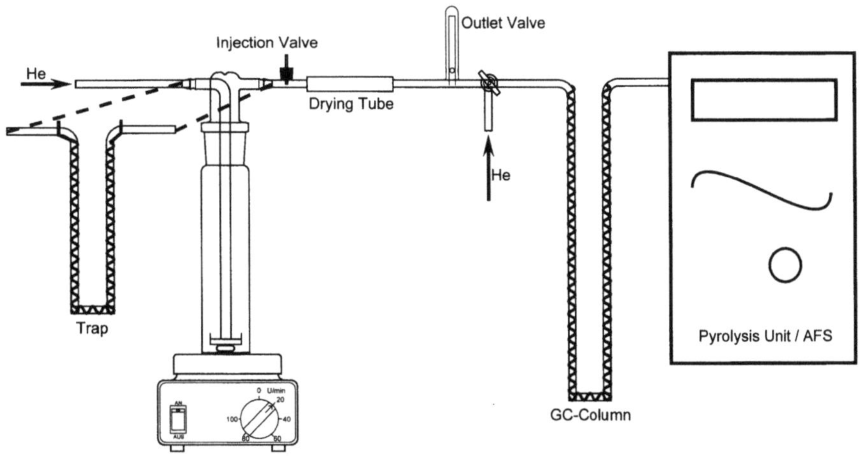

Fig. 2. Schematic diagram of method B (GC/AFS): GC column is packed with Supelcoport (80/100 mesh) (10% Methylsilicone) SP-2100 Helium: 185 ml/min

Gaseous samples were directly measured, i.e. without derivatisation, using atomic fluorescence spectrometry. Depending upon the gas source, 1 to 10 l of the gas was passed though the cryotrap (cooled to -110°C). Trapped mercury species were subsequently thermally desorbed and refocussed on the GC column to remove matrix gases such as methane. In contrast to the gaseous samples, most liquid and all solid samples were subjected to derivatisation with $NaBH_4$ to yield gaseous educts for analysis. Aliquots (1 ml or 0.01 - 1g) of sample were dispensed into the reaction vessel. Buffer was added to this to obtain a volume of 25 ml. Mercury species were hydride generated by the addition of 1 ml $NaBH_4$ (6%) to the buffer-sample mixture. The resulting hydrides were continuously purged out of the reaction vessel, through a drying tube (filled with magnesium perchlorate) to remove water and cryofocussed onto the GC-column (Supelcoport 80/100) (-196°C) using helium. Volatile mercury species were subsequently thermally desorbed from the GC column under helium (heated over a period of 25 minutes to a final temperature of 100°C).

To enable the direct analysis of mercury species with high vapor pressure, a heated injection valve was incorporated.

To detect the mercury species with the AFS, they must first be pyrolysed to elemental mercury. The optimal pyrolysis temperature was found to be 800°C (Fig. 3).

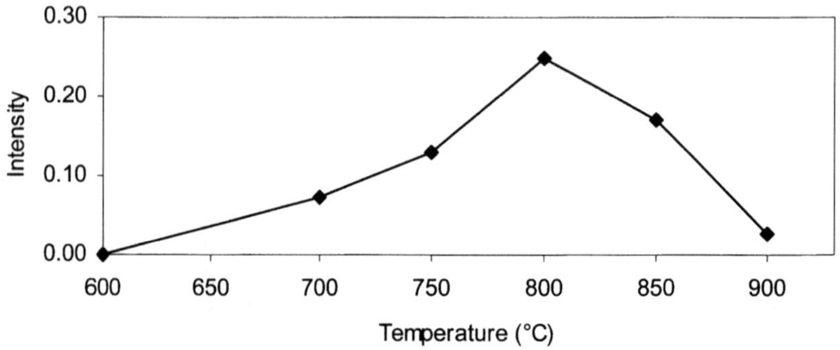

Fig. 3. Optimisation of pyrolysis (GC/AFS) for 20 pg Me_2Hg at different temperatures.

Monomethylmercury was initially hydride generated at pH ~ 1. This however gave rise to dismutation of the monomethylmercury to elemental and dimethylmercury (Fig. 4).

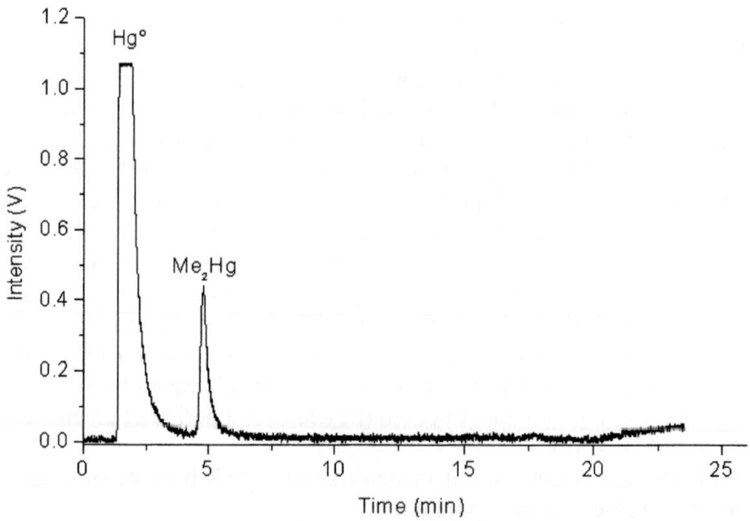

Fig. 4. Typical chromatogram of $NaBH_4$-derivatised MeHgCl (pH ~ 1) showing presence of elemental mercury and dimethylmercury

Through the use of acetate or citrate buffer (pH ~ 4.5), the dismutation of MeHgCl to Hg° and Me_2Hg can be inhibited so that MeHgH is the sole hydride generation product. The remaining Hg° is a contaminant product from $NaBH_4$, and typically amounted to 50 – 200 pg/ml for a 6% $NaBH_4$ solution (Fig. 5).

The hydride generation of mercury and monomethylmercury is pH dependent. The reaction mechanism is described in equations (1) and (2). Craig (1993) has demonstrated the formation of monomethylmercury hydride following hydride generation of monomethylmercury with sodium borohydride. The formation of HgH_2 as a reaction intermediate during the hydride generation of Hg^{2+} to Hg° may be assumed (Craig, 1999).

$$Hg^{2+} + 2\ NaBH_4 + 6\ H_2O \rightarrow Hg^0 + 2\ Na^+ + 2\ H_3BO_3 + 7\ H_2 \quad (1)$$

$$(CH_3)Hg^+ + NaBH_4 + 3\ H_2O \rightarrow (CH_3)HgH + Na^+ + H_3BO_3 + 3\ H_2 \quad (2)$$

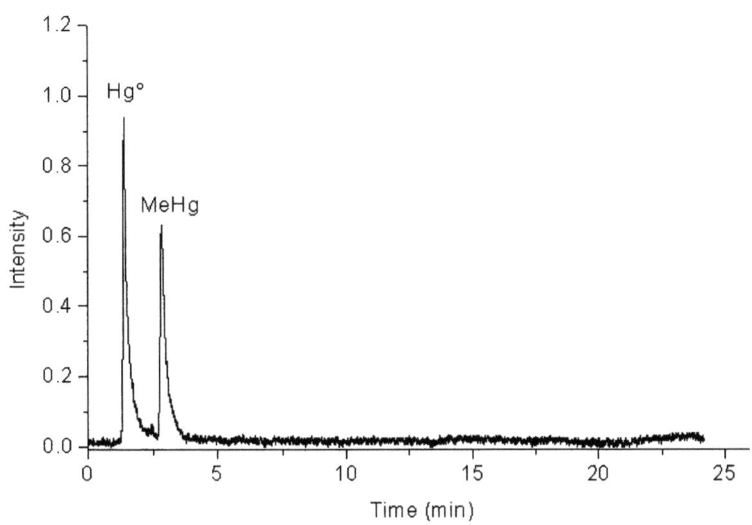

Fig. 5. Typical GC/AFS chromatogram of $NaBH_4$-derivatised MeHgCl (pH ~ 4.5). Hg° occurs as typical contamination of the $NaBH_4$

Ethylation of MeHgCl standard was used to obtain MeEtHg so that retention time comparison could be made with samples. The reaction conditions required for ethylation using $NaBEt_4$ are analogous to those described above for hydride generation. Ethylation of MeHgCl under the same reaction conditions as used for hydride generation resulted in the detection of MeEtHg as well as low quantities (less then 10%) of Me_2Hg and Et_2Hg (Fig. 6).

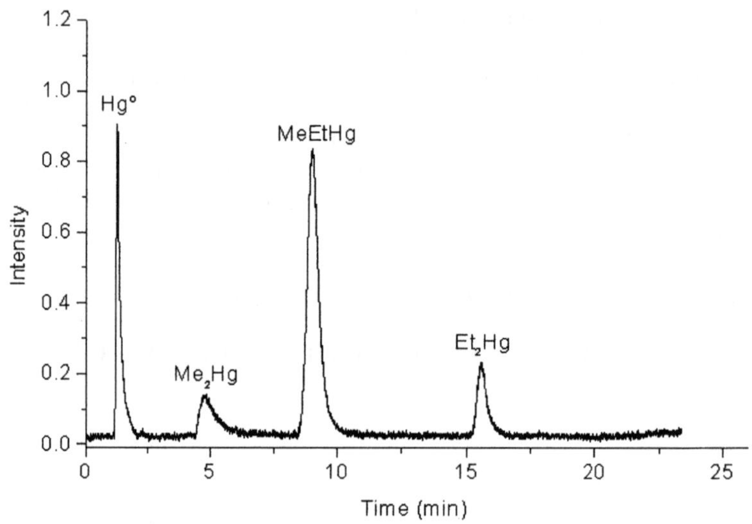

Fig. 6. Typical GC/AFS chromatogram showing species obtained by derivatisation of MeHgCl using $NaEt_4Br$

Using the optimised GC-conditions it was possible to separate a mixture of six mercury species; Hg°, MeHgH, Me_2Hg, EtHgH, MeEtHg, Et_2Hg within 16 minutes (Fig. 7). These volatile mercury standards were injected directly (Hg°, Me_2Hg, Et_2Hg) or obtained by hydride generation or ethylation of MeHgCl.

To obtain optimal resolution of species eluting early from the column, the temperature programme of the GC was changed in that the initial temperature was held for a longer period. With this programme it was possible to resolve early eluting species despite the presence of large quantities of Hg° in the sample (such as occur in natural gas). Standards of the volatile organomercury species MeHgH, Me_2Hg, MeEtHg, EtHgH and EtHg were obtained through hydride generation or ethylation of MeHgCl. To enable the direct analysis of gas condensate, the liquid was injected directly through the heated injection valve.

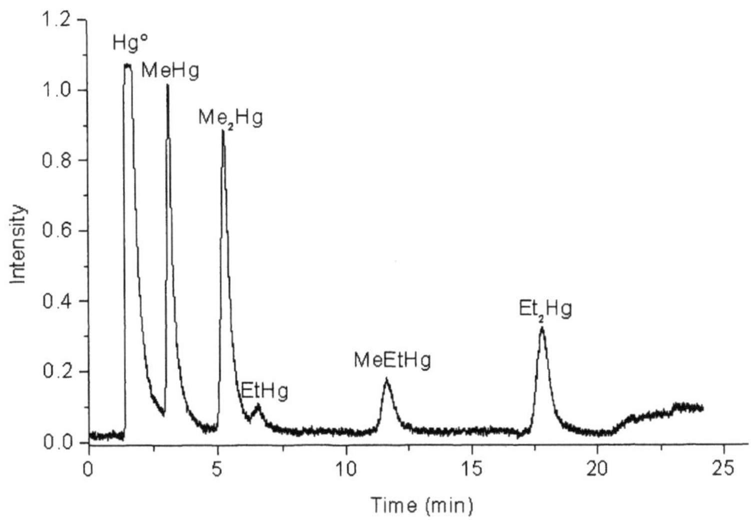

Fig. 7. Typical GC/AFS chromatogram of six mercury species.

Results

Environmental samples

Analysis of natural gas samples by cryotrapping up to 10 l of gas taken directly from the municipal supply, revealed that in addition to elemental mercury, which is generally found in natural gas, up to 234 pg/l Me_2Hg and 48 pg/l Et_2Hg were also detected (Fig. 8A). Samples were passed through a cryotrap, desorbed and re-focused on a GC column as already described in the methods section.

The detected mercury species with a retention time of 8 min, i.e. between Me_2Hg and Et_2Hg, may be MeEtHg. Retention time variation of this species was noted, however this may be due to matrix effects from methane or other constituents of natural gas.

The mean monthly average of hydrocarbon content of natural gas is shown in Table 1.

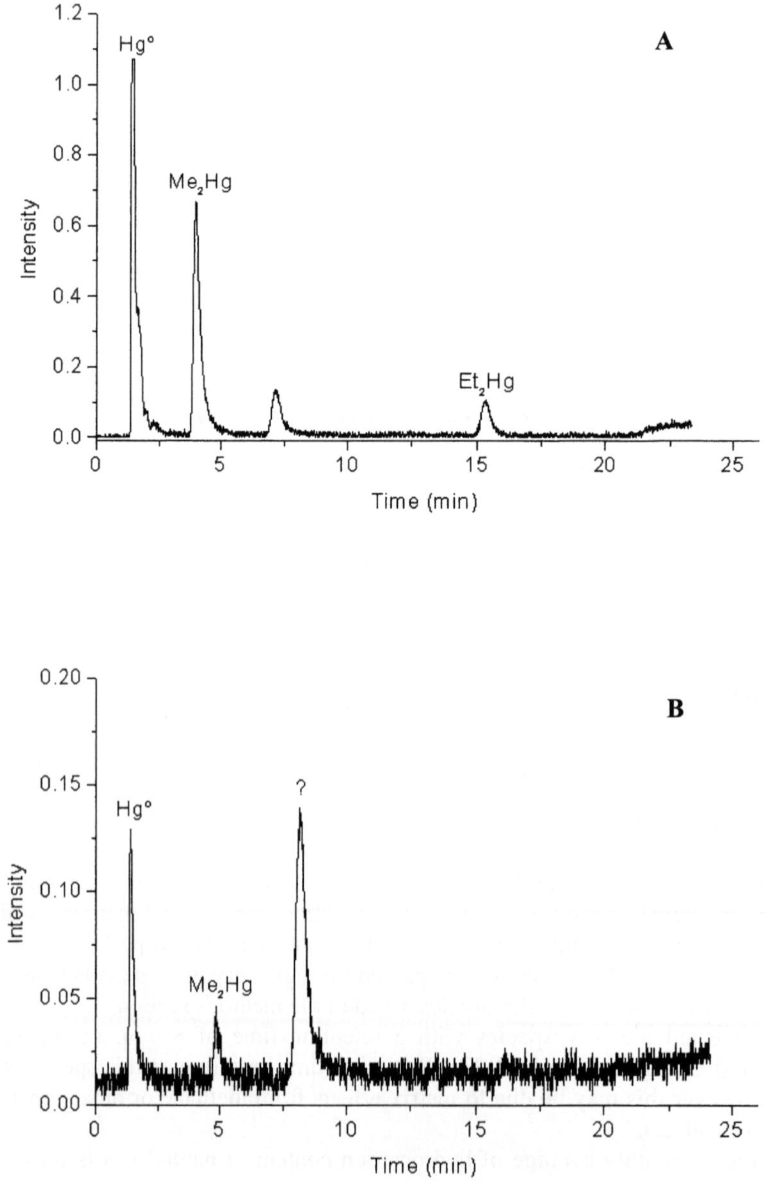

Fig. 8. Typical GC/AFS chromatogram of A) municipal gas B) sewage gas.

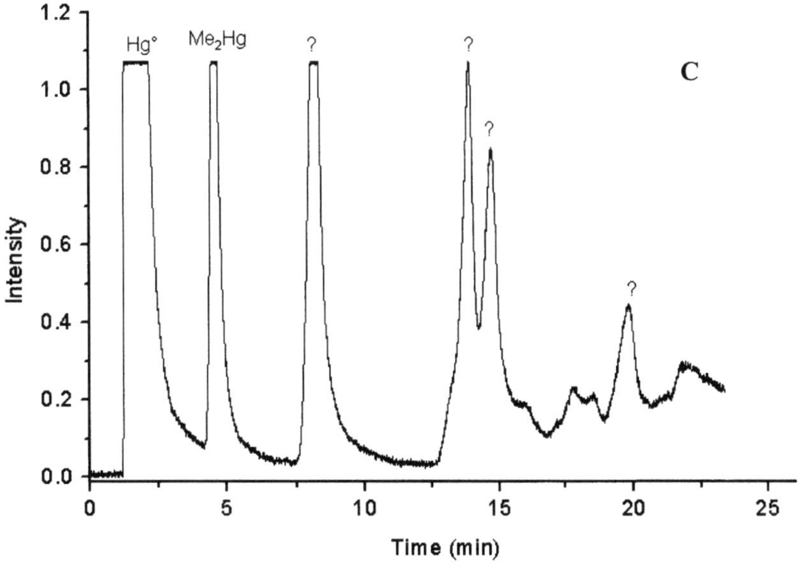

Fig. 8 cont'd. Typical GC/AFS chromatogram of C) gas condensate.

Table 1. Hydrocarbon content of municipal gas.

		mol.-%
Carbon dioxide	CO_2	0.70
Nitrogen	N_2	1.45
Oxygen	O_2	<0.01
Methane	CH_4	90.62
Ethane	C_2H_6	5.50
Propane	C_3H_8	1.10
Iso-Butane	i-C_4H_{10}	0.30
n-Butane	n- C_4H_{10}	0.15
Iso-Pentane	i-C_5H_{12}	0.05
n-Pentane	n- C_5H_{12}	0.03
Hexane and other hydrocarbons	C_6H_{14} + ...	0.10

Me_2Hg was also detected in digester gas from a sewage treatment plant. A typical chromatogram of the mercury species present in 1 l digester gas is shown in Figure 8B. As with the natural gas an unknown mercury species (possibly MeEtHg) was detected.

To purify natural gas from accompanying contaminants such as water and mercury before it enters the municipal supply, the gas passes through a low temperature separation (LTS) system. Through sequential pressure reduction, the temperature of the gas is reduced to around -32 °C (Joule Thompson Effect). This results in the condensation of long-chain hydrocarbons and the mercury, the so called "gas condensate". Through this step, up to a ten-fold reduction of mercury content can be achieved. Analysis of samples of gas condensates revealed high concentrations of elemental mercury (670 mg/l) as well as Me_2Hg (180 µg/l) (Fig. 8C). The unknown peak with retention time 8 min. (possibly MeEtHg) was again detected, as was a wide range of unidentified compounds with higher retention times. It is possible that these unknown species are ethyl- and phenylmercury compounds. The results of spiking experiments were however, inconclusive. The possibility of abiotic ethylation of mercury occurring under the high pressure (650 bar) and temperature (150 °C) conditions found in the LTS system cannot be discounted.

Biological samples

Amalgam containing mercury has been used for many years as a filling material in dentistry. Due to health concerns regarding use of the material, it has widely been replaced by ceramic and synthetic materials, but is still in use. It is known that small amounts of mercury can leach out of the amalgam and be detected in the breath (Berglund 1990). Figure 9 shows the detection of mercury in the breath of two volunteers with 14 and 16 fillings respectively. Samples I and II were taken directly after waking (0.01 - 2 pg/l). Following intensive mastication (with chewing gum), significantly higher concentrations of mercury were noted: 14 to 77 pg/l (samples III and IV). Elevation of the background mercury concentration seen upon waking was also noted after cleaning of the teeth and consumption of a hot drink (4 - 14 pg/l) (Sample V).

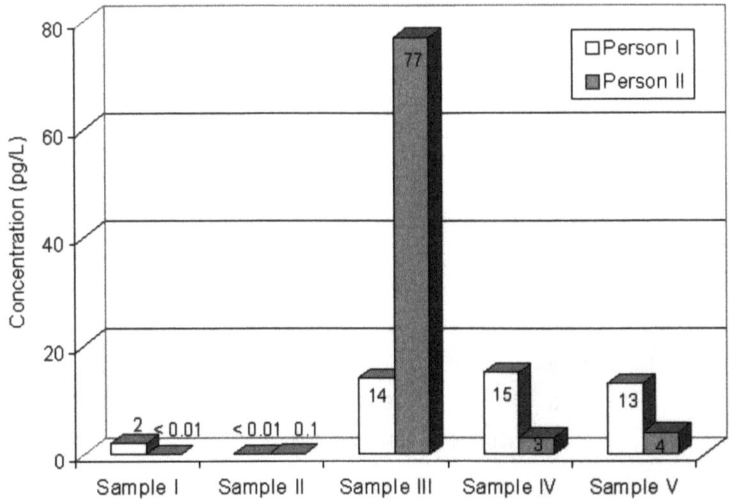

Fig. 9. Elemental mercury in breath. Samples were taken directly after waking (I and II). Following intensive mastication (with chewing gum; samples III and IV) and after cleaning of the teeth and consumption of a hot drink (Sample V).

A summary of mercury content in the breath of subjects with differing number of amalgam fillings is shown in Figure 10. No significant correlation between the number of amalgam fillings and mercury detection could be made; possibly because of the wide variation in age, condition and size of the fillings in the different subjects. A significantly higher concentration of mercury was noted however, in samples from subjects with more than 14 amalgam fillings. In general, these results are consistent with those reported for chewing tests (Mackert and Berglund 1997; Takahashi et al. 2001; Berdouses et al. 1995; Gay et al. 1978).

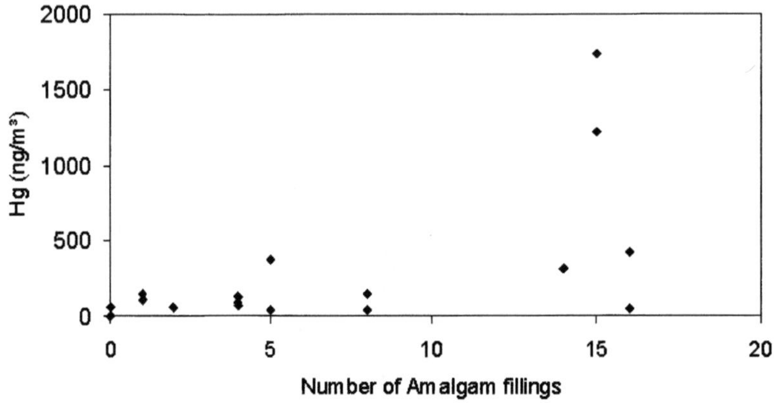

Fig. 10. Relationship between mercury breath, and number of fillings.

Removal of amalgam from the mouth cavity may release mercury, a portion of which can be ingested or inhaled during the procedure. Yannai and Berdicevsky (1991) as well as Huppertz (1998) reported that the oral bioflora of humans is able to biomethylate inorganic mercury. *Streptococcus mitio*, *Streptococcus mutans* and *Streptococcus sangius* have all been demonstrated to produce methylmercury from both pulverised amalgam and $HgCl_2$ during an incubation period of four days (Heintze et al. 1983). The analyses were perfomed by a gas-liquid chromatographic method.

Sputum samples taken from a subject immediately following the removal of four fillings hydride genation revealed the presence of the species MeHg, Me_2Hg as well as EtMeHg with concentrations of 34.9, 3.1 and 16.4 µg/l.

The presence of vitamins C and E has been reported to cause the transformation of mercury to methylated mercury species (Carl 2002). The kinetics of abiotic transformation of mercury by acetic acid has been studied by Gardfeldt et al. (2003).

The algae *Chlorella pyrenoidosa* is a known complexing agent for mercury compounds. It was used in the study here to identify the presence of methylmercury compounds following removal of amalgam fillings. Following amalgam removal, sputum was mixed with 2.5 g of algae *Chlorella pyrenoidosa* for 5 minutes. Subsequent disruption of the algal complex *via* hydride generation revealed the presence of methylmercury compounds which may otherwise be protein-bound or associated with other components of the sputum (Fig. 11).

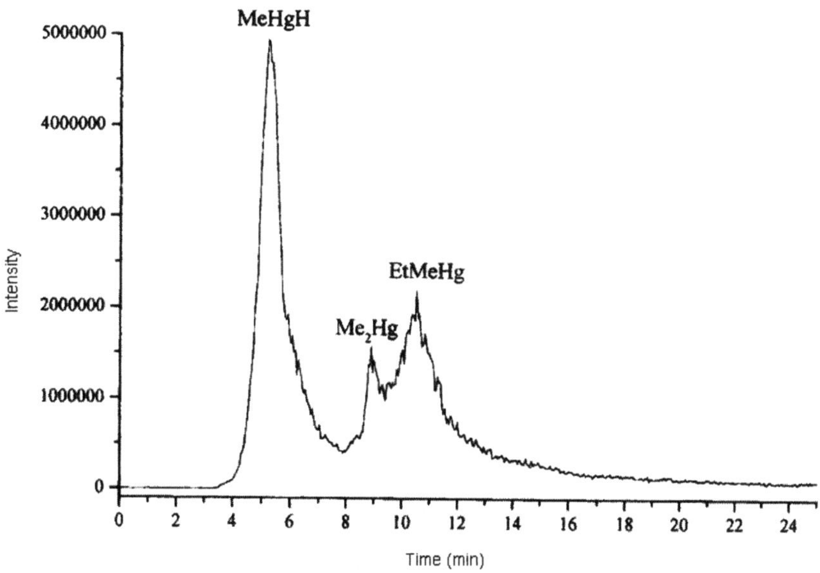

Fig. 11. Typical GC/ICP-MS chromatogram of mercury species in saliva following amalgam removal. Sputum samples were hydride generated with $NaBH_4$ to generate volatile mercury species prior to analysis.

Fish and sea-food constitute a significant source of mercury in our diet. Analysis of tinned tuna-fish (fish and brine analysed separately) *via* hydride generation GC/AFS at pH 4.5 revealed that around 90% of the mercury extracted was in the monomethylated form.

Human excretion of mercury occurs *via* both urine and faeces. The elimination of mercury can be encouraged through the ingestion of the heavy-metal binding agent Asparagus P. The detailed mechanism of metal binding to Asparagus P is not currently known, however it is thought that the mercury compounds are able to bind to available sulfhydryl groups. Figure 12 shows typical chromatograms of urine samples of two volunteers given two days following the ingestion of the metal-binding agent. As was noted with the natural gas measurements, unknown mercury compounds were also detected in urine samples.

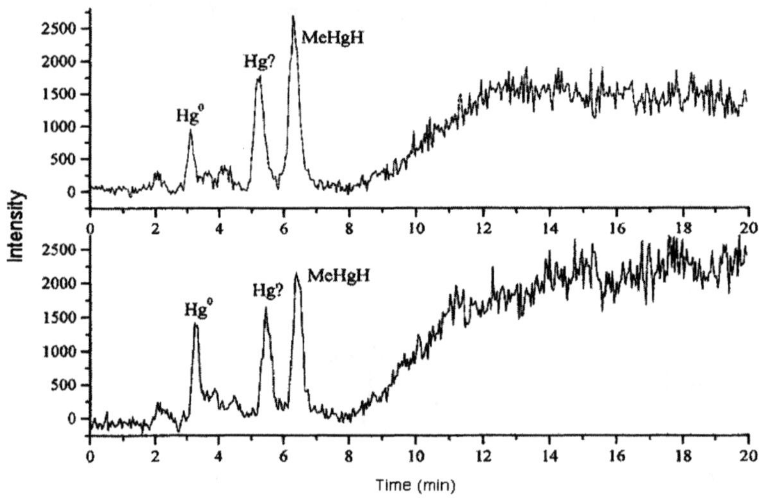

Fig. 12. Mercury species in human urine of two volunteers.

In addition to breath, faeces and urine samples were also monitored over a four-day period following amalgam removal. Inorganic mercury (65 ng/l) was detected in the morning-urine sample on the first day after amalgam removal. The concentration of mercury subsequently fell over the following days. Low levels of methylmercury (1 - 14 ng/l) were detected intermittently in the urine samples. In contrast, up to 43 ng/kg of methylmercury were detected in faeces samples. Neither mono- or dimethylmercury was detected in faeces before amalgam was removed. Immediately following amalgam removal however, the concentration of these species rose to 9 and 25 ng/kg respectively on the first day. The concentration of monomethylmercury rose further to 43 ng/kg by the second day. The concentration of both species decreased over the following days to 16 and 3 ng/kg for mono- and dimethylmercury respectively.

Various organomercury species were detected in the head hair of a subject who typically works with organometallics. Elemental mercury as well as monomethylmercury, monoethylmercury and an unknown mercury species were all detected (Fig. 13).

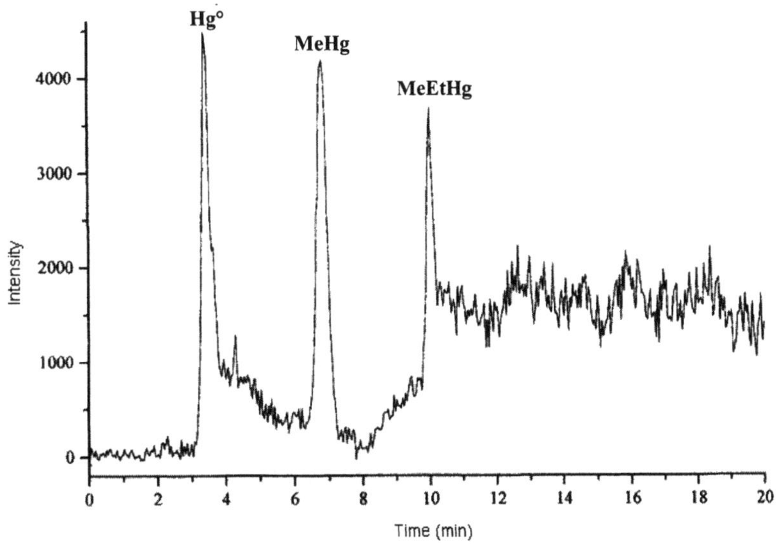

Fig. 13. Mercury species in human hair. Samples were hydride generated and analysed by ICP-MS.

Although Valentino et al. (1995) and Sarmani et al. (1997) demonstrated enrichment of mercury concentrations in hair by factors between 200 and 300 compared to blood levels and Toribara (2001) could monitor dimethylmercury poisoning of a human by means of hair analysis, Nielsen et al. (1994) reported that mercury concentrations in the hair of mice exposed to methylmercury were not correlated. The mechanism of mercury association with hair is not as yet fully understood.

Validation of data

Comparison of data obtained from GC/AFS and GC/ICP-MS analysis of gas samples from a digester at a sewage treatment plant revealed the presence of an unknown (possibly MeEtHg) mercury species with a retention time of around 8 min. As described before, retention time variation of this species was noted.

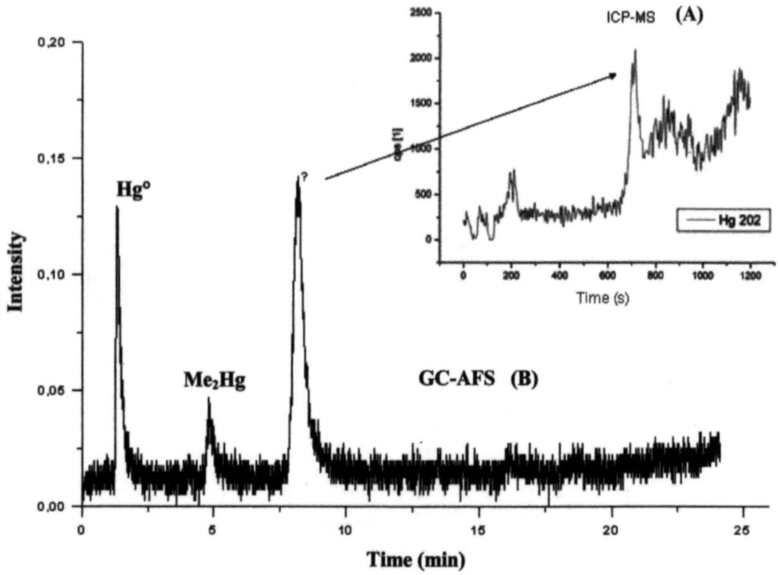

Fig. 14. methods A & B: sewage gas

The two methods (AFS and ICP-MS) were compared using breath samples. Generally, the AFS-method returned higher concentrations of mercury than the ICP-method did (Fig. 15). While the AFS was calibrated with external standards, a semiquantitative interelement calibration was used for the ICP-MS (Feldmann 1995). A possible explanation may be that in the case of volatile Hg species in gases the semi quantitative calibration leads to an underestimation of the species concentrations present. Similar comments regarding the use of semi-quantitative calibration for the ICP-MS have been made by Feldmann (1997).

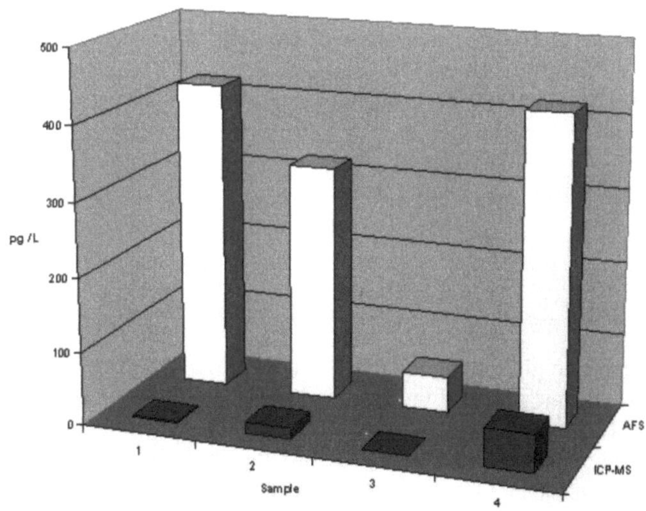

Fig. 15. Comparison of GC/ICP-MS and GC/AFS methods for the analyses of sewage gas

Discussion

Dimethylmercury in environmental gases

The mean concentration of total gaseous mercury over northern Europe (between Berlin and Stockholm) is between 1 and 2 pg/l (Schmolke et al. 1999). The predominant mercury species is known to be elemental mercury; methylmercury chloride can account for up to 5% of total mercury and dimethylmercury up to 0.5%. Based on this data, and similar data for methylated/halogenated species of tin, lead, and arsenic, Mester and Sturgeon (2002) suggested that the release of toxic trace element species to the atmosphere by environmental processes may be underestimated.

Within sediments, conversion of Hg(II) to elemental and organic mercury takes place (Weber et al. 1998); for dimethylmercury concentrations in the µg/kg range have been reported (Wassermann et al. 2002).

After careful separation by vaporization/trapping, volatile mercury species are usually measured by AFS, which ensures low detection limits (Shafawi et al. 1999). Using GC for species separation Pongratz and Heumann (1998) detected dimethylmercury concentrations in surface sea-water samples of the Antarctic and Arctic Ocean from <0.5 to 150 pg/l. We measured similar concentrations in purified natural gas and in the water phase of gas condensates (Table 2).

Table 2. Dimethylmercury concentrations of environmental gases.

Sample	Me_2Hg
Communal gas	136 pg/l
Treated gas	82 pg/l
Raw natural gas	745 pg/l
Gas condensate (water)	31 pg/l
Gas condensate	155 – 668 µg/l

Significantly high dimethylmercury concentrations were present in raw natural gas and the organic phase of the gas condensate; in the latter up to 23 µg/l was reported by Tao et al. (1998), around 100 µg/l for total mercury (II) compounds were determinded by HPLC/AAS (Schickling and Broekaert 1995). Tao et al. (1998) also detect ethylated species in condensates, which correlates with our findings here.

The studies of Bouyssiere et al. (2000) and Shafawi et al. (2000) are compatible with our findings, in that the removal of elemental mercury from natural gas is more efficient than is the removal of organic mercury detected. In industrially contaminated areas, elemental, ionic, methyl, ethyl, and phenylmercury as well as a number of unidentified species have been reported (Hempel et al. 1995; Hintelmann et al. 1995).

Mercury species generated in metabolic processes

Organic mercury species can enter the human body *via* ingestion, through vaccination, e.g. up to 25 µg ethylmercury *via* Thimerosal (Sodium ethylmercurithiosalicylate), or may be generated *in situ* by enzymatic methylation. Methylmercury in the diet is very effectively (90 to 95%) absorbed in the gastrointestinal tract (Bernard and Perdue 1984; Inskip and Piotrowski 1985). After a few days, approximately 10% of the ingested amount has crossed the blood/brain-barrier and entered the brain (Dales 1972); consequently the central nervous system is usually considered to be the most critical target for methylmercury toxicity. Neuronal degeneration by calcium deposition in the nervous system of rats was observed by Mori et al. (2000) after injection of dimethylmercury.

Although methylmercury in the blood is mainly associated with erythrocytes (half live approx. two months), inorganic mercury accumulates in the plasma fraction. Only trace amounts of methylmercury are excreted *via* urine and faeces (Ishihara 2000). Some of the ingested methylmercury may be demethylated in liver and/or kidney as well as in brain and faeces (Edwards and McBride 1975). Norseth and Clarkson (1971) have proposed that protein-bound methylmercury may also be biotransformed to inorganic mercury in the colon. Demethylation may be performed by phagocytic cells as well as by the intestinal microflora (Clarkson 1997). The main elimination route for inorganic mercury is *via* urinary excretion (Carrier et al. 2001a,b; Young et al. 2001). Accumulation of inorganic mercury in the kidneys of Canadian Indians chronically consuming fish containing high concentrations of methylmercury has been reported (Cappon and Smith 1981).

Biosynthesis of methylmercury from ionic mercury in the gastrointestinal tract has been shown to occur in human faeces in the presence of methanogens (Edwards and McBride 1975; Rowland et al. 1975), and in the oral cavity by *Streptococci* (Heintze et al. 1983). 0.05 to 0.26% of the mercury (as $HgCl_2$) injected into the stomach was converted to monomethylmercury in experiments with rats (Ludwicki 1989).

Elimination of mercury from the human body is generally a slow process, and not only restricted to methylmercury, which may stay in blood and tissues from months to years: Inhalation of elemental mercury by volunteers without amalgam fillings led to retention of 69% of the inhaled dose followed by a medium half time of this species in expired breath of two days (Sandborgh-Englund et al. 1998). About 1% of the absorbed mercury was excreted *via* urine during the first three days after exposure, and 8 to 40% during the next month.

Significant amounts of mercury are released from amalgam fillings: Over a one-day study period, up to 125 µg Hg was detected in oral emissions in addition to the detection of up to 19 µg Hg in urine and up to 190 µg Hg in faeces (Skare and Engquist 1994). Lygre et al. (1999) have shown that the concentration of mercury in sal*via* correlates with the number of amalgam fillings, however, there is no direct relationship to mercury concentrations in urine and blood, which more closely represent the body burden (Ganss et al. 2000). 80% of mercury from amalgam particles and mercury bound to sulfhydryl groups is excreted in an oxidised form (Engquist et al. 1998). Compared to non-dental workers, the concentrations of total and inorganic blood mercury are significantly higher in dental workers (Chang et al. 1992). No such correlation for organomercury species in blood has however been reported.

Conclusions

In addition to the presence of elementary mercury as a contaminant, dimethylmercury is a common trace compound found in natural gas (concentration range pg/l), and is also a trace component of ambient air (concentration range fg/l. In contrast to an air matrix, the absence of oxygen in the natural gas matrix ensures the stability of dimethylmercury. Since diethylmercury may be associated with dimethylmercury, and ethane is a minor component of natural gas, organomercury compounds in fossil gases may possibly be formed chemically during catagenesis than by bacteria during diagenesis.

Elemental mercury was the only species detected in human breath, its concentration being roughly correlated with quality and surface area of exposed amalgam fillings. Although monomethylmercury can be found in saliva, it has not been possible yet to reproduce proof for the existence of dimethylmercury. However, following the removal of amalgam fillings, substantial amounts of organic mercury can be found in faeces strengthening a biovolatilisation hypothesis similar to bismuth (see Hirner et al., this volume).

Within the anaerobic environment of the colon, the microflora present may be able to biotransform mercury to the hydride and methylated species, which can subsequently diffuse into the blood stream, and be exhaled or cross the blood/brain barrier. Last but not least it should be mentioned that because of the wide variation in internal/external calibration methods which are used in volatile mercury speciation, international gas standards are urgently needed for validation purposes.

In general it should be mentioned that many of the pilot studies reported here are still waiting for more statistical validation.

Acknowledgements

The authors are grateful to the Deutsche Forschungsgemeinschaft for project funding (Hi276/10).

Literature

Barat RB, Poulos AT (1998) Detection of mercury compounds in the gas phase by Laser Photo-Fragmentation / Emisson Spectroscopy. Appl Spectrosc 52:1360-1363

Berdouses E, Vaidyanathan TK, Dastane A et al. (1995) Mercury release from dental amalgams – an in–vitro study under controlled chewing and brushing in an artificial mouth. J Dent Res 74:1185-1193

Berglund A (1990) Estimation by a 24-hour study of the daily dose of intra-oral mercury vapor inhaled after release from dental amalgam. J Dent Res 69:1646-1651

Bernard SR, Perdue P (1984) Metabolic models for methyl and inorganic mercury. Health Physics 46:695-699

Bouyssiere B, Baco F, Savary L, Lobinski R (2000) Analytical methods for speciation of mercury in gas condensates. Oil & Gas Science and Technology 55:639-648

Cappon CJ, Smith JC (1981) Mercury and selenium content and chemical form in human and animal tissue. J Anal Toxicol 5:90-98

Carl E (2002) Beratungsstelle für Amalgamvergiftungen, Gräflingen

Carrier G, Bouchard M, Brunet RC, Caza M (2001b) A toxicokinetic model for predicting the tissue distribution and elimination of organic and inorganic mercury following exposure to methyl mercury in animals and humans. II. Application and validation of the model in humans. Toxicol Appl Pharmacol 171:50-60

Cai Y, Monsalud S, Furton KG (2000) Determination of methyl- and ethylmercury compounds using gas chromatography atomic fluorescence spectrometry following aqueous derivatization with sodium tetraphenylborate. Chromatographia 52:82-86

Carrier G, Brunet RC, Caza M, Bouchard M (2001a) A toxicokinetic model for predicting the tissue distribution and elimination of organic and inorganic mercury following exposure to methyl mercury in animals and humans. I. Development and validation of the model using experimental data in rats. Toxicol Appl Pharmacol 171:38-49

Chang SB, Siew C, Gruninger SE (1992) Factors affecting blood mercury concentrations in practicing dentists. J Dent Res 71:66-74

Clarkson TW (1997) The toxicity of mercury. Crit Rev Clin Lab Sci 34:369-403

Craig PJ, Mennie D, Needhamm et al (1993) Mass spectroscopic and nuclear-magnetic-resonance evidence confirming the existence of methyl mercury hydride. J Organomet Chem 447:5-8

Craig PJ, Jenkins RO, Stojak GH (1999) The analysis of inorganic and methyl mercury by derivatisation methods: Opportunities and difficulties. Chemosphere 39:1181-1197

Dales LG (1972) The neurotoxicity of alkyl mercury compounds. Am J Med 53:219-232

Dietz C, Madrid Y, Camara C (2001) Mercury speciation using the capillary cold trap coupled with microwave-induced plasma atomic emission spectroscopy. J Anal At Spectrom 16:1397-1402

Dietz C, Madrid Y, Camara C et al. (2000) The capillary cold trap as a suitable instrument for mercury speciation by volatilization, cryogenic trapping, and gas chromatography coupled with atomic absorption spectrometry. Anal Chem 72:4178-4184

Dunemann L, Hajimiragha H, Begerow J (1999) Simultaneous determination of Hg(II) and alkylated Hg, Pb, and Sn species in human body fluids using SPME-GC/MS-MS. Fresenius J Anal Chem 363:466-468

Ebinghaus R, Tripathi RM, Wallschläger D, Lindberg SE (1998) Natural and anthropogenic mercury sources and their impact on the air-surface exchange of mercury on regional and global scales. GKSS 98/E/51, Geesthacht, 50 pp

Edwards T, McBride BC (1975) Biosynthesis and degradation of methylmercury in human faeces. Nature 253:462-464

Engquist A, Colmsjö A, Skare I (1998) Speciation of mercury excreted in feces from individuals with amalgam fillings. Arch Environ Health 53:205-213

Feldmann J (1995) Erfassung flüchtiger Metall- und Metalloidverbindungen in der Umwelt mittels GC/ICP-MS. Cuvillier Verlag, Göttingen

Feldmann J (1997) Summery of a calibration method for the determination of volatile metal(loid) compounds in environmental gas samples by using gas chromatography-inductively coupled plasma mass spectrometry. J Anal At Spectrom 12:1069-1076

Ganss C, Gottwald B, Traenckner I et al. (2000) Relation between mercury concentrations in sal*via*, blood and urine in subjects with amalgam restorations. Clin Oral Invest 4:206-211

Gardfeld K, Munthe J, Stromberg D, Lindqvist O (2003) A kinetic study on the abiotic methylation of divalent mercury in the aqueous phase. Science Tot Environ 304:127-136

Gay DD, Cox RD, Reinhardt JW (1979) Chewing releases mercury from fillings. The Lancet 5:985-986

Grüter UM, Kresimon J, Hirner AV (2000) A new HG/LT-GC/ICP-MS multi-element speciation technique for real samples in different matrices. Fresenius J Anal Chem 368:67-72

Guidotti M, Vitali M (1998) Determination of urinary mercury and methylmercury by solid phase microextraction and GC/MS. J High Res Chrom 21:665-666

Heintze U, Edwardsson S, Derand T, Birkhed D (1983) Methylation of mercury from dental amalgam and mercuric chloride by oral streptococci in vitro. Scand J Dent Res 91:150-152

Hempel M, Wilken R-D, Miess R, Hertwich J, Beyer K (1995) Mercury contaminated sites - behavior of mercury and its species in lysimeter experiments. Water, Air, Soil Pollut 80:1089-1098

Hintelmann H, Hempel M, Wilken R-D (1995) Observation of unusual organic mercury species in soils and sediments of industrially contaminated sites. Environ Sci Technol 29:1845-1850

Hirner AV, Rehage H, Sulkowski M (2000) Umweltgeochemie. Steinkopff Verlag, Darmstadt, 836 S.

Inskip MJ, Piotrowski JK (1985) Review of the health effects of methyl mercury. J Appl Toxicol 5:113-133

Ishihara N (2000) Excretion of methyl mercury in human feces. Arch Environ Health 55:44-47

Ludwicki JK (1989) Studies on the role of gastrointestinal tract contents in the methylation of inorganic mercury compounds. Bull Environ Contam Toxicol 42:283-288

Lygre GB, Hol PJ, Eide R et al. (1999) Mercury and silver in saliva from subjects with symptoms self-related to amalgam fillings. Clin Oral Invest 3:216-218

Mackert JR, Berglund A (1997) Mercury exposure from dental amalgam fillings: Absorbed dose and the potential for adverse health effects. Crit Rev Oral Biol Med 8:410-436

Mester Z, Sturgeon RE (2002) Detection of volatile organometal chloride species in model atmosphere above seawater and sediment. Environ Sci Technol 36:1198-1201

Mori F, Tanji K, Wakabayashi K (2000) Widespread calcium deposits, as detected using the alizarin red S technique, in the nervous system of rats treated with dimethyl mercury. Neuropathology 20:210-215

Mothes S, Wennrich R (1999) Solid phase microextraction and GC-MIP-AED for the speciation analysis of organomercury compounds. J High Resol Chromatogr 22:181-182

Nielsen JB, Andersen O, Grandjean P (1994) Evaluation of mercury in hair, blood and muscle as biomarkers for methylmercury exposure in male and female mice. Arch Toxicol 68:317-321

Norseth T, Clarkson TW (1971) Intestinal transport of ^{203}Hg-labeled methyl mercury chloride. Arch Environ Health 22:568-577

Pongratz R, Heumann KG (1998) Determination of concentration profiles of methyl mercury compounds in surface waters of polar and other remote oceans by GC-AFD. Int J Environ Anal Chem 71:41-56

Puk R, Weber JH (1994a) Critical review of analytical methods for determination of inorganic mercury and methylmercury compounds. Appl Organomet Chem 8:293-302

Puk R, Weber JH (1994b) Determination of mercury(II), monomethylmercury cation, dimethylmercury and diethylmercury by hydride generation, cryogenic trapping and atomic-absorption spectrometric detection. Anal Chim Acta 292:175-183

Rowland IR, Grasso P, Davies MJ (1975) The methylation of mercuric chloride by human intestinal bacteria. Experientia 31:1064-1065

Sandborgh-Englund G, Elinder CG, Johanson G, Lind B, Skare I, Ekstrand J (1998) The absorption, blood levels, and excretion of mercury after a single dose of mercury vapor in humans. Toxicol Appl Pharmacol 150:146-153

Sarmani SB, Hassan RB, Abdullah MP et al. (1997) Determination of mercury and methylmercury in hair samples by neutron activation. J Radioanal Nucl Chem 216:25-27

Schickling C, Broekaert JAC (1995) Determination of mercury species in gas condensates by on-line coupled high-performance liquid chromatography and cold-vapor atomic absorption spectrometry. Appl Organomet Chem 9:29-36

Schmolke SR, Schroeder WH, Kock HH et al. (1999) Simultaneous measurements of total gaseous mercury at four sites on a 800 km transect: spatial distribution and short-time variability of total gaseous mercury over central Europe. Atmos Environ 33:1725-1733

Shafawi A, Ebdon L, Foulkes M et al. (2000) Preliminary evaluation of adsorbent-based mercury removal systems for gas condensate. Anal Chim Acta 415:21-32

Shafawi A, Ebdon L, Foulkes M, et al. (1999) Determination of total mercury in hydrocarbons and natural gas condensate by atomic fluorescence spectrometry. Analyst 124:185-189

Skare I, Engquist A (1994) Human exposure to mercury and silver released from dental amalgam restorations. Arch Environ Health 49:384-394

Takahashi Y, Tsuruta S, Hasegawa J et al. (2001) Release of mercury from dental amalgam fillings in pregnant rats and distribution of mercury in maternal and fetal tissues . Toxicology 163:115-126

Tao H, Murakami T, Tominaga M et al. (1998) Mercury speciation in natural gas condensate by gas chromatography inductively coupled plasma mass spectrometry. J Anal At Spectrom 13:1085-1093

Toribara TY (2001) Analysis of single hair by XRF discloses mercury intake. Hum Exp Toxicol 20:185-188

Välimäki I, Perämäki P (2001) Determination of mercury species by capillary column GC-QTAAS with purge and trap preconcentration technique. Mikrochim Acta 137:191-201

Valentino L, Torregrossa MV, Saliba LJ (1995) Health effects of mercury ingested through consumption of seafood. Water Sci Technol 32:41-47

Wasserman JC, Amouroux D, Wasserman MAV, Donard OFX (2002) Mercury speciation in sediments of a tropical coastal environment. Environ Technol 23:899-910

Weber JH, Evans R, Jones SH et al. (1998) Conversion of mercury(II) into mercury(O), monomethylmercury cation, and dimethylmercury in saltmarsh sediment slurries. Chemosphere 36:1669-1687

Yannai S, Berdicevsky I, Duek L (1991) Transformations of inorganic mercury by Candida albicans and Saccharomyces cerevisiae. Appl Environ Microbiol 57:245-247

Young JF, Wosilait WD, Luecke RH (2001) Analysis of methylmercury disposition in humans utilizing a PBPK model and animal pharmacokinetic data. J Toxicol Environ Health 63:19-52

Chapter 7

Biovolatilisation of metal(loid)s by microorganisms

K. Michalke & R. Hensel

Introduction

The production of volatile metal(loid) compounds through biomethylation and biohydridisation is a significant part of the biogeochemical cycles of metals (e.g. Bi, Hg, Sn), metalloids (e.g. As, Sb) and non-metals (e.g. S, Se, Te). Decisive for the mobility of a certain element species is its affinity to solid matrices, its solubility in solvents, and its volatility. Consequently, physical, chemical and biochemical reactions, which change these properties, will also influence the distribution of the respective elements in the environment. Of particular importance in this regard are various interactions of inorganic compounds with biota resulting in a thorough redistribution of the elements. Since most of these volatile compounds exhibit higher toxicity than their inorganic educts, these reactions lead to an uncontrolled distribution of toxicants in the environment (Craig 1988; Fergusson 1991; Thayer 1984, 1991).

Volatile methyl and hydride derivatives of metals and metalloids have been detected in gases released from anthropogenic environments such as waste water treatment plants or waste deposits (Feldmann et al. 1994, 1995) as well as from environments not significantly affected by anthropological activities such as sediments, wetlands (Wickenheiser et al. 1998) and hydrothermal springs (Hirner et al. 1998). Some of the metal(loid) derivatives are exclusively of anthropogenic origin (e.g. alkylated tin and lead derivatives), whereas some of these compounds are produced by chemical reactions (e.g. trans-alkylation); many however, are formed by biochemical reactions mediated by microorganisms (e.g. methylation and hydride formation).

Up till now, more than thirty different volatile metal(loid) compounds have been identified in different settings in a concentration range of ng to mg m^{-3}. In the case of emissions from waste and wastewater treatment processes, methylated and hydride derivatives of arsenic, antimony and bismuth represent the major portion of the volatiles (Feldmann et al. 1995; Michalke et al. 2000). Despite the high ecological impact of the uncontrolled mobilisation of metal(loid)s *via* biovolatilisation, the responsible organisms of the metal(loid)-metabolising biosphere and the underlying molecular processes of the biotransformations of inorganic metal(loid)s to their volatile derivatives are largely unknown. As a result, intense research activity is necessary to assess the real capacity of metal(loid) biotransformation under various environmental conditions, and to develop strategies for the control of processes leading to this unwelcome metal(loid) mobilisation.

Analytical setup for identifying and quantifying metal(loid)-organic derivatives

Several analytical techniques have been developed to analyse and quantify metal(oid)-organic derivatives in environmental samples such as wastewater, soils and sediments as well as in laboratory assays to study the formation, conversion and degradation of volatile organometal(loid) compounds. Volatile metal(loid) species can be extracted by purging the sample with an inert gas (e.g. N_2, He, Ar) and subsequent enrichment of these compounds on an adsorption trap at low temperatures (cryo-trapping); by enrichment on coated fibres (solid phase-microextraction (SPME); or by extraction with organic solvents. Because of the excellent selectivity and sensitivity, hyphenated techniques have been widely used to separate volatile metal(loid) species by coupling a gas chromatographic system to different detectors, such as atomic absorption spectrometry (Dirkx et al. 1994), microwave induced plasma atomic emission spectrometry (Pereiro et al. 1999; Junyapoon et al. 1999) and mass spectrometry (Morcillo et al. 1995; Dunemann et al. 1999), including inductively coupled plasma mass spectrometry (deSmaele et al. 1999; Wickenheiser et al. 1998).

Non-volatile, ionic species of precursors or degradation products of volatile metal(loid)-organic compounds can be separated by high performance liquid chromatography, but in most cases, matrices of samples from the environment impede a direct application to the liquid chromatographic system. Non-volatile metal(loid) derivatives, such as arsenosugars have to be extracted from the sample matrix prior to the application to the analytical system (Geiszinger et al. 2002). In other cases, a derivatisation of the respective species by *in situ* hydride generation or alkylation of ionic metal(loid) species and a subsequent separation of the resultant derivatives from the sample matrix is recommended (Bouyssiere et al. 2002).

Volatile metal(loid) compounds in the environment

Methyl and hydride derivatives of metal(loid)s in the environment are mainly produced by biological processes, in which organisms of all three domains, Archaea, Bacteria and Eucarya, are involved. The terms "biomethylation" and "biohydridisation" refer to the relevant reactions mediated by biota, i.e. methyl transfer to a metal(loid) (Challenger 1945) or the addition of hydrogen to a metal(loid), respectively (Wickenheiser et al.1999; Michalke et al. 2000).

Recently two reviews appeared concerning biomethylation of metal(loid)s (Thayer 2002, Bentley and Chasteen 2002)

The biogenic production of volatile metal(loid) compounds has been known since the pioneering work of Gosio (1897), who investigated the volatilisation of arsenic-containing pigments by moulds. The gas with a garlic like odour produced by the transformation was later identified by Challenger (1945) as trimethylarsine (Me_3As). Since the pioneering work of Gosio and Challenger, intensive research has been performed with respect to biotransformations of arsenic. The derivatisation of arsenic has been studied extensively with various fungi (Cox and Alexander 1973; Cullen et al. 1995), bacteria (Bachofen et al.1995; Honschopp et al. 1996; Michalke et al. 2000), archaea (Bachofen et al.1995; Michalke et al. 2000) and mammals (Thompson 1993; Zakharyan et al. 1995; Aposhian et al.1997), including humans (Crecelius 1977). In addition to arsenic, Challenger (1945) reported the biomethylation of selenium and tellurium to dimethylselenide and dimethyltellurium by moulds. These studies were confirmed later by other authors (Doran 1982; Gao and Tanji 1995; Van Fleet-Stalder and Chasteen 1998).

As demonstrated for arsenic, the biovolatilisation of the chemically related elements antimony and bismuth also occurs in nature. Although trimethylantimony and trimethylbismuth have been shown to represent the largest fractions of volatile metal(loid) derivatives in gases released from municipal waste deposits and sewage gases (Feldmann et al. 1994, 1995), only recently was the biogenic origin of these compounds in sewage sludge demonstrated (Michalke et al. 2000). Obviously, biomethylation of antimony is mediated by members of all three domains, as shown for fungi (Andrewes et al. 1998; Andrewes et al. 2001; Jenkins et al. 1998), bacterial species (Jenkins et al., 2002; Michalke et al. 2000) and methanoarchaea (Michalke et al. 2000). In contrast, the volatilisation of bismuth was reported only recently for bacterial and methanoarchaeal species (Michalke et al. 2000, 2002; Wickenheiser et al. 2000) and nothing is known about biomethylation of bismuth by eucaryotes. The common sewage-sludge microorganism *Methanobacterium formicicum* seems to be very versatile in producing volatile metal(loid) compounds. It not only methylates arsenic and antimony, but also bismuth to the corresponding trimethyl derivatives. Additionally, the production of the respective hydrides and the partially methylated monomethyl- and dimethylhydrides of these elements could be shown by this organism (Michalke et al. 2000, 2002).

Similar to arsenic, the study of biomethylation of mercury was motivated by tragic poisonings. Between 1941 and 1956 in Japan (Luke and Tedeschi 1982) and in 1972 in Iraq (Bakir et al. 1973) several thousand cases of poisonings - many of them fatal - by organomercury compounds were reported. The intoxications in Japan were caused by the uptake of fish from the Minamata Bay area, which contained high amounts of methylmercury due to microbial conversion in discarded waste waters containing mercury (Luke and Tedeschi 1982).

The biomethylation of inorganic tin has been reported (Hallas et al. 1982; Ashby and Craig 1988). Organotin compounds gained special interest as biocide additives (e.g. tributyltin). These compounds can leach into aquatic environments and undergo dealkylation and methylation (Clark et al. 1988; Errecalde et al. 1995).

Towards an understanding of metal(loid) biotransformation processes

Analyses of the role of the microflora engaged in the synthesis of volatile derivatives of metal(loid)s would not only contribute to a basic understanding of biomethylation and biohydridisation, but would also provide a basis for the rational controlling of these processes, especially in anthropogenic settings such as waste water treatment facilities and waste deposits.
Considerable efforts have been made in the past to define and differentiate the natural biotransformation potential of microorganisms, nevertheless little is known about the identity of the organisms involved in biotransformations under environmental conditions. Recent research strategies generally follow three approaches: (i) Direct monitoring of metal(loid) derivatives in the environment; (ii) monitoring of metal(loid) volatilisation by microbial populations enriched from their original environment; (iii) monitoring of metal(loid) volatilisation by isolated strains. In the following, we will describe the various approaches in brief, and discuss their advantages and limitations.

Direct monitoring of volatile metal(loid) derivatives in the environment

A rewarding object for studying the biotransformation of metal(loid)s is represented by the anthropogenic habitat of anaerobic surplus sewage sludge stabilisation. Direct monitoring of the volatile metal(loid) derivatives is essential for giving insights into the real diversity of these compounds, but can not however, differentiate them with respect to abiotic or biotic origin.

Gases released from stabilisation tanks at sewage treatment plants contained volatile methyl and hydride derivatives of antimony, arsenic, bismuth, lead, mercury, tellurium and tin (Table 1).

Table 1. Volatile metal(loid) compounds in sewage gases released from anaerobic sewage sludge stabilisation tanks (Krupp 1999; Michalke et al. 2000).

volatile metal(loid) compound	concentration [ng m^{-3}]
dimethylmercury (Me_2Hg)	16 – 57
tetraethyllead (Et_4Pb)	7 – 143
tetramethyltin (Me_4Sn)	n.d. – 189
dimethylarsine (Me_2AsH)	n.d. – 511
monomethylarsine ($MeAsH_2$)	6 – 678
arsine (AsH_3)	21 – 761
dimethyltellurium (Me_2Te)	n.d. – 1 859
trimethylarsine (Me_3As)	46 – 3 321
trimethylbismuth (Me_3Bi)	1 665 – 46 234
trimethylantimony (Me_3Sb)	483 – 77 527

n.d. not detected

Table 2. Metal(loid) content of sewage sludge samples and volatile metal(loid)s detected in the headspace of the incubated samples after 1 week at 37 °C under anaerobic conditions (Michalke et al. 2000).

metal(loid)	mg kg^{-1} dry weight [a]	volatile derivative	mg m^{-3} [b]	relative conversion rate [c]
Sb	1.7 ± 0.3	Me_3Sb	77.5 ± 4.6	5980
Bi	1.4 ± 0.3	Me_3Bi	46.2 ± 6.9	4433
Te	0.2 ± 0.1	Me_2Te	1.9 ± 0.15	1247
As	15.2 ± 1.3	AsH_3	0.76 ± 0.21	
		$MeAsH_2$	0.68 ± 0.45	47 [d]
		Me_2AsH	0.51 ± 0.36	
		Me_3As	3.3 ± 1.3	
Hg	3.0 ± 1.2	Hg^0	0.29 ± 0.16	18
Sn	25.4 ± 0.3	Me_4Sn	0.29 ± 0.18	1

[a] Mean values ± absolute errors are listed.
[b] Mean values of three independent measurements ± standard deviations are listed.
[c] conversion rates are normalised to the element with the lowest conversion rate, i.e. tin.
[d] calculated for the sum of all volatile As species.

In most cases, derivatives of arsenic, antimony and bismuth were the dominant volatile metal(loid) compounds detected in the emitted gases. However, the predominance of these volatile derivatives did not correlate with the content of the respective metal(loid)s in the sewage sludge; e.g. the conversion rates of antimony and bismuth to volatile Me_3Sb and Me_3Bi in incubated sludge samples were by orders of magnitude (up to approx. 6000-fold) higher than the conversion rates of arsenic or tin (Table 2). The data could be interpreted in terms of different stabilities of the resultant compounds, and/or in terms of a different bioavailability of the elements for biotransformation and a different inherent susceptibility of the different metal(loid)s towards volatilisation by microorganisms in sewage sludge. This can only be answered by studies under more defined conditions.

Monitoring of metal(loid) volatilisation by microbial populations enriched from original environments

The isolation of specific physiological groups of microorganisms from environmental samples through the application of culture enrichment methods, can be used as a first approximation of the biogenic production of volatile metal(loid) derivatives in the setting investigated: They could provide closer insights into general pre-requisites of biovolatilisation of metal(loid)s and give valuable hints for the *in situ* derivatisation capacity of certain physiological groups of microorganisms.

Since, however, the experiments are performed under conditions excluding – either intentionally or non-intentionally - certain members of the original microbial community and thus disturbing the complex interactions of the natural population, the observed activities are certainly not directly comparable with those obtained under *in situ* conditions and must therefore be interpreted with some caution: The disintegration results in a disruption of the metabolic co-operation within the microbial population and consequently in a respective change of the metal(loid) transformation activity of their members. An example of the close relationship between metabolic co-operation and methylation activity has been demonstrated for a co-culture of *Desulfovibrio desulfuricans* and *Methanococcus maripaludis* by Pak and Bartha (1998). Although only *D. desulfuricans* is able to methylate mercury, the inhibition of the non-methylating *M. maripaludis* in a co-culture of both organisms led to an inhibition of organometal(loid) production, because *D. desulfuricans* was inhibited by the hydrogen which would otherwise be utilised and removed by the methanoarchaeum.

As an example, to define groups of microorganisms that are involved in the production of the volatile metal(loid) compounds in sewage sludges, we added several growth inhibitors to sewage sludge samples, which are known to be effective against various groups of prokaryotes (ampicillin against bacteria in general, sodium molybdate against sulfate reducing bacteria, bromoethane sulfonate, monensin, and lasalocide against methanogenic archaea; iodopropane mainly, but not exclusively, against methanogenic archaea). In addition to monitoring the formation of volatile metal(loid) derivatives, methane production was followed as an indicator for the metabolic activity of methanogens present in the samples.

As shown in Table 3, the drugs bromoethane sulfonate, monensin, lasalocide, and iodopropane, which are effective against methanogens, inhibit the methane production from 47 % to almost 100% as compared to the control. On the other hand, samples incubated with ampicillin or sodium molybdate, which are known to inhibit either bacteria in general (ampicillin) or sulphate reducing bacteria in particular (sodium molybdate) produced 23 % or 12 % more methane than the control, respectively. The increased methane production in such samples may be due to a decreased competition for nutrients, or is possibly favoured by nutrients, which are released by the lysis of the inhibited members of the population.

Table 3. Production of volatile metal(loid) derivatives and methane in sewage sludge samples in the presence of different biocides incubated for one week anaerobicaly at 37 °C in the dark. Due to the high variance the numbers are given in a range as percentage of control samples. Increased amounts are printed in bold, decreased amounts are printed in plain text. n=3.

	control	ampicillin (200 µg/ml)		sodium molybdate (4 mM)		bromoethane sulfonate (1 mM)		monensin (100 µg/ml)		iodopropane (10 µM)		lasalocide (100 µg/ml)	
expected activity against		bacteria		sulfate reducing bacteria		methanogens		methanogens		methanogens		methanogens	
methane	21.2 ± 1.4 mM	**23% ± 7%**		**12% ± 5%**		-47% ± 7%		-64% ± 1%		-79% ± 15%		-98% ± 1%	
	[mg m^{-3}]	from	to	from	to	from	to	from	to	from	to	from	to
AsH_3	0.76 ± 0.21	-3%	**22%**	n.d.	n.d.	-32%	0,2%	-59%	-43%	n.d.	n.d.	n.d.	n.d.
$MeAsH_2$	0.68 ± 0.45	-8%	**16%**	n.d.	n.d.	n.d.	n.d.	-2%	**24%**	n.d.	n.d.	n.d.	n.d.
Me_2AsH	0.51 ± 0.36	n.d.	n.d.	n.d.	n.d.	-52%	**11%**	**13%**	**206%**	n.d.	n.d.	**2450%**	**3980%**
Me_3As	3.3 ± 1.3	-82%	-54%	**418%**	**732%**	-71%	**238%**	**455%**	**1400%**	-37%	-29%	**890%**	**1900%**
Me_2Te	1.9 ± 0.15	-24%	-16%	-76%	-64%	-89%	-49%	-84%	-78%	-35%	-13%	-91%	-86%
Me_3Bi	46.2 ± 6.9	-97%	-48%	-33%	**4%**	-98%	-17%	-99%	**97%**	-54%	-44%	**67%**	**141%**
Me_3Sb	77.5 ± 4.6	-78%	-20%	-70%	-57%	**290%**	**891%**	**124%**	**234%**	-8%	**139%**	**34%**	**330%**

n.d. not detected

In contrast to the methane production, the influence of the various drugs on the production of volatile metal(loid) compounds is not uniform showing considerable variations in replicate experiments (Table 3). Despite the rather heterogeneous data set, however, some trends can be observed.

A significant promoting effect on the production of Me_3As, Me_2AsH, Me_3Bi and Me_3Sb (up to a 40-fold increase compared to the control) was noted in monensin and lasalocide incubations. These compounds are expected to mainly inhibit methanogens by their permeability promoting effect on membranes (Hilpert and Dimroth, 1982). The mode of action of these drugs is supported by the observation of decreased methane production in our experiment and may indicate that non-methanogenic microorganisms are responsible for the production of Me_3As, Me_2AsH, Me_3Bi and Me_3Sb. The enormous increase of the volatilised metal(loid) derivatives could be explained by the permeability promoting activity of these biocides facilitating the uptake of metal(loid)s into the cell by overcoming the diffusion barrier of the cell-membrane thus resulting in an increased production of the respective volatile metal(loid) compounds suggesting that the penetration of the inorganic metal(loid) through the cell membrane promotes biomethylation. Furthermore, the exit of volatile biomethylation products through the cell membrane may also be increased, causing an apparent increase in the biomethylation capability. These explanations, however, would implicate that the methylation is mainly limited by the availability of the metal(loid)s and does not depend necessarily on the integrity of the cell. In contrast to monensin and lasalocide, compounds such as ampicillin, sodium molybdate and bromoethane sulfonate, inhibit the production of volatile metal(loid) derivatives for the most part – with the exception of Me_3As and Me_3Sb - suggesting that bacteria in general, the sulfate reducing bacteria and the methanoarchaea contribute all to the biovolatilisation process of metal(loid)s in sewage sludges. Although the general inhibitor iodopropane inhibits mainly methanogenic microorganisms – as also shown by its negative effect on the methane production - by mimicking the essential co-factor of methanogenesis co-enzyme M, the decreased production of volatile metal(loid) compounds can not be assigned to a specific group of organisms because iodopropane also deactivates corrinoid dependent reactions (Keneally and Zeikus, 1981), which are essential cofactors of the C_1-metabolism of all organisms.

Despite the inherent ambiguity of the obtained data set, the observed effects clearly account for a biogenic origin of the various volatile metal(loid) derivatives found in the emitted gases of anaerobic sludge stabilisation plants and support the assumption that members of various physiological groups are engaged in the biovolatilisation process.

Table 4. Production of volatile metal(loid) derivatives by anaerobic sewage sludge microorganisms (Michalke et al. 2000 and 2002).

		As	Sb	Bi	Se	Te	Hg	Sn
archaea	*Methanobacterium formicicum*	AsH_3 $MeAsH_2$ Me_2AsH Me_3As X^a	SbH_3 $MeSbH_2$ Me_2SbH Me_3Sb	BiH_3, $MeBiH_2$ Me_2BiH Me_3Bi	Me_2Se Me_2Se_2	Me_2Te	Hg^0	n.d.
	Methanosarcina barkeri	AsH_3, X^a	Me_3Sb	n.d.	Me_2Se Me_2Se_2	n.d.	Hg^0	n.d.
	Methanothermobacter thermautotrophicus	AsH_3	Me_3Sb	n.d.	n.d.	n.d.	Hg^0	n.d.
peptolytic bacteria	*Clostridium collagenovorans*	Me_3As	Me_3Sb	Me_3Bi	Me_2Se Me_2Se_2	Me_2Te	Hg^0	n.d.
sulfate reducing bacteria	*Desulfovibrio vulgaris*	Me_3As	Me_3Sb	n.d.	Me_2Se Me_2Se_2	n.d.	Hg^0	n.d.
	Desulfovibrio gigas	Me_3As	n.d.	n.d.	Me_2Se Me_2Se_2	Me_2Te	Hg^0	n.d.
	sterile control	n.d.	n.d.	n.d.	n.d.	n.d.	Hg^0	n.d.

n.d. not detected

X^a unidentified volatile arsenic species

Monitoring of metal(loid) volatilisation by isolated strains

Pure culture experiments with various species confirmed that the ability of biomethylation and biohydridisation is widely spread within different groups of microorganism. Table 4 lists the volatile metal(loid) species detected in the headspace of pure cultures of different common sewage sludge microorganisms (Michalke et al. 2000 and 2002).

All microorganisms tested were able to transform arsenic, and most volatilised antimony and selenium species. The presence of volatile elemental mercury in all samples however, is not an evidence for its biogenic origin (e.g. from the reduction of Hg^{2+} or the demethylation of Me_2Hg), since the formation of elemental mercury could be also followed in sterile controls, due likely to the reductive conditions of the assay. No microbial species in our assays was able to volatilise tin or mercury by methylation. Among the microorganisms tested, the methanogen *M. formicicum* showed the highest productivity in terms of the biotransformation of metal(loid)s with both number and amount of organometal(loid) species produced being higher than for other microorganisms (Michalke et al. 2000 and 2002). *M. formicicum* not only produced the volatile permethylated compounds of arsenic, antimony and bismuth, but also partially methylated hydrides and the corresponding hydrides of these elements.

Strikingly, the capability for producing the various organometal(loid) derivatives is rather diverse – even amongst closely related species. Thus, neither *D. vulgaris* nor *D. gigas* were able to methylate mercury, although the close relative *Desulfovibrio desulfuricans* strain LS being a well known methylator of mercury (Bermann et al., 1990; Baldi et al. 1993 and 1995). The observation is all the more surprising since generally sulfate reducers are generally assumed to be the key methylating organisms of mercury in lake sediments (Compeau and Bartha, 1985). Impressive in this regard is also the striking difference between the excessive methylation capacity of *M. formicicum,* which exceeds that of the methanogens *M. thermautotrophicus* and *M. barkeri* in quantity and diversity of the products (Michalke et al. 2000 and 2002). The obvious diversity in biomethylation and biohydridisation capacities - even within groups of related organisms – may be explained by a generally high variability of the methylation and hydridisation phenotypes in nature.

As a further disadvantage, the pure culture techniques allow limited insights into the productivity of the respective strain within its original habitat, which is characterised by a combination of specified parameters and conditions such as pH, temperature, metal(loid) ion concentration, carbon source, redox potential, cell density, growth phase or interactions with other members of the consortium, which can hardly be simulated *in vitro* or - more probably – by non-comparable, mostly suboptimal reaction conditions of the pure culture assays, which only allow the recognition of different and rather narrow sections of the whole band-width of the methylation capacity of a certain organism in its environmental setting. An example is the common sewage sludge organism *Methanosarcina barkeri* which only methylates bismuth in the presence of polydimethylsiloxanes (Wickenheiser et al. 2000) which are not constituents of its growth medium, but are present in high concentrations in sewage sludges (Gruemping et al. 1999).

Thus, also the missing production of organometal(loid) derivatives as observed under pure culture conditions does not necessarily account for the general inability of the organism to produce metal(loid)-organic compounds in an environmental setting, but rather, may be due to differing growth conditions *ex situ* and *in situ*.

Studies of axenic cultures are indispensable to analyse exemplarily the molecular mechanism of methylation or hydridisation in detail, however a comprehensive assessment of the ecologically relevant methylation and hydridisation activity of a micoorganism can only be explored with *in situ* methods, which enable both monitoring of the process of methylation and hydridisation, as well as the identification of responsible organisms directly in the habitat. Whereas for the former (*in situ* determination of methylation and hydridisation activities) respective developmental work has to be invested, for the latter (*in situ* identification of microorganisms) an established methodology is already available, namely fluorescent *in situ* hybridisation (FISH) (Amann et al. 1990).

To get an insight into the biodiversity of methanoarchaea in sewage sludge and to determine the abundance of *M. formicicum* in the surplus sewage sludge the microflora was analyzed by the use of fluorescent *in-situ* hybridisation with indocarbocyanine (CY3) labeled specific 16S rRNA targeted oligonucleotide probes. For these studies 16 S rRNA probes with different specificities were used: (i) ARCH915, a specific probe for archaea (Stahl and Amann1991), (ii) MB1174, a probe specific for the family *Methanobacteriacea* (Raskin et al. 1994), (iii) MBF60, a strain specific probe for *M. formicicum* strain DMSZ 1535 (1312) (this work), (iv) MBF262, a group specific probe for the *M. formicicum* strains DSMZ 1535^T, DSMZ 1312, DSMZ 3636, Fcam and *Methanobacterium subterraneum*, *Methanobacterium palustre* (this work). Epifluorescence microscopy showed that 12.8 ± 3.2 % of the DAPI-stained cells ($5.38 * 10^9$ cells $ml^{-1} \pm 10.2$ %) hybridised with the ARCH915 probe. Assuming that all cells that hybridised with probe ARCH915 belong to methanogenic archaea, 8.3 ± 2.8 % thereof hybridised with the family-specific probe for *Methanobacteriaceae* (MB1174), 13.6 ± 6.4 % hybridised with the group-specific probe of *M. formicicum*-related organisms (MBF262), and 9.6 ± 7.8 % hybridised with the *M. formicicum* strain-specific probe (MBF66), respectively (Fig. 2). Based on these results, *M. formicicum* represents nearly 1 % of the population of the methanogenic archaea in the sewage sludge investigated.

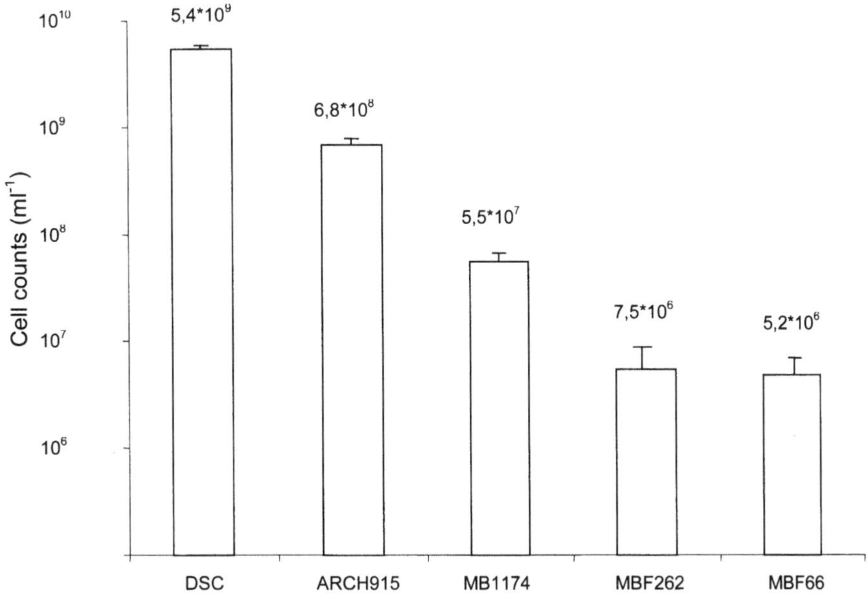

Fig. 1. Biodiversity of methanoarchaea in sewage sludge. Cell counts of FISH for the probes specific for archaea (ARCH915), Methanobacteriaceae (MB1174), *M. formicicum*-related organisms (MBF262 5'-TTACTGGCTTGGTGGGCATT–3') and *M. formicicum* strains DSMZ 1535 and DSMZ 1312 (MBF66 5'-CGCCACGACCCCGAAGGATCG-3') compared to the DAPI stained cell counts (DSC). Hybridisations were performed on polycarbonate filters for 4 h at 48°C with 20 % formamide following a washing step at 50 °C for 30 min according to Amann et al. (1990). The data shown are mean values of three independent samplings and hybridisation experiments ± standard deviations.

Although *M. formicicum* represents only a small fraction of the whole sewage sludge microbial community its cell counts of approx. $5 * 10^6$ ml^{-1} and its high biomethylation and biohydridisation capacity - as determined by pure culture experiments - may contribute to the volatilisation of metal(loid)s in the sewage sludge. However, the *ex situ* capabilities of this organism does not necessarily account for its function *in situ*.

Conclusion and perspectives

Although the first reports about the ability of microorganisms to transform metal(loid)s to volatile derivatives appeared at least 120 years ago, there is still little known about the mechanisms defining the specificity of the enzymatic reactions concerning inorganic educts and emitted products, about the responsible organisms of biovolatilisation and about the conditions governing the production of these compounds, despite the high ecotoxicological relevance of these processes. Thus, the research on biovolatilisation of metal(loid)s, a partial aspect of biotransformation of these elements, is still in its infancy.

The present data about biovolatilisation deduced from emissions of volatile metal(loid) compounds from various settings, their production in enrichment and pure culture experiments can only give a rough impression of the real extent of the underlying transformations, mainly because these experiments do not consider the *in situ* conditions adequately. Therefore, a successful strategy to obtain more insight into these ecologically important processes must aim at a methodology, which is able to monitor the process of methylation and hydridisation directly in the habitat by following the intracellular synthesis of respective metal(loid) products and their precursors. Our future research strategy will focus therefore, on a methodology combining both *in situ* analytical speciation of metal(loid)s and identification of microorganisms by FISH.

Literature

Amann RI, Krumholz L, Stahl DA (1990) Fluorescent-oligonucleotide probing of whole cells for determinative phylogenetic and environmental studies in microbiology. J Bacteriol 172:762-70

Andrewes P, Cullen WR, Feldmann J, Koch I (1998) Methylantimony compound formation in the medium of *Scopulariopsis brevicaulis* cultures: 3CD3-L-methionine as a source of the methyl group. Appl Organomet Chem 13:681–688

Andrewes P, Cullen WR, Polishchuk E, Reimer KJ (2001) Antimony biomethylation by the wood rotting fungus *Phaeolus schweinitzii.* Appl Organomet Chem 15:473–480

Aposhian HV (1997) Enzymatic methylation of arsenic species and other new approaches to arsenic toxicity. Annu Rev Pharmacol Toxicol 37:397-419

Ashby J R and P J Craig 1988 Environmental methylation of tin: an assessment Sci Tot Environ 73:127-133

Bachofen R, Birch L, Buchs U, Ferloni P, Flynn I, Jud G, Tahedel H, Chasteen TG (1995) Volatilization of arsenic compounds by microorganisms. *In* RE Hinchee (ed) Bioremediation of Inorganics. Batelle Press, Columbus

Bakir FS, Damluji F, Amin-Zaki L, Murtadha M, Khalidi A, Al-Rawi NY, Tikriti S, Dhahir HI, Clarkson TW, Smith JC, Doherty RA (1973) Methylmercury Poisoning in Iraq. Science 181:230-241

Baldi F, Parati F, Filippelli M (1995) Dimethylmercury and dimethylmercury-sulfide of microbial origin in the biogeochemical cycle of Hg. Water Air Soil Pol 80:805-815

Bentley R, Chasteen TG (2002) Microbial methylation of metalloids: Arsenic, antimony, and bismuth. Microbiol Mol Biol R 66:250-271

Bermann M, Chase T, Bartha R (1990) Carbon flow in mercury biomethylation by *Desulfovibrio desulfuricans.* Appl Environ Microbiol 56:298-300

Bouyssiere B, Szpunar J, Lobinski R (2002) Gas chromatography with inductively coupled plasma mass spectrometric detection in speciation analysis. Spectrochim Acta B 57:805-828

Challenger F (1945) Biological methylation. Chem Rev 36:315-318

Clark EA, Sterritt RM, Lester JN (1988) The fate of tributyltin in the aquatic environment. Environ Sci Technol 22:600-604

Compeau CG, Bartha R (1985) Sulfate-reducing Bacteria: Principal methylators of mercury in anoxic estuarine sediments. Appl Environ Microbiol 50:498-502

Cox DP, Alexander M (1973) Production of trimethylarsine gas from various arsenic compounds by three sewage fungi. Bull Environ Contam Toxicol 9:84-88

Craig P J (ed) (1986) Organometallic Compounds in the Environment: Principles and Reactions. John Wiley & Sons Inc, New York

Craig PJ, Glockling F (Eds) (1988) The Biological Alkylation of Heavy Elements. The Royal Society of Chemistry, London

Crecelius EA (1977) Changes in the chemical speciation of arsenic following ingestion by man. Environ Health Perspect 19:147-150

Cullen WR, Li H, Pergantis SA, Eigendorf GK, Mosi AA (1995) Arsenic biomethylation by the microorganism *Apiotrichum humicola* in the presence of L-methionine-*methyl*-d3. Appl Organomet Chem 9:507-515

De Smaele T, Moens L, Sandra P, Dams R (1999) Determination of organometallic compounds in surface water and sediment samples with SPME-CGC-ICPMS. Mikrochimica Acta 130:241-251

Dirx WMR, Lobinski R, Adams FC (1994) Speciation analysis of organotin in water and sediments by gaschromatography with optical spectrometric detection after extraction seperation. Anal Chim Acta 286:309-318

Doran JW (1982) Microorganisms and the biological cycling of selenium. Adv Microbiol Ecol 6:1-32

Dunemann L, Hajimiragha H, Begerow J (1999) Simultaneous determination of Hg(II) and alkylated Hg, Pb, and Sn species in human body fluids using SPME-GC/MS-MS. Fresen J Anal Chem 363:466-468

Errécalde O, Astruc M, Maury G, Pinel R (1995) Biotransformation of butyltin compounds using pure strains of microorganisms. Appl Organomet Chem 9:23-28

Feldmann J, Grümping R, Hirner AV (1994) Determination of volatile metal and metalloid compounds in gases from domestic waste deposits with GC/ICP-MS. Fresenius J Anal Chem 350:228-234

Feldmann J, Grümping R, Hirner AV (1994) Determination of volatile metal and metalloid compounds in gases from domestic waste deposits with GC/ICP-MS. Fresenius J Anal Chem 350:228-234

Feldmann J, Hirner AV (1995) Occurrence of volatile metal and metalloid species in landfill and sewage gases. Int J Environ Anal Chem 60:339-359

Feldmann J, Hirner AV (1995) Occurrence of volatile metal and metalloid species in landfill and sewage gases. Int J Environ Anal Chem 60:339-359

Fergusson JE (1991) The Heavy Elements Chemistry Environmental Impact and Health Effects. Pergamon Press, Oxford

Gao S, Tanji KK (1995) Model for biomethylation and volatilization of selenium from agricultural evaporation ponds. J Environ Qual 24:191-197

Geiszinger AE, Goessler W, Kosmus W (2002) An arsenosugar as the major extractable arsenical in the earthworm Lumbricus terrestris. Appl Organomet Chem 16:473-476

Gosio B (1897) Zur Frage wodurch die Giftigkeit arsenhaltiger Tapeten bedingt wird. Ber Deutsch Chem Ges 30:1024-1026

Grumping R, Michalke K, Hirner AV, Hensel R (1999) Microbial degradation of octamethylcyclotetrasiloxane. Appl Environ Microbiol. 65:2276-2278

Hallas LE, Means JC, Cooney JJ (1982) Methylation of tin by estuarine microorganisms. Science 215:1505

Hilpert W, Dimroth P (1982) Conversion of the chemical energy of methylmalonyl-CoA decarboxylation into a Na^+ gradient. Nature 296:584-585

Hirner AV, Feldmann J, Krupp E, Grümping R, Goguel R, Cullen WR (1998) Metal(loid)organic compounds in geothermal gases and waters. Organic Geochemistry 29:1765-1778

Honschopp S, Brunken N, Nehrkorn A, Breunig HJ (1996) Isolation and characterization of a new arsenic methylating bacterium from soil. Microbiol Res 151:37-41

Huysmans KD, Frankenberger WT, (1991) Evolution of trimethylarsine by a *Penicillium* sp isolated from agricultural evaporation pond water. Sci Total Environ 105:13-28

Jenkins RO, Craig PJ, Goessler W, Miller D, Ostah N, Irgolic KJ (1998) Biomethylation of inorganic antimony compounds by an aerobic fungus: *Scopulariopsis brevicaulis* Environ Sci Technol 32:882–885

Jenkins RO, Forster SN, Craig PJ (2002) Formation of methylantimony species by an aerobic prokaryote: *Flavobacterium* sp. Arch Microbiol 178: 274-278

Junyapoon S, Ross AB, Bartle KD, Frere B (1999) Injection by programmed temperature vaporization injection (PTV) of gaseous samples for gas chromatography atomic emission spectrometry (GC-AED). J High Res Chrom 22 (1):47-51

Keneally W, Zeikus JG (1981) Influence of corrinoid antagonists on Methanogen metabolism. J Bact 146:133-140

Krupp E (1999) Analytik umweltrelevanter Metall(oid)-Spezies mittels gaschromatographischer Trennmethoden. Ph.D. thesis University of Essen

Lee N, Nielsen PH, Andreasen KH, Juretschko S, Nielsen JL, Schleifer KH, Wagner M (1999) Combination of fluorescent in situ hybridization and microautoradiography-a new tool for structure-function analyses in microbial ecology. Appl Environ Microbiol 65:1289-1297

Luke GT, Tedeschi MD (1982) The Minamata disease. Am J For Med Pat 3:335-338

Michalke K, Meyer J, Hirner AV, Hensel R (2002) Biomethylation of bismuth by the methanogen *Methanobacterium formicicum.* Appl Organomet Chem 16:221-227

Michalke K, Wickenheiser EB, Mehring M, Hirner AV, Hensel R (2000) Production of volatile derivatives of metal(oid)s by the microflora involved in the anaerobic digestion of sewage sludge. Appl Env Microbiol 66:2791-2796

Morcillo Y, Cai Y, Bayona JM (1995) Rapid determination of methyltin compounds in aqueous samples using solid phase microextraction and capillary gas chromatography following in-situ derivatization with sodium tetraethylborate. J High Res Chrom 18:767-770

Pak K, Bartha R (1998) Mercury methylation by interspecies hydrogen and acetate transfer between sulfidogens and methanogens. Appl Environ Microbiol 6:1987-1990

Pereiro IR, Wasik A, Lobinski R (1999) Speciation of organotin in sediments by multicapillary gas chromatography with atomic emission detection after microwave-assisted leaching and solvent extraction-derivatization. Fresen J Anal Chem 363:460-465

Raskin L, Stromley JM, Rittmann BE, Stahl DA (1994) Group-specific 16S rRNA hybridization probes to describe natural communities of methanogens. Appl. Environ. Microbiol. 60:1232-1240

Stahl DA, Amann R (1991) Development and application of nucleic acid probes in bacterial systematics. p. 205 – 248. In E. Stackebrandt and M. Goodfellow (eds), Nucleic acid techniques in bacterial systematics. John Wiley Sons Ltd. Chichester, England.

Thayer J S (1984) Organometallic compounds and living Organisms. Academic Press Inc, Orlando

Thayer JS (1995) Environmental Chemistry of the Heavy Elements. VCH, Weinheim

Thayer JS (2002) Biological methylation of less-studied elements. Appl Organomet Chem 16:677-691

Thompson DJ (1993) A chemical hypothesis for arsenic methylation in mammals. Chem Biol Interactions 88:89-114

Van Fleet-Stalder V, Chasteen TG (1998) Using fluorine-induced chemiluminescence to detect organo-metalloids in the headspace of phototrophic bacterial cultures amended with selenium and tellurium. J Photochem Photobiol B: Biology 43:193-203

Wickenheiser EB, Michalke K, Hensel R, Drescher C, Hirner AV, Brutishauser B, Bachofen R (1998) Volatile compounds in gases emitted from the wetland bogs near Lake Cadagno. *In* Peduzzi R, Bachofen R, Tonolla M (eds) Documenta dell'Instituto Italiano di Idrobiologia No. 63, Lake Cadagno

Wickenheiser EB, Michalke K, Hirner AV, Hensel R, Flassbeck D (2000) The biological methylation of bismuth; evidence for the involvement of polydimetylsiloxanes in the biologically-mediated methylation of metals. In: Centeno JA Collery Ph Vernet G Finkelman RB Gibb H Etienne JC (eds) Metal ions in biology and medicine. John Libbey Eurotext Paris 6:120-123pp

Zakharyan R, Wu Y, Bogdan GM, Aposhian HV (1995) Enzymatic methylation of arsenic compounds: assay partial purification and properties of arsenite methyltransferase and monomethylarsonic acid methyltransferase of rabbit liver. Chem Res Toxicol 8:1029-1038

Chapter 8

The Effect of Phosphate on the Bioaccumulation and Biotransformation of Arsenic(V) by the Marine Alga *Fucus gardneri*

S. C. R. Granchinho, W. R. Cullen, E. Polishchuk, and K. J. Reimer

Introduction

Arsenic is present in the natural environment and living organisms in different chemical forms (Cullen and Reimer 1989). In the marine environment, high concentrations of arsenic occur in macroalgae with concentration factors of 1000 to 10,000 reported (Cullen and Reimer 1989; Francesconi and Edmonds 1993; Francesconi and Edmonds 1997), and algae seem to be able to accumulate arsenic more efficiently than the higher members of the food web. Few studies have been reported, however, on either the mechanism of arsenic accumulation, or on the algae pathways through which the arsenic is accumulated and biotransformed (Francesconi and Edmonds 1997).

It has been suggested that arsenate [As(V)] enters living cells via an active phosphate transport system that is not able to distinguish between the arsenate and phosphate species (Sanders and Windom 1980). In oxygenated seawater, $H_2PO_4^-$ and $H_2AsO_4^-$ would be the predominant species of phosphorus and arsenic. However, the situation with regard to algal uptake is not clear, with several apparently conflicting reports in the literature (Francesconi and Edmonds 1993; Francesconi and Edmonds 1997). For example, although arsenate and phosphate were shown to compete for uptake by unicellular algae (Bottino et al. 1978), a related study with two species of brown macroalgae provided no evidence for a common uptake mechanism (Klumpp 1980). At low phosphate levels, arsenate uptake by unicellular algae increased with increasing phosphate concentrations, presumably as a consequence of phosphate stimulated algal metabolism (Andrae and Klump 1979). In a study of several natural algal populations, the algae readily assimilated arsenate even though phosphate concentrations remained high, indicating that arsenate uptake was not contingent with phosphate depletion (Francesconi and Edmonds 1997). Thus it was of interest to determine the effect of phosphate on arsenate uptake by *Fucus gardneri*, a species that we have studied for a number of years (Harrington et al. 1997; Lai et al. 1998; Granchinho et al. 2001; Granchinho et al. 2002).

Experimental

Reagents and Chemicals

As described previously (Granchinho et al. 2001) all chemicals used were of analytical grade or better and obtained from commercial sources. Glassware and plasticware were cleaned by soaking in 2 % Extran solution for at least one night. They were rinsed with deionized water, and then soaked in 0.1 M HNO_3 solution for at least one night. They were then rinsed with deionized water and air-dried. All glassware was autoclaved prior to use to store the standard solutions or for the algal studies.

Standard solutions of arsenite (from arsenic trioxide, As_2O_3, Fisher), arsenate (from sodium arsenate heptahydrate, $Na_2HAsO_4{\bullet}7H_2O$, Sigma), monomethylarsonic acid (Pfalz & Bauer), dimethylarsinic acid (Aldrich), and arsenobetaine (synthesized as described by Edmonds et al. (1977) were made up in deionized water. Standard samples of oyster tissue SRM (NIST-1566a), Fucus sample (IAEA-140/TM), and kelp powder (laboratory standard-Galloway's naturally kelp powder, Richmond, B.C., Canada) were also available to confirm the retention times of arsenosugars obtained from the HPLC/ICP-MS chromatograms.

Medium and Antibiotics

The artificial seawater that was used in the arsenic exposure experiments, ASP6 F2, was identical as that previously used by Fries (Fries 1977) to culture members of the family *Fucaceae*. The background phosphate concentration in the medium was 0.2 ppb. Extra phosphate as K_2HPO_4 was added to the medium (0.5, 1.0, 1.5, 2.0 or 4.0 μg/ml) for both the acclimation and uptake stages. The pH of the artificial seawater was adjusted from 9.72 to 7.76, the pH of the seawater that was collected with the algae, by the addition of HCl. [The pH reported for the ASP6 F2 artificial seawater is 8.3 (Fries 1980)]. The seawater was prepared without the vitamins, autoclaved, and stored at room temperature. The vitamin solution and the vitamin B_{12} solution were sterilized by using 0.22 μm sterile filters prior to addition to the autoclaved seawater. The artificial seawater with the vitamins was then stored at 4 °C.

The antibiotic/antimycotic solution was prepared according to the recipe for A3-antibiotics as described by Xuewu and Kloareg (1992), and was filter sterilized before being used. The antibiotic/antimycotic solution was added to the media to maintain axenic conditions.

Fucus gardneri Samples

Whole young brown algae, *F. gardneri*, were collected at Brocton point, Stanley Park, Vancouver, BC. All the samples were collected at low tide levels. Seawater collected at the site was later sterilized by autoclaving, and then used for washing the samples. All steps involving the alga after the collection were performed under sterile conditions in a laminar flow hood.

Samples of young *F. gardneri* were rinsed in a sterile beaker approximately 5 times with sterile seawater, followed by 4 washings with distilled water. The samples were then covered with a mixture of antibiotic/antimycotic solution and seawater (~2 ml per 100 ml seawater) for approximately 5 to 10 minutes. The mixture was removed and the *F. gardneri* washed with sterile seawater. The *F. gardneri* was separated and placed into 1-l Erlenmeyer flasks, each flask containing 400 ml artificial seawater, 4 ml antibiotic/antimycotic solution, an amount of K_2HPO_4 (0.5, 1.0, 1.5, 2.0 or 4.0 ug/ml) and between 26.75 and 30.05 g of *F. gardneri*. The flasks were placed in an incubator (Conviron Environmental Chamber) at a temperature of 15 °C/7 °C on a 12 hour day/night cycle: the light strength was 100 lux for the day cycle. The flasks were shaken once every few days to ensure adequate dissolved oxygen in the media.

No bacterial growth was evident in the cultures after 14 days of acclimation. The algae were rinsed separately 3 times with artificial medium and placed into sterile flasks, which contained 200 ml medium, 2 ml antibiotics, 100 µg arsenate, the same level of K_2HPO_4 that was used for the acclimation, and between 14.90 and 26.65 g of *F. gardneri*. The flasks were placed into the incubator at 15 °C/7 °C on a 12 hour cycle for the exposure experiment. The exposure experiment was run for 28 days and 1 ml samples of the medium were taken twice a week during that period. After the exposure experiment, the *F. gardneri* samples were collected, frozen and freeze-dried.

Sample Preparation and Analysis

All biomass samples collected were weighed, frozen (-20 °C) immediately, and then freeze-dried. All freeze-dried samples were kept at -20 °C until they were extracted by using a procedure similar to that described by Shibata and Morita (1992). Kelp powder (a laboratory standard), oyster tissue SRM (NIST-1566a) and *Fucus* sample (IAEA-140/TM) were similarly extracted as reference materials. The extract solutions were frozen until just before analysis. The samples were analyzed by using the ion-pairing HPLC/ICP-MS method as described previously (Lai et al. 1998).

All the liquid samples taken during the experiments were frozen immediately to preserve sample integrity until the time of the analysis. The liquid samples were analyzed by using anion-exchange HPLC/ICP-MS again as described previously (Lai et al. 1998).

Results and discussion

Whole young *F. gardneri* samples were chosen for the exposure experiments instead of the tips because, as indicated by Fries (1980), the tips of *F. gardneri* grow very slowly and consequently several months would be needed to see any growth. The *F. gardneri* samples that were collected from the sampling sites underwent a rigorous treatment to minimize microbial contamination. No bacterial contamination was seen during the experiments. Other more rigorous techniques for cleaning the algae could have been used, but many of these would have resulted in the destruction of the algal cells.

The *F. gardneri* remained intact and healthy looking throughout most of the experiment, although at the end some wilting was evident and the color was darker (brownish/green) than when freshly collected. The media were not changed during either the acclimation period or during the arsenic exposure period, so the algae samples were most likely experiencing starvation conditions at the end of each period. The culture media also became increasingly colored (ranging from light orange to dark orange) over the course of the experiment, and this was particularly noticeable for the media containing the As(V) species.

The major arsenicals found in the *F. gardneri* extracts before the acclimation period were arsenosugar 1c and 1d (Figures 1 and 2). Arsenite, dimethylarsinic acid, As(V) and arsenosugars 1a and 1b (Figures 2 and 3) were also detected.

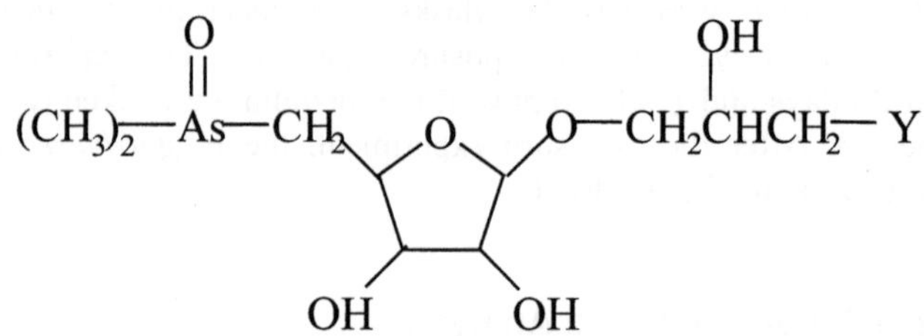

	Y
1 a	-OH
b	$-OPO_3HCH_2CH(OH)CH_2OH$
c	$-SO_3H$
d	$-OSO_3H$

Fig. 1. The structures of some of the arsenosugars found in the marine environment

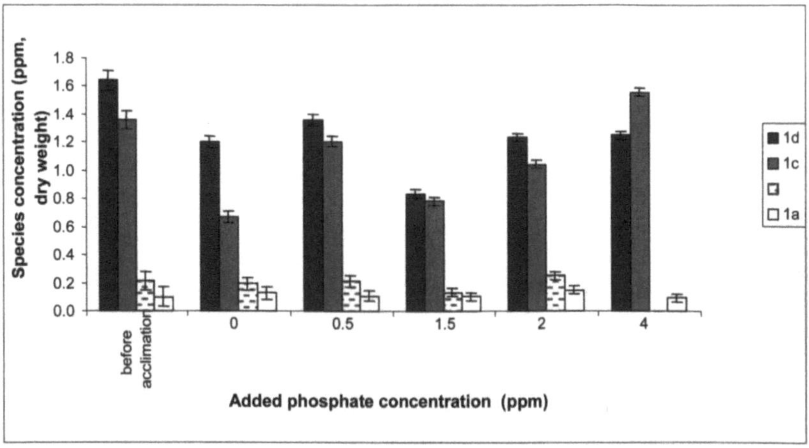

Fig. 2: Arsenosugars found in *F. gardneri* extracts before and after 14 days of acclimation

After acclimation, the algal extracts showed a slight decrease in Me_2AsOOH and in arsenosugars 1c and 1d (Figures 2 and 3). A slight increase in 1c concentration was seen from high phosphate media. The concentration of the other arsenic species in the extracts stayed relatively constant during the acclimation period and did not vary with the phosphate concentrations. The loss of arsenic species from the alga has been noted previously (Granchinho et al. 2001). In those experiments the loss was substantial; however, the *Fucus* samples had been collected from a different location and during a different season so the results cannot be compared with those from the present work. The arsenicals are probably lost by passive diffusion because the rate is independent of the phosphate concentration. A related study revealed that the sugars are released unchanged into the culture media where they can be detected by using HPLC/ICP-MS (Bellman 2001).

After 14 days of acclimation, the *F. gardneri* samples were washed and exposed to 500 ppb of arsenate and the appropriate level of added K_2HPO_4 (0.5, 1.0, 1.5, 2.0 or 4.0 µg/ml). The extracts of the samples obtained at the end of the arsenate exposure period showed an increase in the concentration of all arsenic compounds (arsenite, $MeAsO(OH)_2$, Me_2AsOOH, arsenate and arsenosugars 1a through 1c) except for arsenosugar 1d which decreased slightly relative to the concentrations at the beginning of the uptake experiment (Tables 1 and 2, Figures 4 and 5). The concentration of arsenosugar 1c in the algae produced at low phosphate concentration was higher than that found in the freshly collected sample: the concentration decreased as the concentration of phosphate in the media increased. Similar results were also seen for arsenosugars 1a and 1b (Figure 4). There is considerable variation in the concentration of the other arsenicals which are increased after arsenate exposure and the results do not correlate with the changes in the phosphate concentration (Figure 5) although the dramatically increased concentration of arsenite found at low phosphate concentration indicates enhanced uptake of arsenic.

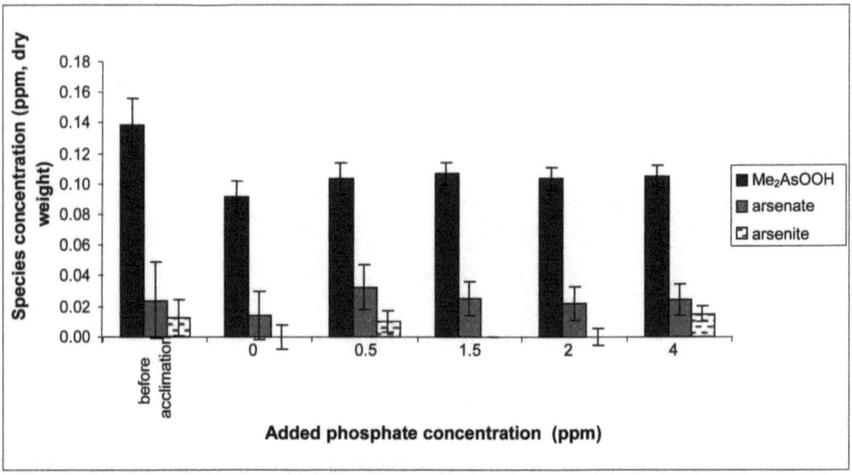

Fig. 3. Me_2AsOOH and inorganic arsenic species found in *F. gardneri* extracts before and after 14 days of acclimation

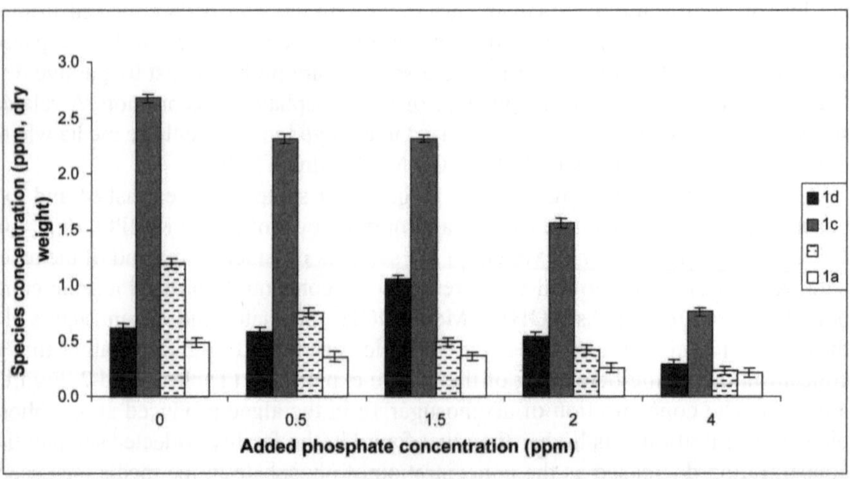

Fig. 4. Arsenosugars found in *F. gardneri* samples following exposure to arsenate in phosphate amended media.

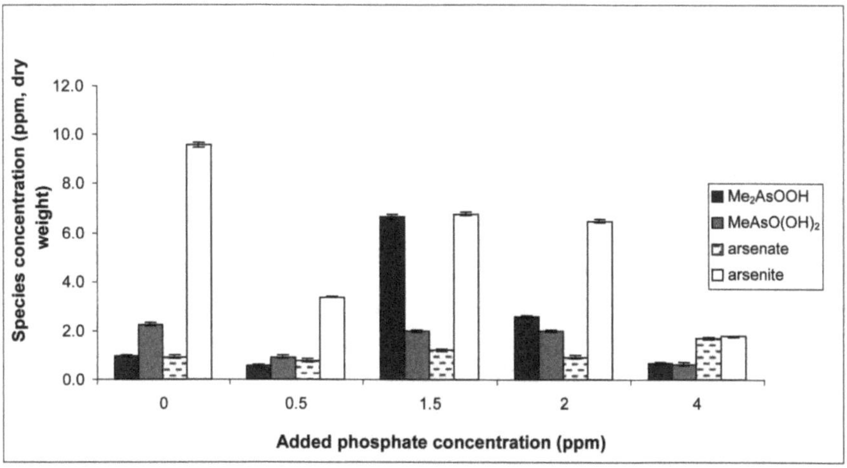

Fig. 5. $MeAsO(OH)_2$, Me_2AsOOH and inorganic arsenic species found in *F. gardneri* samples following exposure to arsenate phosphate amended media

Table 1. Arsenic speciation of *F. gardneri* extracts (ppm, dry weight) following acclimation in phosphate amended media.

Algae sample	Arsenic species found						
	arsenite	Me_2AsOOH	arsenate	1a	1b	1c	1d
before acclimation	0.01	0.14	0.02	0.10	0.21	1.36	1.64
algae + 0.0 μg/ml K_2HPO_4	trace[a]	0.09	0.01	0.13	0.19	0.67	1.20
algae + 0.5 μg/ml K_2HPO_4	0.01	0.10	0.03	0.11	0.21	1.21	1.36
algae + 1.5 μg/ml K_2HPO_4	trace	0.11	0.03	0.10	0.13	0.78	0.84
algae + 2.0 μg/ml K_2HPO_4	trace	0.10	0.02	0.15	0.25	1.04	1.23
algae + 4.0 μg/ml K_2HPO_4	0.01	0.11	0.02	0.10	0.00	1.56	1.25

[a] "trace" amounts are greater than or equal to the detection limit

Table 2. Arsenic speciation of *F. gardneri* extracts (ppm, dry weight) following arsenate exposure in phosphate amended media.

Algae sample	Arsenic species found							
	arsenite	$MeAsO(OH)_2$	Me_2AsOOH	arsenate	1a	1b	1c	1d
algae + 0.0 µg/ml K_2HPO_4	9.59	2.24	0.95	0.93	0.48	1.19	2.67	0.62
algae + 0.5 µg/ml K_2HPO_4	3.37	0.95	0.59	0.80	0.36	0.75	2.31	0.58
algae + 1.5 µg/ml K_2HPO_4	6.77	2.01	6.65	1.22	0.36	0.49	2.31	1.05
algae + 2.0 µg/ml K_2HPO_4	6.49	1.99	2.59	0.94	0.26	0.42	1.56	0.54
algae + 4.0 µg/ml K_2HPO_4	1.77	0.65	0.69	1.71	0.21	0.23	0.76	0.29

Samples of the media were collected and were analyzed by using anion-exchange HPLC/ICP-MS (Figures 6-8). The two species that are formed presumably as a result of uptake and transformation of the arsenate were arsenite and Me_2AsOOH (Figures 7,8) and their rate of formation depends on the phosphate concentration. At lower added phosphate concentrations (0.0 and 0.5 µg/ml), higher amounts of As(III) and Me_2AsOOH are seen in the medium, while at higher concentrations (2.0 and 4.0 µg/ml), lower amounts of these species are produced. These results are consistent with the hypothesis that at low phosphate concentration more arsenate is taken up by the algae, to be biotransformed into arsenite and Me_2AsOOH, because the uptake of arsenate and phosphate is competitive. The inhibition of arsenate uptake and transformation at higher phosphate concentrations (Figure 6) supports this postulate of a common uptake mechanism for arsenate and phosphate.

These results from *F gardneri* are very similar to those obtained from axenic cultures of the unicellular alga *Polyphysa peniculus* grown in artificial seawater (Cullen et al.1994). Arsenate is taken up and metabolized to arsenite and dimethylarsinate. Arsenic species are also lost from the cells when they are transferred to arsenic free media. However, arsenosugar are not found in *P peniculus*.

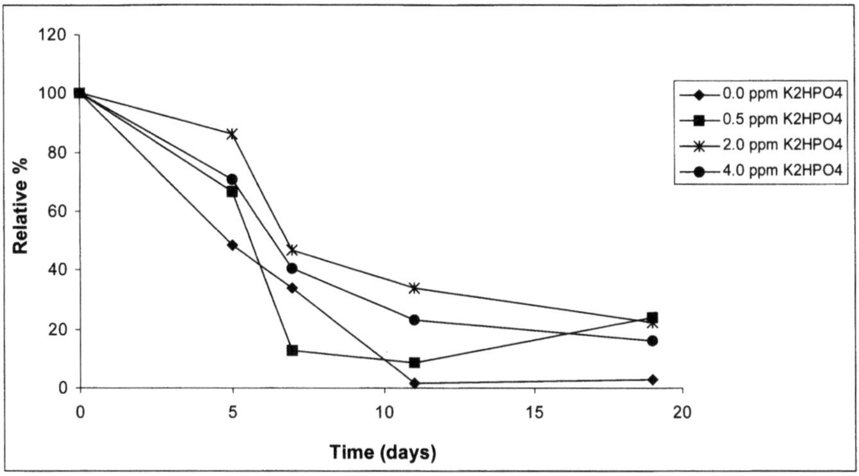

Fig. 6. Arsenate (As(V)) variation in medium samples collected during uptake of arsenate in phosphate amended media: added phosphate concentrations given in the box.

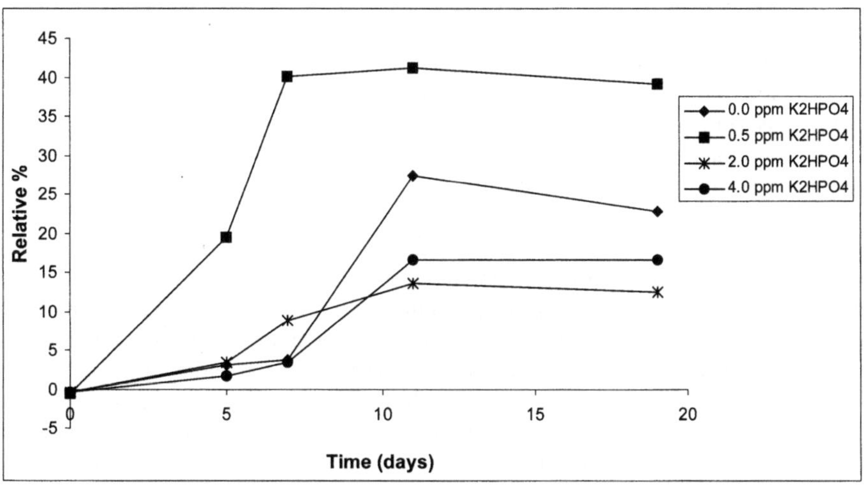

Fig. 7. Arsenite (As(III)) variation in media samples collected during the uptake of arsenate from phosphate amended media: added phosphate concentrations given in the box.

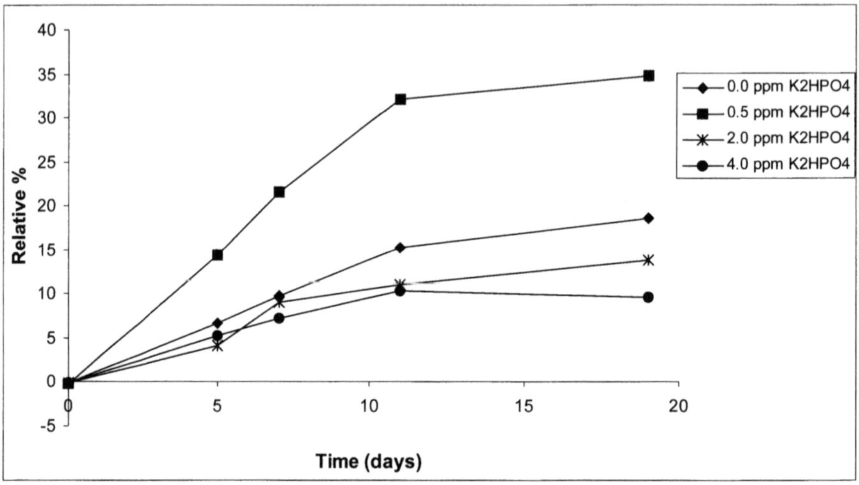

Fig. 8. Dimethylarsinic acid variation in media samples collected during uptake of arsenate from phosphate amended media: added phosphate concentrations given in the box.

A biogenic pathway has been proposed to account for the formation of the arsenosugars from arsenate in seawater that follows the sequence: arsenate, arsenite, monomethylarsenic acid, dimethylarsinic acid, dimethylarsenoyladenosine, to arsenosugars. Only one study to date on the unicellular alga *Chaetoceros concavicornis* has demonstrated that arsenate can be a precursor to at least one arsenosugar (Edmonds et al. 1997). Our early work with the macroalga *Fucus gardneri* revealed that there is a seasonal dependence on the arsenic concentration and arsenic speciation in the samples and that the distribution of the arsenosugars differed from growing tip to the rest of the plant (Lai et al. 1998). However, this study did not reveal any clues about the biochemical origin of the arsenosugars. When the macroalga *Fucus serratus* was grown in aquaria in arsenate amended raw seawater (approximately 0 ppb, 20 ppb, 50 ppb and 100 ppb arsenate) the arsenic concentration and speciation in the algal extracts, particularly arsenite, monomethylarsenic acid, dimethylarsinic acid, varied with time of exposure and seawater arsenic concentration (Geiszinger et al. 2001). Some increases with time were seen in the concentrations of the arsenosugars 1a, 1b and 1d but the concentration of 1c, the major arsenical in *Fucus serratus*, did not increase. The results were interpreted as providing some support for the postulated biogenic pathway.

Our initial work on the uptake and metabolism of arsenate by *axenic* cultures of *Fucus gardneri* (Granchinho et al. 2001) was initiated in an attempt to eliminate all biochemical processes that might be associated with bacteria and fungi, thus ensuring that, say, any arsenosugars would be produced only by the *Fucus*. As reported above, a major loss of arsenic species was observed during the acclimation period when the artificial seawater contained approximately 2 ppb phosphate. There was a further decrease in the concentration of two arsenosugars on exposure of the *Fucus* samples to seawater containing 500 ppb arsenate, and a slight increase in the concentration of 1a, a minor species in the natural macro alga. The experiment provided little evidence for the hypothesis that arsenosugars are produced by macroalgae from arsenate.

In the course of these experiments a very resilient fungus identified as *Fusarium oxysporum* was observed to grow on or in the *Fucus*. Because this was a possible source of the arsenosugars found in the macroalga the fungus was grown in pure culture in the presence of 500 ppb arsenate. The usual metabolites, arsenite, and dimethylarsinic acid were produced, but no arsenosugars. Thus the fungus was not a source of the arsenosugars found in *Fucus gardneri.*

In the present paper the phosphate concentration was varied to ascertain if the production of arsenosugars by *Fucus* could be stimulated by a deficit of phosphate, which would encourage uptake of arsenate, or by an excess of phosphate, which would possibly increase metabolism. The results indicate that at low phosphate concentration arsenate uptake is enhanced and as a result the concentration of the usual metabolites arsenite and dimethylarsinic acid are increased in the *Fucus* and the seawater. There is also an increase in the concentration of one arsenosugar, 1c, providing some evidence that the sugar(s) is a macroalgal metabolite. It should be noted that Benson and Summons (1981) discussed the probable importance of environmental phosphate in regulating arsenic metabolism some time ago.

The results of Geiszinger et al. (2001) suggest that studies at lower arsenate concentrations over a longer time period would be rewarding. High arsenate concentrations can be toxic and metabolic processes may be slow. We leave this for others to follow up.

Acknowledgments

The authors are grateful to the Natural Sciences and Engineering Research Council of Canada for financial assistance and to Mr. Bert Mueller for help with the ICP-MS analysis. The referees' comments are greatly appreciated.

Literature

Andreae MO, Klumpp D (1979) Biosynthesis and release of organoarsenic compounds by marine algae. Environ Sci Tech 13:738-741

Bellman K (2001) Arsenic excretion by a marine alga, Fucus gardneri. B.Sc Thesis. University of British Columbia Benson AA, Summons RE (1981) Arsenic accumulation in Great Barrier Reef invertebrates. Science 211:482-483

Bottino NR, Newman RD, Cox ER, Stockton R, Ioban M, Zingaro RA, Irgolic KJ (1978) The effects of arsenate and arsenite on the growth and morphology of the marine unicellular algae *Tetraselmis chui* (Chlorophyta) and *Hymenomonas carterae* (Chrysophyta). J Exp Mar Biol Ecol 33:153-168.

Cullen WR, Reimer KJ (1989) Arsenic speciation in the environment. Chem Rev 89:713-764

Cullen WR, Harrison LG, Li H, Hewitt G (1994) Bioaccumulation and excretion of arsenic compounds by a marine unicellular alga, *Polyphysa peniculus*. Appl Organomet Chem 8:313-324

Edmonds JS, Francesconi KA, Cannon JR, Raston, CL, Skelton, BW, White AH (1977) Isolation, crystal structure and synthesis of arsenobetaine, the essential constituent of the western rock lobster *Panulilrus longipes cygnus* George. Tetrahedron Lett 18:1543-1546

Edmonds JS, Shibata Y, Francesconi KA, Rippingdale RJ, Morita M (1997) Arsenic transformations in short marine food chains studied by HPLC-ICP MS Appl Organomet Chem 11:281-287

Francesconi KA, Edmonds JS (1993) Arsenic in the sea. In: Ansell RN, Gibson and Barnes M (eds) Oceanography and Marine Biology: An Annual Review 31, UCL Press, pp 111-151

Francesconi KA, Edmonds JS (1997) Arsenic and marine organisms. Adv Inorg Chem 44:147-185

Fries L (1977) Growth regulating effects of phenylacetic acid and p-hydroxy phenylacetic acid on *Fucus spiralis* L. (Phaeophyceae, Fucales) in axenic culture. Phycol 16:451-455

Fries L (1982) Selenium stimulates growth of marine macroalgae in axenic culture. J Phycol 18:328-331

Geiszinger A, Goessler W, Pedersen SN, Francesconi KA (2001) Arsenic biotransformation by the brown macroalga *Fucus* serratus. Environ Toxicol Chem 20:2255-2262

Granchinho SCR, Polishchuk E, Cullen WR, Reimer KJ (2001) Biomethylation and bioaccumulation of arsenic(V) by marine alga *Fucus* gardneri. Appl Organomet Chem 15:553-560

Granchinho SCR, Franz C, Polishchuk E, Cullen, WR, Reimer KJ (2002) Transformation of arsenic(V) by the fungus Fusarium oxysporum melonis isolated from the alga *Fucus* gardneri. Appl Organomet Chem 16:721-726

Harrington CF, Ojo AA, Lai VW-M, Reimer KJ, Cullen WR (1997) The identification of some water-soluble arsenic species in the marine brown algae *Fucus* distichus. Appl Organomet Chem 11:931-940

Klumpp DW (1980) Characteristics of arsenic accumulation by the seaweeds Fucus spiralis and Ascophyllum nodosum. Mar Biol 58:257-64

Lai VW-M, Cullen WR, Harrington CF, Reimer KJ (1998) Seasonal changes in arsenic speciation in Fucus species. Appl Organomet Chem 12:243-251

Liu X, Kloareg B (1992) Explant axenization for tissue culture in marine macroalgae. Chin J Oceanol Limnol 10:268-275

Sanders JG, Windom HL (1980) The uptake and reduction of arsenic species by marine algae. Estuar Coast Mar Sci 10:555-567

Shibata Y, Morita M (1992) Characterization of organic arsenic compounds in bivalves. Appl Organomet Chem 6:343-349

Chapter 9

Molecular modeling studies of specific interactions between organometallic compounds and DNA

R. Yonchev, H. Rehage, H. Kuhn

Introduction

During the past 40 years, organometallic chemistry has been developed into a large and important branch of chemistry linking the fields of organic and inorganic chemistry. Organometallic compounds have practical applications as catalysts for industrial syntheses, as antiknock additives for gasoline, and as biocides. In addition to this, during the late 1970s, the discovery of the antitumour activities of titanocene dichloride and certain diorganotin derivatives stimulated much interest in the research of organometallic compounds as antitumour agents. Titanocene dichloride has been proven to be a potent agent against breast, lung and intestinal (colon) cancer tissues. In contrast to the serious nephrotoxicity, myelotoxicity, peripheral neurathy of the well known inorganic antitumour agent, diaminedichloroplatinum (II) (cisplatin), titanocene dichloride only exhibits slight side effects with regard to the liver when used in therapeutically necessary amounts. Thus, the organometallic compounds constitute a potent new class of antitumour agents. One of the main problems is the prediction of the structures of resulting interactions between DNA and organometallic compounds and possible biological and physiological impact of these compounds.

The main goal of this preliminary study is to determine the suitability of computer alignment methods of ligands to DNA by using firstly known structures at ligands docked to DNA, before we investigate unknown structures.

Theory and methods

In order to make predictions we must know how to calculate the energy of the system and how to align the ligand to the main structure. Usually calculation of energy is difficult, but the energy of the system could be parameterized. In this way we can calculate a set of parameters for each atom in order to perform further energy calculations. Such set of parameters is called "force field". There are a lot of force fields, treating organic molecules, but the main problem is they do not have parameters for heavy metals. We have used Extensible Systematic Force Field (ESFF), which is appropriate for organometallic compounds.

Extensible Systematic Force Field (ESFF)

ESFF (Biosym/MSI 1995) was derived using a mixture of Density Functional Theory (DFT) calculations on dressed atoms to obtain polarizabilities, gas-phase and crystal structures, etc. The training set included primarily organic and organometallic compounds and a few inorganic compounds. The focus was on crystal structures and sublimation energies. The training set included models containing each element in the first 6 periods up to lead (Z = 82) (except for the inert gases), Sr, Y, Tc, La, and the lanthanides (except for Yb).

Parameters and charges are generated on the fly, based on the model configuration, the local environment, and the derived rules.

The analytic energy expressions for the ESFF are provided in equation 1. Only diagonal terms are included.

$$
\begin{aligned}
E_{pot} &= \underbrace{\sum_b D_b \left[1 - e^{\left(-\alpha\left(r_b - r_b^0\right)\right)}\right]^2}_{(1)} \\
&+ \underbrace{\begin{cases}
\sum_a \dfrac{K_a}{\sin^2 \theta_a^0}\left(\cos\theta_a - \cos\theta_a^0\right)^2 & (normal) \\
\sum_a 2K_a\left(\cos\theta_a + 1\right) & (linear) \\
\sum_a K_a^{\theta_a} \cos^2 \theta_a & (perpendicular) \\
\sum_a \dfrac{2K_a}{n^2}\left(1 - \cos\left(n\theta_a\right)\right) + 2K_a^{\left(-\beta\left(r_{13} - \rho_a\right)\right)} & (equatorial)
\end{cases}}_{(2)} \\
&+ \underbrace{\sum_\tau D_\tau \left(\frac{\sin^2\theta_1 \sin^2\theta_2}{\sin^2\theta_1^0 \sin^2\theta_2^0} + sign \frac{\sin^n\theta_1 \sin^2\theta_2}{\sin^n\theta_1^0 \sin^n\theta_2^0} \cos[\pi\tau] \right)}_{(3)} \\
&+ \underbrace{\sum_o D_o \chi^2}_{(4)} + \underbrace{\sum_{nb} \left(\frac{A_i B_j + A_j B_i}{r_{nb}^9} - 3\frac{B_i B_j}{r_{nb}^6} \right)}_{(5)} + \underbrace{\sum_{nb} \frac{q_i q_j}{r_{nb}}}_{(6)}
\end{aligned}
\qquad (1)
$$

where D and K are force constants, θ are bond angles, A and B are van der Waals parameters, χ is electronegativity, q are partial charges of atoms and r are distances between atoms.

The bond energy is represented by a Morse functional form, where the bond dissociation energy D, the reference bond length r^0, and the anharmonicity parameters are needed. In constructing these parameters from atomic parameters, the force field utilizes not only the atom types and bond orders, but also considers whether the bond is endo or exo to 3-, 4-, or 5-membered rings.

The rules themselves depend on the electronegativity, hardness, and ionization of the atoms as well as atomic anharmonicities and the covalent radii and well depths. The latter quantities are fit parameters, and the former three are calculated.

The ESFF angle types are classified according to ring, symmetry, and π-bonding information into five groups:

The *normal* class includes unconstrained angles as well as those associated with 3-, 4-, and 5-membered rings. The ring angles are further classified based on whether one (exo) or both bonds (endo) are in the ring. Additionally, angles with only central atoms in a ring are also differentiated.

- The *linear* class includes angles with central atoms having sp hybridization, as well as angles between two axial ligands in a metal complex.

- The *perpendicular* class is restricted to metal centers and includes angles between axial and equatorial ligands around a metal center.

- The *equatorial* class includes angles between equatorial ligands of square planar (sqp), trigonal bipyramidal (tbp), octahedral (oct), pentagonal bipyramidal (pbp), and hexagonal bipyramidal (hbp) systems.

- The *π–system* class includes angles between pseudo atoms. This class is further differentiated in terms of normal, linear, perpendicular, and equatorial types.

The rules that determine the parameters in the functional forms depend on the ionization potential and, for equatorial angles, the periodicity. In addition to these calculated quantities, the parameters are functions of the atomic radii and well depths of the central and end atoms of the angle, and, for planar angles, two overlap quantities and the 1-3 equilibrium distances.

To avoid the discontinuities that occur in the commonly used cosine torsional potential when one of the valence angles approaches 180°, ESFF uses a functional form that includes the sine of the valence angles in the torsion. These terms ensure that the function goes smoothly to zero as either valence angle approaches 0° or 180°, as it should. The rules associated with this expression depend on the central bond order, ring size of the angles, hybridization of the atoms, and two atomic parameters for the central atom, which is fit. The functional form of the out-of-plane energy is the same as in CFF91, where the coordinate (ϕ) is an average of the three possible angles associated with the out-of-plane center. The single parameter that is associated with the central atom is a fit quantity.

The charges are determined by minimizing the electrostatic energy with respect to the charges under the constraint that the sum of the charges is equal to the net charge on the molecule. This is equivalent to equalization of electronegativities.

The derivation of the rule begins with the following equation for the electrostatic energy:

$$E = \sum_{i}\left(E_i^0 + \chi_i q_i + \frac{1}{2}\eta_i q_i^2 \right) + \sum_{i>j} B\frac{q_i q_j}{R_{ij}} \tag{2}$$

where χ is the electronegativity and η the hardness. The first term is just a Taylor series expansion of the energy of each atom as a function of charge, and the second is the Coulomb interaction law between charges. The Coulomb law term introduces a geometry dependence that ESFF for the time being ignores, by considering only topological neighbors at effectively idealized geometries.

Minimizing the energy with respect to the charges leads to the following expression for the charge on atom *i*:

$$q_i = \frac{\lambda - \chi_i - \Delta\chi_i}{\eta_i} \tag{3}$$

where λ is the Lagrange multiplier for the constraint on the total charge, which physically is the equalized electronegativity of all the atoms. The $\Delta\chi$ term contains the geometry-independent remnant of the full Coulomb summation.

Equations 2 and 3 give a totally delocalized picture of the charges in a relatively severe approximation. To obtain reasonable charges as judged by, for example, crystal packing calculations, some modifications to the above picture have been made. Metals and their immediate ligands are treated with the above prescription, summing their formal charges to get a net fragment charge. Delocalized π–systems are treated in an analogous fashion. σ–systems are treated using a localized approach in which the charges of an atom depend simply on its neighbors. Note that this approach, unlike the straightforward implementations based on the equalization of electronegativity, *does* include some resonance effects in the π–system.

The electronegativity and hardness in the above equations must be determined. In earlier force fields they were often determined from experimental ionization potentials and electron affinities; however, these spectroscopic states do not correspond to the valence states involved in molecules. For this reason, ESFF is based on electronegativities and hardnesses, calculated using density functional theory. The orbitals are (fractionally) occupied in ratios appropriate for the desired hybridization state, and calculations are performed on the neutral atom as well as on positive and negative ions.

ESFF uses the 6-9 potential for the van der Waals interactions. Since the van der Waals parameters must be consistent with the charges, they are derived using rules that are consistent with the charges.

Starting with the London formula:

$$\left(B_i \sim \alpha_i^2 \cdot IP\right) \tag{4}$$

where α is the polarizability and *IP* the ionization potential of the atoms, the polarizability, in a simple harmonic approximation, is proportional to *n*/*IP* where *n* is the number of electrons. Across any one row of the periodic table, the core electrons remain unchanged, so that the following form is reasonable:

$$\alpha = \frac{a'}{IP} + \frac{b' n_{eff}}{IP} \tag{5}$$

where a' and b' are adjustable parameters that should depend on just the period, and n_{eff} is the effective number of (valence) electrons. Further assuming that α is proportional to R^3 and that another equivalent expression to that in equation 4 is:

$$B_i \sim \varepsilon R^6 \tag{6}$$

where ε is a well depth, the following forms are deduced for the rules for van der Waals parameters:

$$R_i = \frac{a}{(IP)^{1/3}} + \frac{b \cdot n_{eff}^{1/3}}{(IP)^{1/3}} \quad \text{and} \quad \varepsilon_i = c(IP) \tag{7}$$

The van der Waals parameters are affected by the charge of the atom.
In ESFF we found it sufficient to modify the ionization potential (*IP*) of metal atoms according to their formal charge and hardness:

$$IP = (IP)_0 + q\eta_i \tag{8}$$

and for nonmetals to account for the partial charges when calculating the effective number of electrons.

ESFF atom types are determined by hybridization, formal charge, and symmetry rules. In addition, the rules may involve bond order, ring size, and whether bonds are endo or exo to rings. For metal ligands the *cis-trans* and axial-equatorial positionings are also considered. The addition of these latter types affects only certain parameters (for example, bond order influences only bond parameters) and thus are not as powerful as complete atom types. In one sense they provide a further refinement of typing beyond atom types.

The ESFF has been parameterized to handle all elements in the periodic table up to radon. It is recommended for organometallic systems and other systems for which other force fields do not have parameters. ESFF is designed primarily for predicting reasonable structures (both intra- and intermolecular structures and crystals) and should give reasonable structures for organic, biological, organometallic and some ceramic and silicate models. It has been used with some success for studying interactions of molecules with metal surfaces. Predicted intermolecular binding energies should be considered approximate.

Ligand aligning algorithm

In order to understand how proteins and DNA carry out certain biological functions, how they recognize ligands and form protein-protein or protein-ligand complexes, it is essential to identify protein or DNA functional residues and interaction interfaces such as active sites or binding sites. The functional interfaces can serve as targets for structural based drug design or to guide the site-directed mutagenesis in studying the protein structure-function relationship.

The site search algorithm is shown as a 2D graph in Figure 1 (Biosym/MSI 2001). First, the protein or DNA is mapped onto a grid, which covers the complete protein or DNA space. The grid points are then defined as free points and protein points. The protein points are grid points, within 2 Å from a hydrogen atom or 2.5 Å from a heavy atom. Then, a cubic eraser moves from the outside of the protein toward the center to remove the free points until the opening is too small for it to move forward. Those free points not reached by the eraser will be defined as site points. If a smaller eraser is used, sites with smaller openings will be identified. To find the shallow cleft on the protein surface, a larger eraser should be used. A larger eraser sometimes joins into one site several sites defined by a smaller eraser.

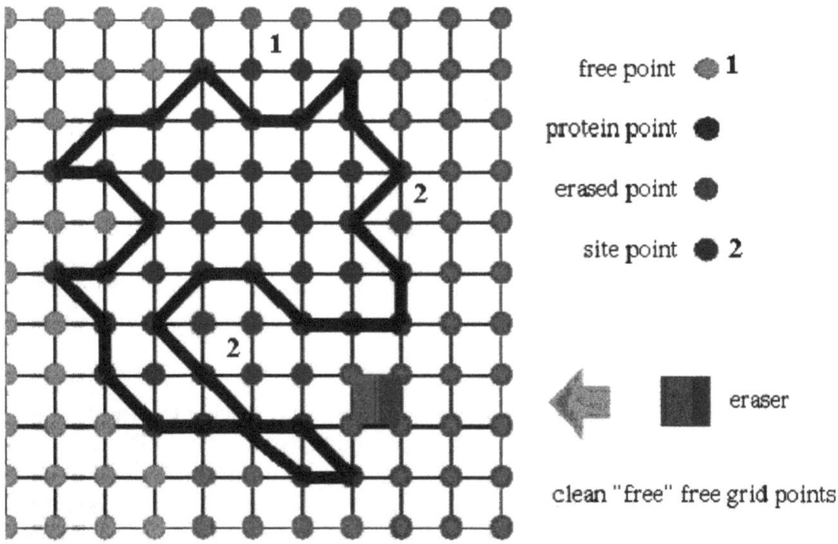

Fig. 1. A 2D representation of the site search algorithm. The eraser is moving from the left towards the protein, and the eraser has processed half of the protein. The dots marked with 1 are the free points in the current setting and are marked differently from the other free points to illustrate a potential binding site if the eraser size is increased.

After a site is located, expanding or contracting the site can modify it. One layer of grid points at the cavity opening site will be added or removed by each expand or contract operation, respectively.

The Monte Carlo method is employed in the conformational search of the ligand. During the search, bond lengths and bond angles are untouched, only torsion angles (excepted those in a ring) are randomized. Therefore, the ligand molecule(s) should be energy minimized to ensure correct bond lengths and bond angles before ligand fitting. Multiple changes may happen at the same time during conformational search. The upper limit of random dihedral perturbation is the same for all torsions (180°). The lower limit of random dihedral perturbation depends on the number of rotating atoms – the more rotating atoms are attached to the bond, the smaller the size of the dihedral perturbation window.

After a new conformation is generated, the fitting is carried out in two steps. First, the non mass-weighted principle moment of inertia (PMI) of the binding site is compared with the non mass-weighted PMI of the ligand according to the following equations. If the value (Fit_{PMI}) is above the threshold or not better than fitting results previously saved, no further docking process will be performed. Another ligand or another conformation of the same ligand will be examined.

PMI: Principal moment of inertia (P_x, P_y, P_z):

$$\text{ratio}_{xy} = P_x/P_y;\ \text{ratio}_{xz} = P_x/P_z;\ \text{ratio}_{yz} = P_y/P_z$$

$$Fit_{PMI} = \sqrt{\Delta ratio_{xy}^2 + \Delta ratio_{xz}^2 + \Delta ratio_{yz}^2}$$

On the other hand, if Fit_{PMI} is better than previously saved results, the ligand is positioned into the binding site according to the PMI. Because PMI is a scalar property, there are four possible positions for the ligand to orient in the binding site. For each position, the corresponding docking score is computed. In addition, an optional rigid body minimization can be applied to each position to optimize the docking score.

The docking score is the negative value of the non-bonded inter-molecular energy between the ligand and the protein. In molecular mechanics, the most time-consuming calculation is the nonbonding interactions. If no cutoffs are used (i.e. in exact calculations), the computational time grows quadratically with the number of atoms in the system. For a typical protein system with a few thousand atoms, this is not practical for docking which requires enormous number of energy evaluations. However, if the protein can be assumed to be rigid, one may use a grid to evaluate interactions. Electrostatic and van der Waals grids are constructed for the calculation.

After the docking score is calculated for each orientation of the ligand, it is compared with the results saved previously. If the new one is better, it is saved. Finally, rigid body minimization is applied to the saved conformations of the ligand to optimize their positions and docking scores.

Results and discussion

DNA dodecamer duplex d(CCTCTGGTCTCC) + d(GGAGACCAGAGG) (Yang et al. 1996) was built into Insight II version 97.0 (MSI, San Diego, California) using the Biopolymer module. This DNA duplex was chosen, because previous studies show that cisplatin interacts either intrastrand (65%) with both N7 atoms of the 2 consecutive guanine bases or interstrand (25%) with guanine-adenine bases (Pilch et al. 2000).

The ligands – cis-diaminedichloroplatinum (II) (or cisplatin), diethyltin dichloride (Et_2SnCl_2) and dimethyltin dichloride (Me_2SnCl_2) were built using the Builder module of Insight II, and energy minimized using the Discover 3 module. Ligands rigid alignment to DNA was performed using the Drug Discovery module of Cerius^2 version 4.2 MatSci (MSI, San Diego, California). Used align method is rigid because we have one constrain – DNA must remain with the same structure.

In aqueous solution both alkyl ligands and chloride ligands can be dissociated depending on pH and concentration. The active forms of cisplatin, diethyltin dichloride and dimethyltin dichloride are with one or both chloride ligands dissociated.

In the DNA dodecamer duplex used there are two types of binding sites. One is the base nitrogen ring of DNA (especially nitrogen rings in two consequent guanine bases) and the other is the phosphate group of DNA. Figure 2 shows the result of rigid alignment of $[(NH_3)_2PtCl(OH_2)]^+$ with DNA.

a)

b)

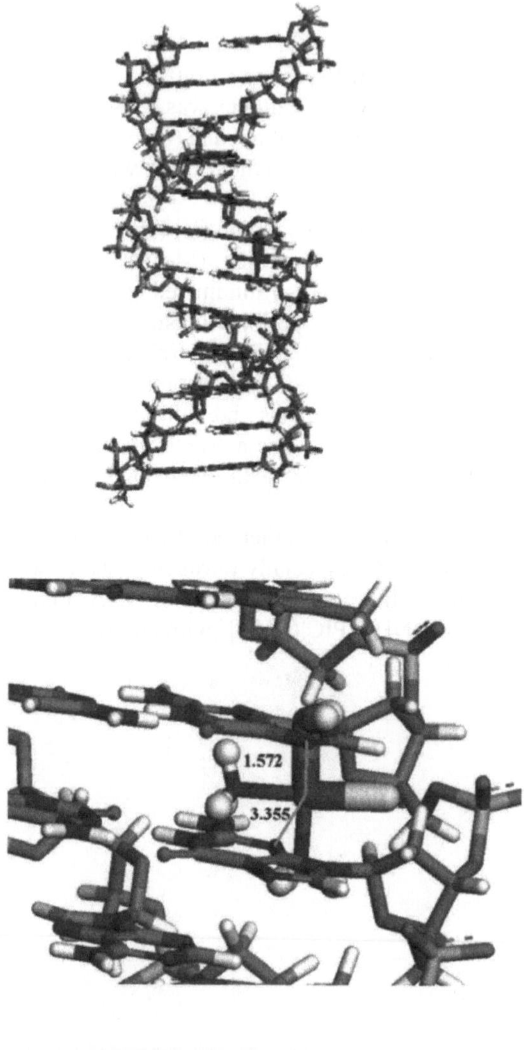

Fig. 2. a) Aligned structure of DNA and $[(NH_3)_2PtCl(OH_2)]^+$ and b) enlargedComputer simulated structure is similar to the published structure. The distance between N 7 atom of guanine base and the platinum atom is 1.572 Å. This distance is of the order of the bond, which means possible bond formation. After some ligand re-orientation (rotation), in order to keep the planar conformation of platinum, it is possible a new bond formation between platinum and N7 atom of the second guanine base, which will deform DNA strand. Figure 3 represents the structure published in protein data bank (Yang et al. 1996).

a)

b)

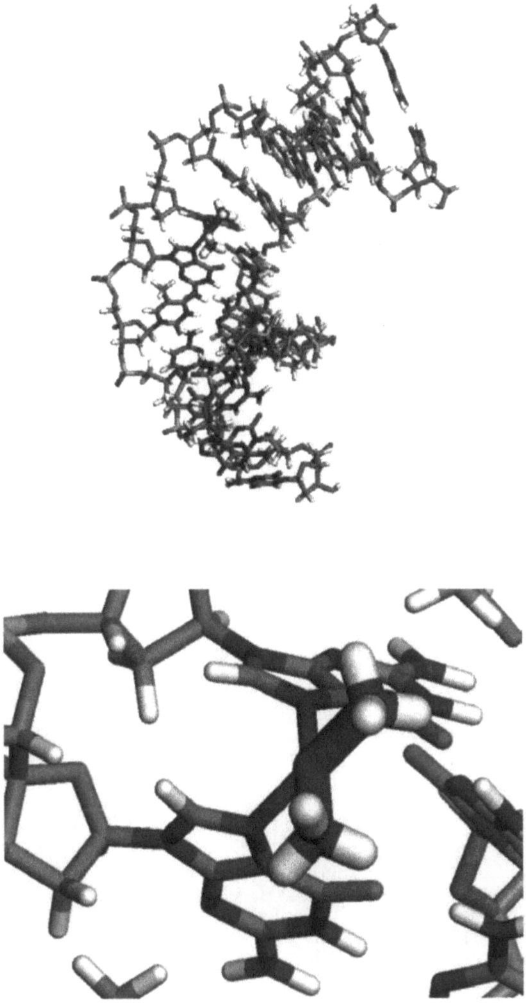

Fig. 3. a) Crystal structure of cisplatin bonded to DNA (Yang et al. 1996) and b) enlarged-
Figures 4 and 5 show the results of rigid alignment of $[(C_2H_5)_2SnCl(OH_2)]^+$ and $[(CH_3)_2SnCl(OH_2)]^+$ with DNA respectively.

a)

b)

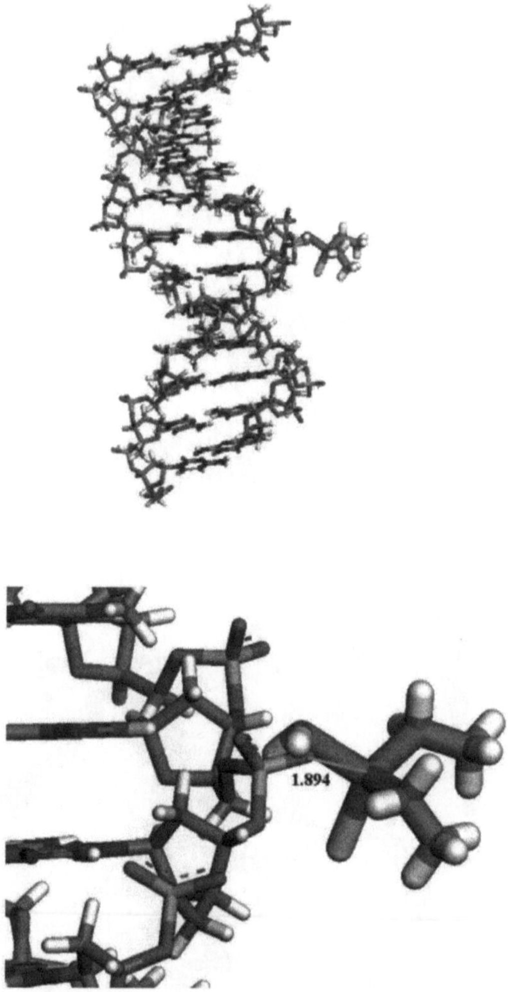

Figure 4. a) Aligned structure of DNA and $[(C_2H_5)_2SnCl(OH_2)]^+$ and b) enlarged

a)

b)

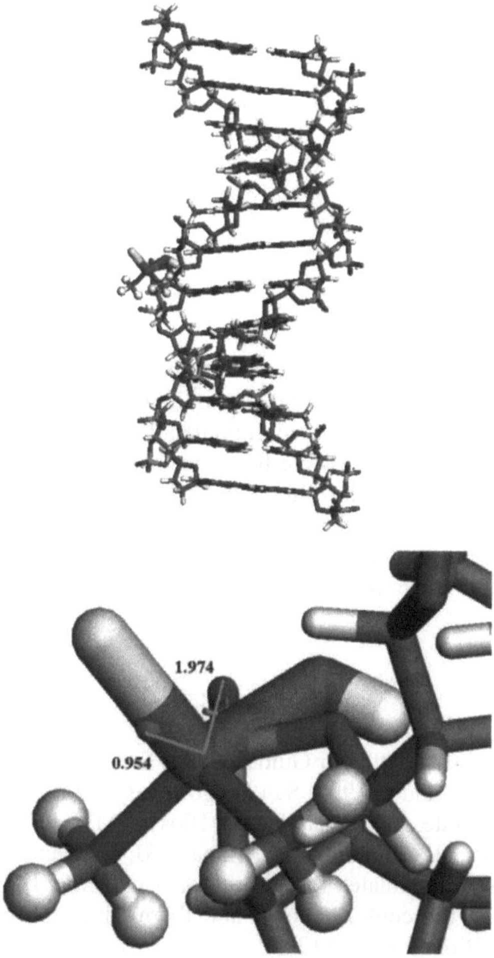

Fig. 5. a) Aligned structure of DNA and $[(CH_3)_2SnCl(OH_2)]^+$ and b) enlarged

Previous studies (Yang et al. 1999) show that organotin (IV) compounds interact with phosphate groups of DNA. Obtained structures show the same trends. The distances between the oxygen atoms and the tin atom are 1.894 Å and 0.954 Å for $[(C_2H_5)_2SnCl(OH_2)]^+$ and $[(CH_3)_2SnCl(OH_2)]^+$ respectively. These distances are of the order of the bond and this could lead to bond formation. Further its possible inter- or intra- strand interaction with phosphate group from the same DNA strand or with phosphate group from another DNA strand.

Conclusions

In this study we have shown that structures predicted with computer alignment of different ligands to DNA are close to experimentally examined structures. Therefore, we can conclude, that at this first stage, the computer methods used are adequate for such predictions. Additional refining of the predictions is possible with using of simulations based on the quantum mechanic methods.

Future work

In the future we will perform hybrid quantum mechanics and molecular mechanics method in order to obtain information on the final structure, including the observation of bond formation between the organometallic compound and the DNA molecule. The same approach is possible for water box simulation in order to include solvent effect during interactions. In the hybrid quantum mechanics and molecular mechanics approach, a molecular system treated by quantum mechanics is embedded into a molecular system treated by molecular mechanics. The latter system can be finite or periodic in three dimensions.

Literature

Discover 2.9.7 / 95.0 / 3.0.0 User Guide. (1995) San Diego: Biosym/MSI

Insight II 95.0 User Guide. (1995) San Diego: Biosym/MSI

Cerius^2 4.6 User Guide. (2001) San Diego: Biosym/MSI

Pilch DS, Dunham SU, Jamieson ER, Lippard SJ, Breslauer KJ (2000) DNA sequence context modulates the impact of a cisplatin 1,2-d(GpG) intrastrand cross-link and the conformational and thermodynamic properties of duplex DNA. J Mol Biol 296:803-812

Yang DZ, Wang AHJ (1996) Structural studies of interactions between anticancer platinum drugs and DNA. Prog Biophys Mol Biol 66:81-111

Yang P, Guo ML (1999) Interactions of organometallic anticancer agents with nucleotides and DNA. Coordination Chem Rev 186:189-211

Chapter 10

Organometal(loid) compounds associated with human metabolism

A.V. Hirner, L. M. Hartmann, J. Hippler, J. Kresimon, J. Koesters, K. Michalke, M. Sulkowski, A. W. Rettenmeier

Introduction

Biomethylation of metals and metalloids is a well-known process ubiquitously occurring in the environment, which leads to the formation of chemical species with significantly higher mobility and altered toxicology. There are only a few historical reports, e.g. about "bismuth breath" or "Gosio gas" dealing with the association of humans with methylated metal(loid)s. Although the toxicity of the latter [later identified as trimethyl arsine (Challenger 1945)] has not been conclusively demonstrated, this gas produced by fungi in wet wallpaper was considered to be the reason for the illness of people living there (Gosio 1897). Amongst other observations, dimethyltellurium in "bismuth breath" of mine workers, dimethylselenium in the upper ng/m^3 range in human breath, as well as the detection of at least twenty-two different organometal(loid) species in human urine are indications for the methylation of metal(loid)s occurring in humans (Cai et al. 1995; Feldmann et al. 1996; Kresimon et al. 2001).

Biomethylation of ingested metal(loid)s can occur in both the liver and kidney, with transferred metabolites being released into the bloodstream, or additionally in the case of the liver, directly into the bile. The liver is generally accepted to be the main site of human arsenic biomethylation (Styblo et al. 2000). Recently, human renal cells (HK-2) have also been shown to possess the ability to biotransform inorganic arsenic (Peraza et al. 2002). With regard to selenium, the enzyme responsible for biomethylation of the selenite metabolite hydrogen selenide is primarily located in the cytosol of lung and liver cells (Mozier et al. 1988; Nakamuro et al. 2000).

Urinary excretion is the major route of arsenic elimination from the body. Arsenic species from seafood sources, including the non-toxic arsenobetaine, are normally excreted within a three day period after ingestion; inorganic arsenic is methylated and excreted mainly as dimethylarsinic acid in urine (Le et al. 1996). The trivalent methylated arsenic species monomethylarsonous acid and dimethylarsinous acid, which were postulated by Challenger (1945) to be intermediates in the biomethylation pathway of inorganic arsenic, have recently been detected in urine (Le et al. 2000; Aposhian et al. 2000). Both of these species have been shown to be more cytotoxic than the inorganic arsenite (Petrick et al. 2000; Vega et al. 2001; Styblo et al. 2000).

Bismuth salts are often used in cosmetic and pharmaceutical preparations; bismuth citrate for example has been a common treatment for peptic ulcers and irritable bowel syndrome (IBS). This element is thought to be excreted both through urine and faeces, but the proportion excreted by each route has not as yet been elucidated. Methylmercury in thc human diet (e.g. in fish) is almost completely absorbed into the bloodstream. The nervous system is the principal target tissue affected by methylmercury in humans, whilst the kidney is the critical organ following the ingestion of inorganic mercury salts (Jonnalagadda and Rao 1993). Selenium is an essential trace element in the human body, forming part of antioxidant enzymes that protect against the effects of oxygen free radicals. It is also essential for the normal functioning of the thyroid and immune systems (Corvilain et al. 1993). Nevertheless, certain forms of selenium are highly toxic and the detection of the detoxified degradation product trimethylselenonium in urine is indicative of exposure to such species (Quijano et al. 1999; Zheng et al. 1998; Li et al. 1999).

As well as biomethylation occuring in the liver and kidneys, the possibility of biomethylation by host microbiological flora e.g. in the intestinal tract cannot be excluded. A wide range of bacterial species commonly isolated from clinical as well as food samples have been demonstrated to possess the ability to biomethylate metal(loid)s such as arsenic, antimony, selenium and bismuth (Shariatpanahi et al. 1981; Shariatpahani et al. 1983; Michalke et al. 2000; Jenkins et al. 2002; Smith et al. 2002). With the exception of *Flavobacterium* sp. (Jenkins et al. 2002) the biomethylating bacteria are facultative or strict anaerobes. While much is known about the interaction of methylated metal(loid)s with bacteria in culture, there is little knowledge with respect to respective processes inside the human body.

Monitoring studies (mainly for arsenic and mercury speciation) following ingestion of elevated levels or metal(loid)s have been performed using human urine or blood (Apostili 1999; Ganss et al. 2000; Kresimon et al. 2001). As yet, no studies regarding the holistic speciation and analysis of organometal(loid)s in humans, i.e. the analysis of breath, sputum, urine, blood and faeces have been performed. We report here on a pilot study on the human metabolism of the metal(loid)s selenium, mercury, arsenic and bismuth.

Experimental

Ingestion experiment

After avoiding intake of fish or other seafood the day before and abstaining from breakfast on the day of the experiment, three volunteers (named X, Y, and Z) ingested a meal consisting of tuna fish, bread and a mixed salad supplemented by Selemun®, Telen®, and Heidelberger Chlorella tablets; the masses of arsenic, bismuth and mercury contained in the meal are listed in Table 1.

Table 1 Total arsenic, bismuth and mercury ingested per volunteer

Species	*via* Tuna (μg)	*via* Tablets (μg)
Hg_{tot}	60	-
MeHg	10	-
Se	195	228
As_{org}	270	1
Bi	-	215000
Pb	25	-

Selenium in tablets was supplied as sodium selenite, bismuth in tablets was supplied as bismuth citrate hydroxide complex. Tuna was analysed by microwave-digestion/ICP-MS and GC/AFS, elemental data for tablets were provided by the manufacturer.

Sampling of breath and saliva was performed before and 2 hours after the meal; blood was sampled before and at 1, 2, and 4 hours after the meal. Urine and faeces samples were given before the meal, and 24 hours later; additional urine samples were delivered by each volunteer during the study period according to urinary urge. Exhaled air was collected in 10 l Tedlar bags, and immediately transferred to traps at -80 °C. Saliva samples were collected in 125 ml polyproylene sampling containers; no mouth wash procedure was employed before samples were given to avoid any unnecessary dilution of the saliva samples. Breath samples were analysed immediately, urine, saliva and whole blood samples were analysed within 24 hours; samples were stored at 4 °C until analysis. Faeces samples were stored at -20 °C until analysis was performed (within 4 weeks). With the exception of breath for which two sub-samples were taken, all other samples were measured in triplicate.

Analytical methods

Faeces samples (20-40 g wet weight) were anaerobically incubated over a 12-day period under N_2-atmosphere at 37 °C in the dark. Elemental species present in the gaseous phase above faeces samples and elemental species present in breath samples were determined by low temperature gas chromatography coupled with plasma mass spectrometry (GC/ICP-MS) as described elsewhere (Wickenheiser et al. 1998). Organometal(loid) species in liquid and solid samples were determined by GC/ICP-MS after derivatisation with $NaBH_4$ (Grüter et al. 2000). In addition, mercury compounds were analysed using GC-atomic fluorescence spectrometry (GC/AFS) (see Hippler et al. in this volume). To test for the presence of highly toxic trivalent organic arsenic species in urine HPLC/AFS was used (Le et al. 1996).

Elemental concentrations in urine and blood were directly measured by ICP-MS after dilution 1:10, while in the case of food and faeces the samples were microwave-digested before analysis by a mixture of HNO_3 (65%, suprapure) and H_2O_2 (35%) 5:1 in closed vessels (CEM Mars 5; ca. 0.3 g faeces or 0.5 g food samples + 10 ml digestion mixture, max. temperature 180 °C; after digestion dilution to final volume of 100 ml). Total elemental concentrations (Ga, Ge, As, Se, Sn, Sb, Te, Hg, Bi) in the urine and whole blood samples, and the food and faeces digests were detected by inductively coupled plasma-mass spectrometry (ICP-MS) (Agilent 7500a, Agilent Technologies, Germany). The ICP-MS was operated at 1260 W rf-power, with argon flows of 15 l min^{-1} (plasma gas), 0.98 l min^{-1} (carrier gas) and 0.9 l min^{-1} (auxiliary gas). Solutions (up to 1 in 100 dilutions) were delivered at 0.3 rps to a Babington nebuliser and routed through a double-pass Scott-type spray chamber cooled at 2 °C. The signal at m/z 77 was monitored in order to control chloride interference upon the ^{75}As signal. Quantitation was performed by external calibration and validated by analysing SERO B2 (whole blood) and SERO 201205 (urine).

With the exception of faeces (total content 10 to 160 μg Pb/kg in irregular distribution) the element lead could not be detected in any of the biological samples, hence it will not be further considered in this paper.

Results

Gaseous samples

In breath, only methylated species of Se, I and Bi could be detected in all samples; trace levels of Me_4Sn and Hg^0 were found only sporadically in just a few samples. There was no clear trend observable between dimethylselenium in pre- and post-meal samples (Table 2), indicating that the selenium metabolism of the volunteers was not significantly influenced by the additionally ingested selenium. This applies likewise to methyl iodide (MeI) (concentrations from 4 to 14 ng/m^3).

Table 2 Organometal(loid) species detected by GC/ICP-MS in breath samples before and after ingestion of a metal-enriched meal

	X		Y		Z	
Species	pre- (ng/m^3)	post- (ng/m^3)	pre- (ng/m^3)	post- (ng/m^3)	pre- (ng/m^3)	post- (ng/m^3)
Me_2Se	1.5	2.0	4.7	1.2	0.7	0.4
Me_4Sn	0.3	n.d.	n.d.	n.d.	n.d.	n.d.
MeI	14.0	12.0	10.0	14.0	4.0	4.0
Hg^0	3.0	1.0	n.d.	n.d.	n.d.	1.0
BiH_3	n.d.	0.8	n.d.	0.6	n.d.	n.d.
$MeBiH_2$	0.2	2.8	0.6	0.3	n.d.	0.3
Me_3Bi	n.d.	n.d.	n.d.	0.1	n.d.	n.d.

n.d. = not detected (<0.03 ng/m^3)

Three different bismuth species could be seen in breath samples (Fig. 1). The species distribution was not uniform among the volunteers (Table 2). In samples X and Y there were indications that methylation to monomethyl species may even be possible at trace substrate concentrations.

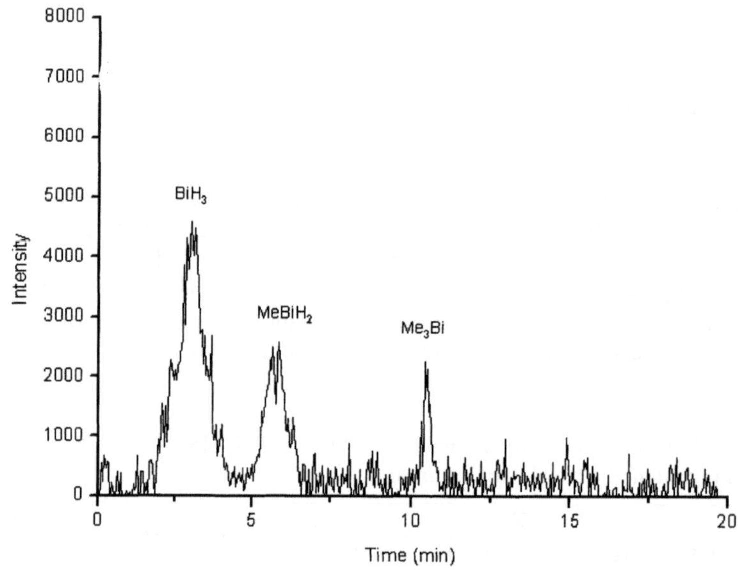

Fig. 1 GC/ICP-MS analysis of breath sample after ingestion of a bismuth enriched meal (volunteer Y). Peaks represent 0.8, 0.3 and 0.1 ng/m^3 for BiH_3, $MeBiH_2$ and Me_3Bi respectively. Identification of BiH_3, $MeBiH_2$ and Me_3Bi is based upon boiling point/retention time correlation (Grüter et al. 2000)

Saliva

Table 3 presents the results of saliva speciation analysis. No systematic trends between pre- and post-meal samples were recognised and so these data cannot be used to evaluate metabolic processes with regard to the metals. Mercury leaching from amalgam fillings of volunteers X and Z as well as residues of the bismuth tablets in the oral cavity of volunteers Y and Z was noted.

Table 3 Elemental species detected in saliva samples (pre- and post-meal).

Species	X		Y		Z	
	pre- (ng/l)	post- (ng/l)	pre- (ng/l)	post- (ng/l)	pre- (ng/l)	post- (ng/l)
$MeAsH_2$	n.d.	n.d.	n.d.	n.d.	91	n.d.
Me_2AsH	n.d.	n.d.	96	n.d.	1324	n.d.
Me_3As	n.d.	n.d.	n.d.	n.d.	412	n.d.
Hg^0	3700	4300	n.d.	208	5100	4500
BiH_3	n.d.	n.d.	n.d.	108601	n.d.	15451
Me_3Bi	n.d.	n.d.	n.d.	n.d.	n.d.	n.d.

n.d. = not detected (<0.08 ng/l)

Urine

Based on total elemental analysis, only the concentrations of arsenic and bismuth in urine were significantly increased during the digestion phase following meal consumption. Mercury and selenium, in contrast, showed no change in concentration during the 24 hour study period (Fig. 2). Mercury concentrations within the range 0.2 to 9.2 µg/l, as well as selenium concentrations of 8 to 86 µg/l were found. The normal mercury concentration in urine typically lies between 0 and 10 µg/l (ARUP 2002) and can increase with the consumption of fish and seafood up to a maximum of 50 µg/l (Eley and Cox 1993). In Germany, the average urinary mercury level is in comparison around 0.9 µg/l (Becker et al. 1998). No increased level of selenium was detected in urine samples given after meal consumption, and the detection range of 8 to 86 µg/l compares favourably with the literature (12.4 to 97.6 µg/l) (Gammelgard and Jons 2000).

For arsenic and bismuth, an increase in metal(loid) concentration in the urine samples was observed following the meal (As: up to 220 µg/l; Bi: up to 3400 µg/l). In Europe, the average arsenic concentration in urine is 11 to 17 µg/l (Le et al. 2000, Buchet et al. 1996). For Germany the average arsenic concentration in the urine is 6.4 µg/l with maximum values up to 160 µg/l (Becker et al. 1998). However, the arsenic concentration in urine depends on the nutrition; e.g. a high fish consumption can lead to far higher urinary arsenic concentrations. The concentration of bismuth in the urine should be low (< 2 µg/l) (ARUP 2002), and indeed in the study here, only low levels of bismuth were detected in initial (t = 0) urine samples. In contrast to arsenic and mercury however, literature evidence regarding bismuth analysis of urine is scarce.

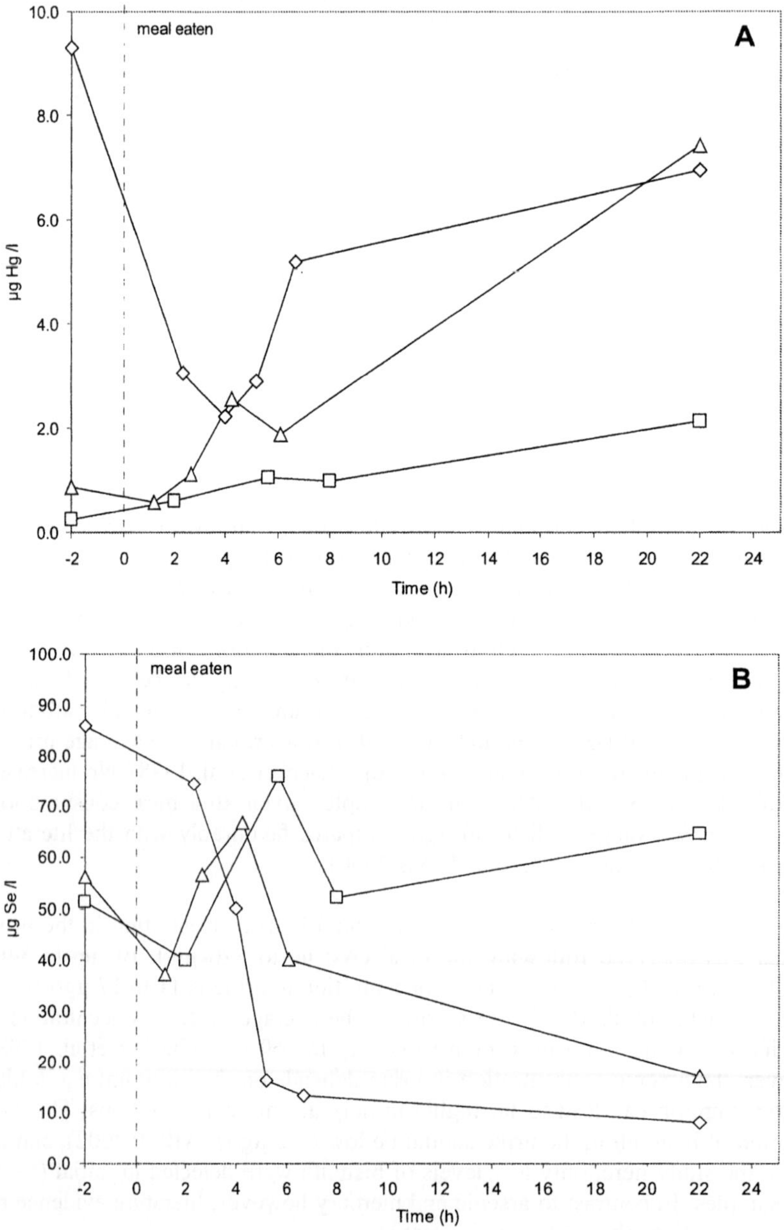

Fig. 2: Total mercury (A) and selenium (B) content of urine over a 24 hour period following ingestion of a metal-enriched meal. Subject X (◇), Y (□) and Z (△).

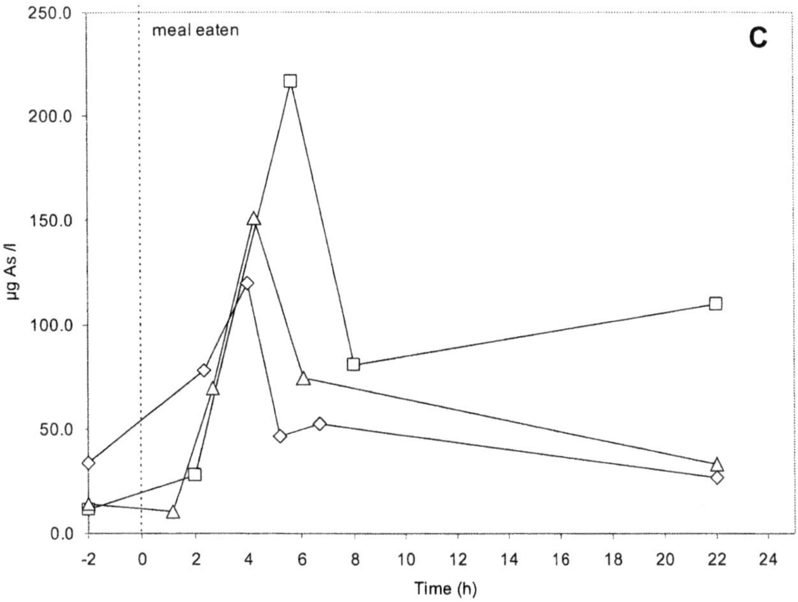

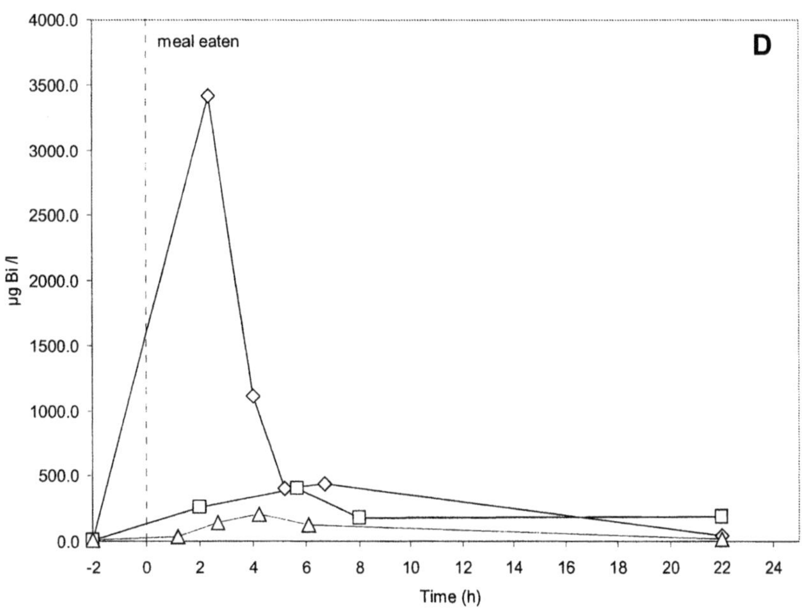

Fig. 2 cont´d: Total arsenic (C) and bismuth (D) content of urine over a 24 hour period following ingestion of a metal-enriched meal. Subject X (◇), Y (□) and Z (△).

The distribution of mono-, di-, and trimethylated arsenic species in urine samples is illustrated in Fig. 3. Of the species tested, Me_2AsOOH (note arsenobetaine and arsenocholine were not analysed) was the most abundant, and trimethylated arsenic could be detected in post-meal samples only. Generally, the distribution of the methylated arsenic species was somewhat irregular; it should however be considered that the derivatisation technique applied is often characterised by matrix-dependent low recovery rates. The highly toxic species $MeAs(OH)_2$ and Me_2AsOH were not detected during HPLC/AFS analysis.

Although the total bismuth concentration in urine exhibits a similar profile and higher concentration compared to arsenic, organic bismuth is much lower than organic arsenic, and could be detected in 5 (out of 51) samples only: $MeBiH_2$ (1.4 to 12 ng/l) and Me_2BiH (21 ng/l).

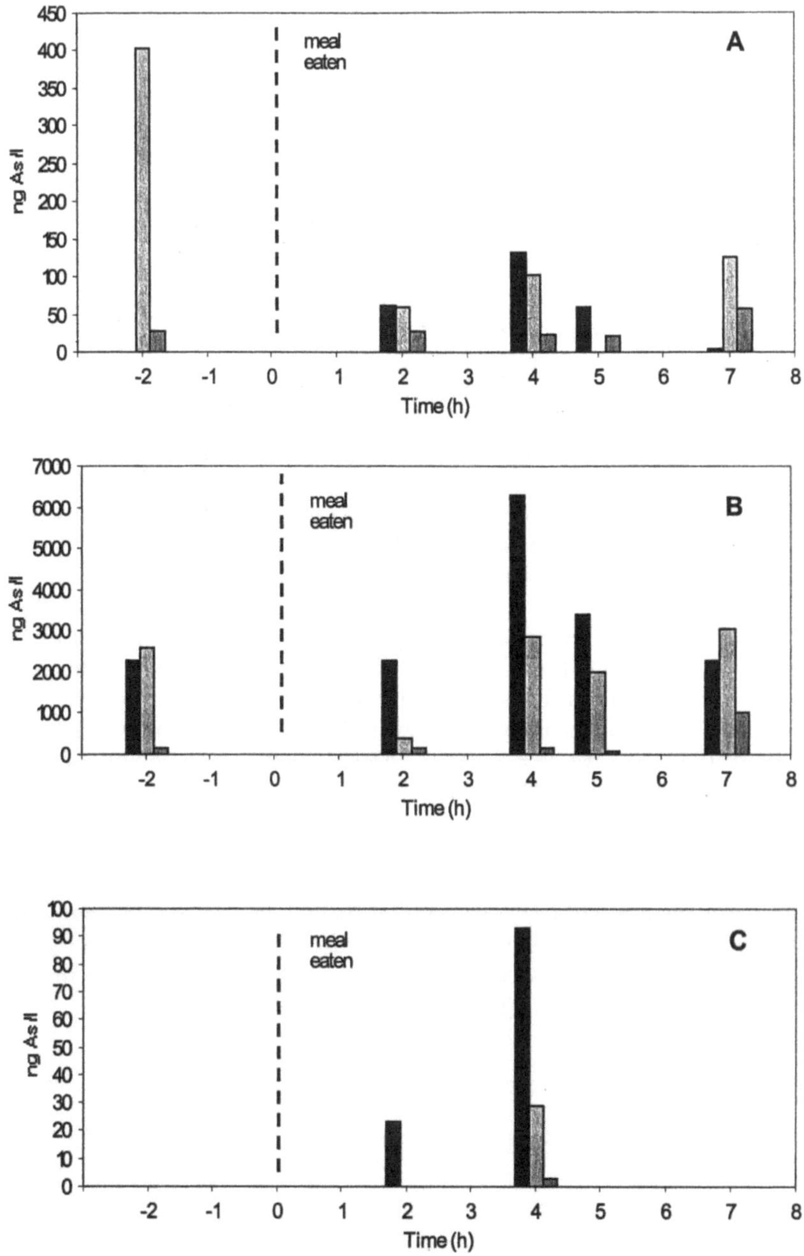

Fig. 3 Arsenic species in urine samples after ingestion of a metal-enriched meal: $MeAsO(OH)_2$ (**A**), Me_2AsOOH (**B**), and Me_3AsO (**C**). Subject X (■), Y (■) and Z (■)

Blood

The selenium concentration in the blood samples, 150 to 200 µg/l, remained relatively constant throughout the study period, and represents a slight increase in comparison to the average values of the German adult population. The normal mean selenium concentrations in plasma and in full blood are 70 and 80 µg/l respectively, within a reference range of 50-120 µg/l plasma/serum (Bundesgesundheitsblatt 2002).

The range of arsenic concentration in blood is relatively large and highly dependant upon diet. For groups exposed to low levels of arsenic, the blood concentration of arsenic ranges from 0.3 to 2 µg/l. In areas where fish (contains high levels of the non-toxic arsenobetaine) is eaten regularly, the average arsenic concentration in the blood increases to 5-10 µg/l (National Research Council 1999). Immediately after consumption of such a meal the blood arsenic concentration rises. Most of the absorbed inorganic and organic arsenic species have a relatively short half-life in the blood (National Research Council 1999), consequently, the concentration of arsenic rapidly decreases again to background levels. This typical profile was observed here in this study and is depicted in Fig. 4.

As with literature evidence concerning bismuth in urine, the number of reports regarding the detection of bismuth in blood is extremely limited in number. As the occurrence in bismuth in foodstuffs is normally very low, it can be expected that the bismuth burden on the body will likewise be very low. Blood bismuth concentrations of 0-5 µg/l have been reported (ARUP 2002) which compare favourably with the detection range of 0-1 µg/l that we found in initial blood samples taken before meal consumption.

While the selenium concentration in blood was fairly constant during the ingestion experiment, a significant increase in elemental content accompanying the digestion of the meal could be observed in the case of arsenic, and more notably for bismuth (Fig. 4); a similar scenario was observed with respect to the methylated species of these two elements (Fig. 5).

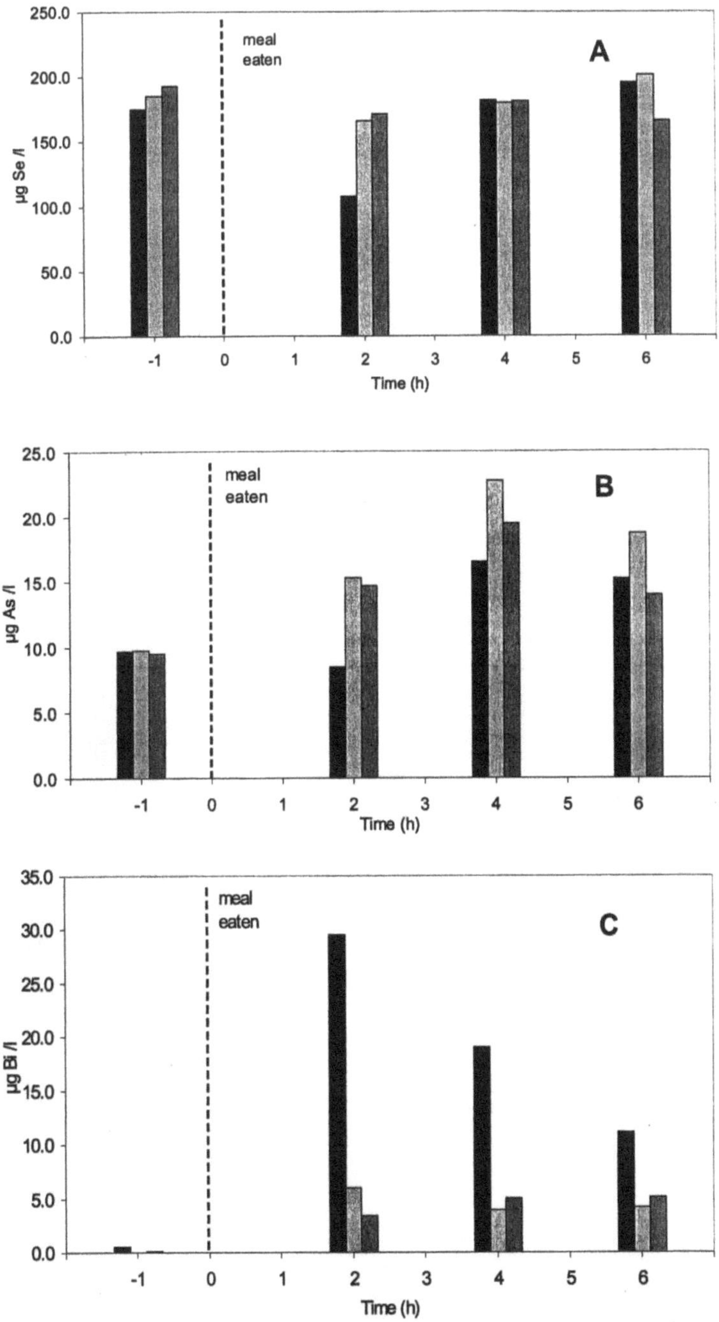

Fig. 4 Total content of selenium (**A**), arsenic (**B**) and bismuth (**C**) in blood during the ingestion experiment. Subject X (■), Y (■) and Z (■)

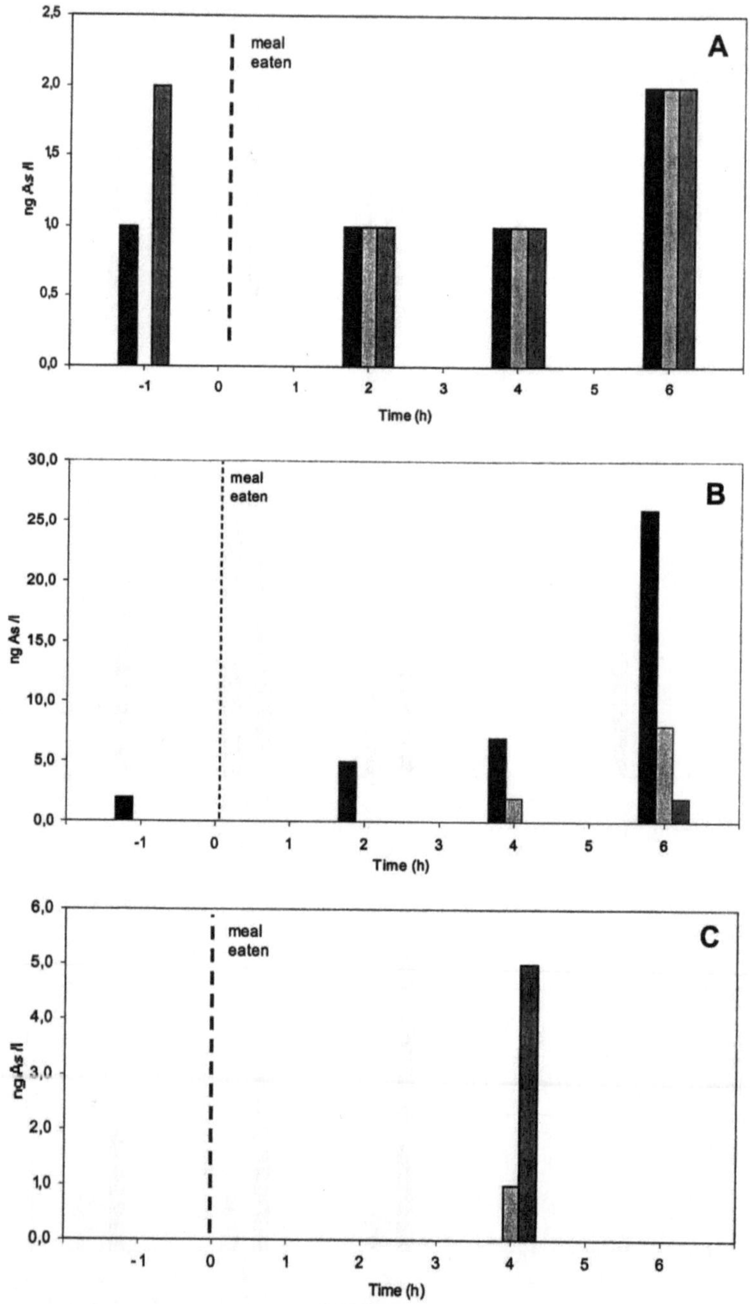

Fig. 5 Arsenic species in blood samples: $MeAsO(OH)_2$ (**A**) and Me_2AsOOH (**B**), Me_3AsO (**C**). Subject X (■), Y (■) and Z (■)

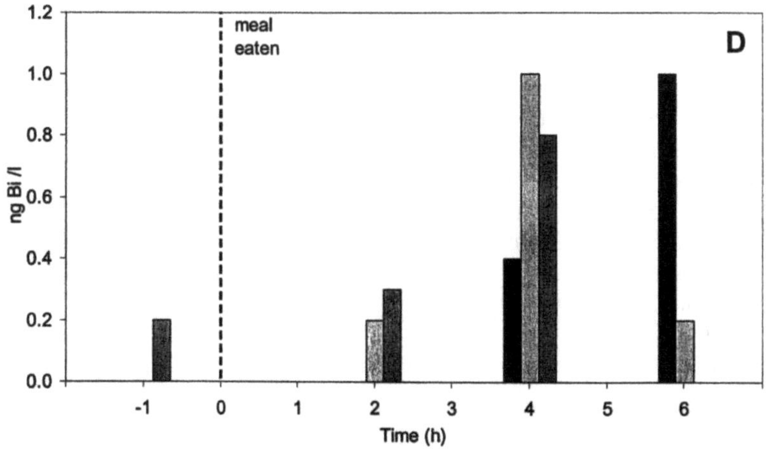

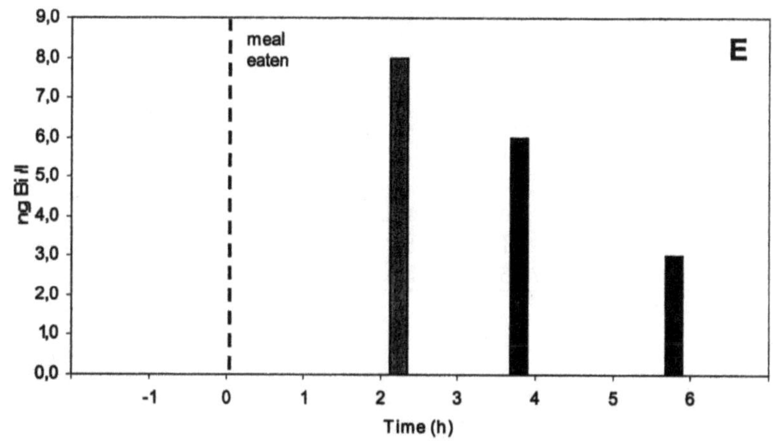

Fig. 5 cont'd Bismuth species in blood samples: $MeBiH_2$ (**D**), and Me_3Bi (**E**). Subject X (■), Y (■) and Z (■)

Faeces

While bismuth was found in faeces only after the meal had been taken (Table 4, Fig. 6), for most of the other elements (Hg, Se, Sb, Pb, Sn, Ga, Ge, In) irregular distributions among the volunteers were observed irrespective of whether the sample had been taken before or after the meal; the concentrations of As and Te were generally found to be lower than the detection limit. Interestingly, mercury was found to be the most abundant metal in faeces samples taken before the meal (Table 4). Unusually high tin concentrations were noted in some samples from volunteer X (range 30 to 60 mg/kg).

Table 4 Total metal content of faeces samples before and after ingestion of a metal-enriched meal

	X		Y		Z	
Element	pre- (mg/kg)	post- (mg/kg)	pre- (mg/kg)	post- (mg/kg)	pre- (mg/kg)	post- (mg/kg)
Ga	0.3	0.2	0.3	0.2	0.3	0.4
Ge	n.d.	n.d.	n.d.	n.d.	n.d.	n.d.
As	n.d.	n.d.	n.d.	n.d.	n.d.	n.d.
Se	0.3	0.2	0.3	0.3	0.2	0.4
Sn	55.6	0.1	0.1	43.6	0.2	0.1
Sb	n.d.	n.d.	n.d.	n.d.	n.d.	n.d.
Te	n.d.	n.d.	n.d.	n.d.	n.d.	n.d.
Hg	0.2	0.1	1.9	0.2	0.1	0.3
Bi	1.0	74.5	1.0	168.1	1.0	240.9

n.d. = not detected (< 30 µg/kg)

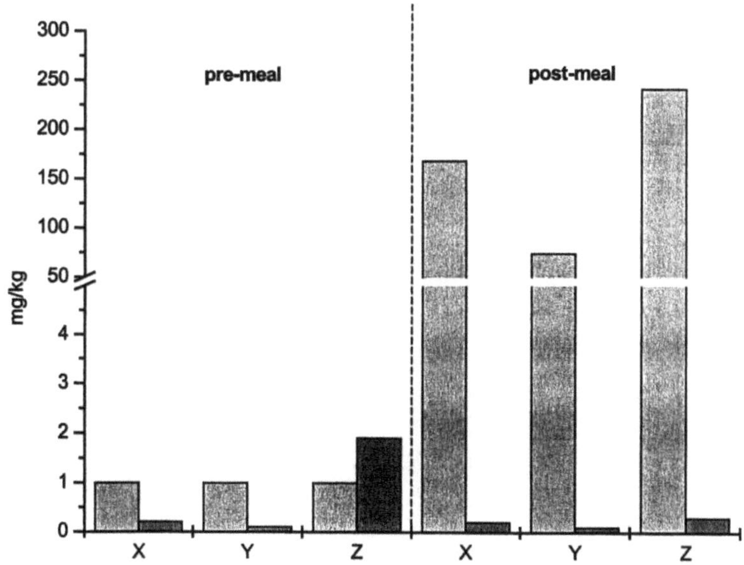

Fig. 6 Bismuth (▒) and mercury (■) concentrations in faeces taken before and after ingestion of the meal

The time profile of bismuth species volatilisation in the headspace of the incubation experiments is graphically illustrated using faeces samples taken after the meal (Fig. 7). Bismuth volatilization apparently works even at trace substrate concentrations, since low levels of volatile BiH_3, MeBi, Me_2Bi and Me_3Bi [up to 1.3 ng Bi/kg (wet weight)] were all detected in the headspace gases of faeces samples given before ingestion of the meal despite low total bismuth concentrations of faeces (ca. 1 mg Bi/kg (wet weight)). The total bismuth in faeces samples increased by up to 200-fold after ingestion (Fig. 6). That the quantity of volatile species did not increase by a comparable proportion (ca. ten-fold increase was observed) suggests that this bismuth may not all be bioavailable. The biovolatilisation of antimony species could also be demonstrated by incubation experiments (data not given here). Since antimony was not one of the elements ingested by the volunteers, this indicates that existing background antimony concentrations in faeces are being volatilised.

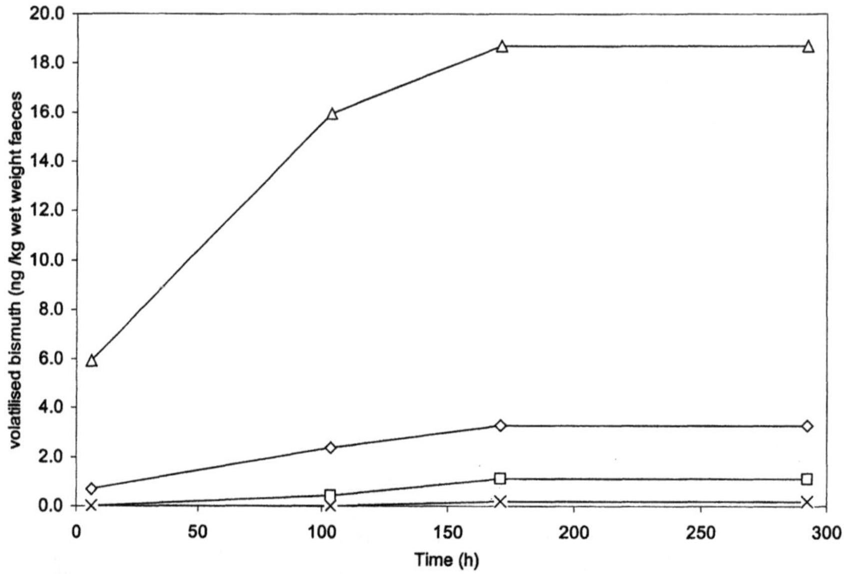

Fig. 7 GC/ICP-MS analysis of headspace gases evolved from faeces following ingestion of a metal-enriched meal: BiH_3 (◇), $MeBiH_2$(□), Me_2BiH (×) and Me_3Bi (△)

Discussion

Metal(loid)s ingested are absorbed in the gastrointestinal tract or arrive in the colon to be excreted via faeces. The absorbed part is transported by blood and lymph to other organs, including the kidney and liver, where enzymatic methylation may occur. Compounds, such as many of the fully methylated/hydride species, have high vapour pressures and may escape from the bloodstream during its lung passage, and thus can be found in breath. Lipophilic substances, for example organic mercury compounds, are able to cross the blood/brain-barrier and eventually find their way into the brain. Thus, only a holistic analytical approach as taken in the present study will provide the data necessary for the overall evaluation of exposure to the metal(loid)s and its potential toxicological significance. The present pilot study has shown that, if both parent compound and metabolites are only analysed in specific compartments such as blood or urine, the formation of potentially harmful degradation products may remain unrecognized.

In the anaerobic environment of the colon, biomethylation may occur because of the presence of microorganisms with biomethylating capability, leading to the formation of organometal(loid) species. Many of these organometal(loid) species are volatile and able to diffuse into blood, thus participating in the processes described above and schematically illustrated in Fig.8. Further evidence for the occurrence of biomethylation in the human gut was obtained from the observations in the present study that bismuth was eliminated predominantly *via* faeces and volatile bismuth compounds appeared in the breath after the ingestion of the meal. As the toxicity of the volatile bismuth species have not yet been elucidated, the toxicological implications of the formation of these species are still unknown. When other elements, e.g. mercury, are processed in a similar manner, highly toxic species with long half-lives are formed.

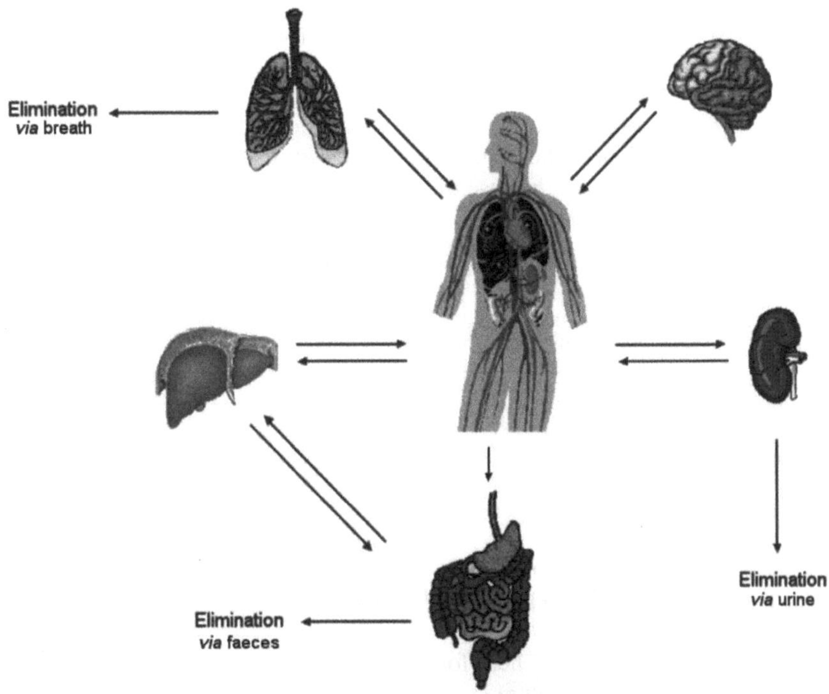

Fig. 8 Metabolic pathways of ingested metal(loid)s in the human body

When discussing the analytical results of our ingestion experiment in the light of the metabolic pathway model presented, specific conclusions may be drawn for three elements:

- Arsenic is absorbed in the small intestine and is likely methylated in the liver and/or the kidney, and is excreted *via* urine mainly in the form of Me_2AsOOH, thus it cannot be detected in the faeces. Liver and kidney might therefore be the primary targets for arsenic toxicity originating from methylated tri- and pentavalent methylated arsenicals.

- Bismuth absorption in the small intestine is very low, thus nearly all of this element arrives in the colon. There, hydride formation and methylation takes place leading to educts diffusing into the blood and eventually escaping *via* breath. Depending on the characteristics of an individual's intestinal flora the composition of the formed volatile bismuth species may vary from person to person, which in turn may effect the toxicological outcome.

- Mercury may in principle be able to follow both pathways, as high levels of mercury were detected both in urine (Table 3) and in faeces (Table 4). As mentioned above, biomethylation of mercury in the gut leads to the formation of the highly neurotoxic monomethylated and dimethylated species.

The different pathways of arsenic, bismuth and mercury can further be illustrated by Table 5 where the percentage of urinary excretion (related to ingested amounts) is calculated. In the case of volunteer X this calculation cannot be done for mercury as the initial mercury concentration in urine was higher than post-meal levels (Fig. 2).

Table 5 % Mass excretion of arsenic, bismuth and mercury *via* urine

Element	X	Y	Z
Hg	n.c.	29.0	96.0
As_{total}	25.0	78.0	39.0
As_{org}	4.1	2.4	0.9
$As_{Me2AsOOH}$	4.0	2.3	0.8
Bi	0.5	0.2	0.1

n.c. = not calculated (pre-meal Hg concentration was higher than post-meal). Total voided urine was 1280 ml (X), 1200 ml (Y) and 1340 ml (Z)

While the case discussed here for arsenic is in accordance with what is published in the pertinent literature (Le et al. 1996; Lintschinger et al. 1998), the results in respect to bismuth are new. No relationship was found in this study between ingestion of methyl mercury and mercury concentrations in urine (Fig. 3a) and faeces (Table 4). With the analytical techniques used, more complex forms of organomercury could not be detected; further method optimisation in relation to the blood matrix will be necessary.

Conclusions

Although the results of this pilot study still have to wait for statistical validation, it could be demonstrated that some differences exist in the metabolism of metal(loid)s in the human body:

- In contrast to selenium, ingestion of arsenic and bismuth compounds increased the total amount and speciation of these elements in urine and blood.
- Most of the ingested bismuth ends up in the colon, where hydride formation and methylation occurs, eventually leading to the presence of bismuthine and methylated bismuth species in blood and breath. The toxicological consequences of this formation are not yet known.
- The process of biological element volatilization in the colon can also be demonstrated for antimony, and may be similar for mercury. In the latter case, highly neurotoxic methylated mercurials will be formed.

Acknowledgements

The authors are grateful to Prof. H. Emons for participating in this project. This work was funded by the Deutsche Forschungsgemeinschaft (Project FOR 415).

Literature

Aposhian VH, Gurzau ES, Le XC, Gurzau A, Healy SM, Lu X, Ma M, Yip L, Zakharyan RA, Maiorino RM, Dart RC, Tirus MG, Gonzalez-Ramirez D, Morgan DL, Avram D, Aposhian MM (2000) Occurrence of monomethylarsonous acid in urine of humans exposed to inorganic arsenic. Chem Res Toxicol 13:693-697

Apostili P (1999) The role of elemental speciation in environmental and occupational medicine. Fresenius J Anal Chem 363:499-504

ARUP Laboratories (2002) ARUPs Guide to Clinical Laboratory Testing. http://www.ARUPlab.com/guides/ug/tests/0099007.htm

Becker K, Kaus S, Krause C, Lepom P, Schulz C, Seiwert M, Seifert B(1998) Umwelt-Survey Band III: Human-Biomonitoring. Stoffgehalte in Blut und Urin der Bevölkerung in Deutschland. Umweltbundesamt, Berlin

Buchet JP, Lison D, Ruggeri M, Foa V, Elia G (1996) Assessment of exposure to inorganic arsenic, a human carcinogen, due to the consumption of seafood. Arch Toxicol 70:773-778

Bundesgesundheitsblatt (2002) Gesundheitsforschung – Gesundheitsschutz, 45(2), 190-195

Cai X-J, Block E, Uden PC, Quimby BD, Sullivan JJ (1995) Allium chemistry - identification of natural-abundance organoselenium compounds in human breath after ingestion of garlic using gas-chromatography with atomic-emission detection. J Agr Food Chem 43:1751-1753

Challenger F (1945) Biological methylation. Chem Rev 36:315-361

Corvilain B, Contempre B, Longombe AO, Goyens P, Gervydecoster C, Lamy F, Vanderpas JB, Dumont JE (1993) Selenium and the thyroid - How the relationship was established. Am J Clin Nutr 57:244-248

Eley BM, Cox SW (1993) The release, absorption, and possible health effects of mercury from dental amalgam: A review of recent findings (erratum to the original review published on September 11, 1993). Br Dent J 175:355-362

Feldmann J, Riechmann T, Hirner AV (1996) Determination of organometallics in intraoral air by LT-GC/ICP-MS. Fresenius J Anal Chem 354:620-623

Gammelgaard B, Jons O (2000) Determination of selenite and selenate in human urine by ion chromatography and inductively coupled plasma mass spectrometry. J Anal Atomic Spectrom 15:945-949

Ganss C, Gottwald B, Traenckner I, Kupfer J, Eis D, Mönch J, Gieler U, Klimek J (2000) Relation between mercury concentrations in saliva, blood, and urine in subjects with amalgam restorations. Clin Oral Invest 4:206-211

Gosio B (1897) Zur Frage, wodurch die Giftigkeit arsenhaltiger Tapeten bedingt wird. Ber Deutsch Chem Ges 30:1024-1026

Grüter UM, Kresimon J, Hirner AV (2000) A new HG/LT-GC/ICP-MS multi-element speciation technique for real samples in different matrices. Fresenius J Anal Chem 368:67-72

Jenkins RO, Forster SN, Craig PJ (2002) Formation of methylantimony species by an aerobic prokaryote: Flavobacterium sp. Arch Microbiol 178:274-278

Jonnalagadda SB, Rao PV (1993) Toxicity, bioavailability and metal speciation. Comp Biochem Physiol C 106:585-595

Kresimon J, Grüter UM, Hirner AV (2001) HG/LT-GC/ICP-MS coupling for identification of metal(loid) species in human urine after fish consumption. Fresenius J Anal Chem 317:586-590

Le XC, Lu XF, Ma MS, Cullen WR, Aposhian HV, Zheng BS (2000) Speciation of key arsenic metabolic intermediates in human urine. Anal Chem 72:5172-5177

Le XC, Ma MS, Wong NA (1996) Speciation of arsenic compounds using high-performance liquid chromatography at elevated temperature and selective hydride generation atomic fluorescence detection. Anal Chem 68:4501-4506

Li FS, Goessler W, Irgolic KJ (1999) Determination of trimethylselenonium iodide, selenomethionine, selenious acid, and selenic acid using high-performance liquid chromatography with on-line detection by inductively coupled plasma mass spectrometry or flame atomic absorption spectrometry. J Chromatogr A 830:159-176.

Lintschinger J, Schramel P, Hatalak-Rauscher A, Wendler I, Michalke B (1998) A new method for the analysis of arsenic species in urine by using the HPLC-ICP-MS. Fresenius J Anal Chem 362:313-318

Michalke K, Wickenheiser EB, Mehring M, Hirner AV, Hensel R (2000) Production of volatile derivatives of metal(loid)s by microflora involved in anaerobic digestion of sewage sludge. Appl Environ Microbiol 66:2791-2796

Mozier NM, McConnell KP, Hoffmann JL (1988) S-Adenosyl-L-methionine:Thioether S-methyltransferase, a new enzyme in sulfur and selenium metabolism. J Biol Chem 10:4527-4531

Nakamuro K, Okuno T, Hasegawa T (2000) Metabolism of selenoamino acids and contribution of selenium methylation to their toxicity. J Health Sci 46:418-421

National Research Council (1999) Arsenic in Drinking Water; Subcommittee on Arsenic in Drinking Water; Washington, DC: National Academy Press

Peraza MA, Kopplin MJ, Carter DE, Gandolfi AJ (2002) Inorganic arsenic biotransformation and mitochondrial toxicity in HK-2 human proximal tubular cells. The Toxicologist 66:A403

Petrick JS, Ayala-Fierro F, Cullen WR, Carter DE, Aposhian HV (2000) Monomethylarsonous acid (MMA^{III}) is more toxic than arsenite in Chang human hepatocytes. Toxicol Appl Pharmacol 163:203-207

Quijano MA, Gutierrez AM, Perez-Conde MC, Camara C (1999) Determination of selenium species in human urine by high performance liquid chromatography and inductively coupled plasma mass spectrometry. Talanta 50:165-173

Shariatpahani MAC, Anderson AC, Abdelghani AA, Englande AJ (1983) Microbial metabolism of an organic arsenical herbicide. Biodeterioration 5:268-277

Shariatpanahi M, Anderson AC, Abdelghani AA, Englande AJ, Hughes J, Wilkinson RF (1981) Biotransformation of the pesticide sodium arsenate. J Environ Sci Health B 16:35-47

Smith LM, Craig PJ, Jenkins RO (2002) Formation of involatile methylantimony species by Clostridium spp. Chemosphere 47:401-407

Styblo M, DelRazo LM, Vega L, Germolec DR, LeCluyse EL, Hamilton GA, Reed W, Wang C, Cullen WR, Thomas DJ (2000) Comparative toxicity of trivalent and pentavalent inorganic and methylated arsenicals in rat and human cells. Arch Toxicol 74:289-299

Vega L, Styblo M, Patterson R, Cullen W, Wang CQ, Germolec D (2001) Differential effects of trivalent and pentavalent arsenicals on cell proliferation and cytokine secretion in normal human epidermal keratinocytes. Toxicol Appl Pharmacol 172:225-232

Wickenheiser EB, Michalke K, Drescher C, Hirner AV, Hensel R (1998) Development and application of liquid and gas-chromatographic speciation techniques with element specific (ICP-MS) detection to the study of anaerobic arsenic metabolism. Fresenius J Anal Chem 362:498-501

Zheng J, Goessler W, Kosmus W (1998) The chemical forms of selenium in selenium nutritional supplements: an investigation by using HPLC/ICP/MS and GF/AAS. Trace Elem Electroly 15:70-75

Chapter 11

Genotoxicity of organometallic species

A.-M. Florea, E. Dopp, G. Obe, A.W. Rettenmeier

Human exposure to organometallic species

The modification of metals and metalloids by formation of volatile metal hydrides and alkylated species (volatile and non-volatile) has a major impact on the processing of these elements in the environment. In general, the formation of such species increases the mobility of the respective element and can result in its accumulation in biological systems (Craig and Glockling 1988).

Many studies have shown that the production of organometal(loid) species is possible and likely whenever metal(loid)s in the presence of methyl donors are exposed to specific microorganisms under anaerobic conditions, at least on a microscale basis (Brinckman and Bellama 1978). These conditions may exist both in natural environments (e.g. wetlands, pond sediments) and in anthropogenic systems (e.g. waste disposal sites and sewage treatment plants). Other sources of organometal(loid)s are industrially produced compounds such as biocides and catalysts.

Organometal(loid)s can be released in the gaseous state or as aerosols into the atmosphere; *via* the solved state, they can enter the hydrosphere and soil. There is a steady exchange of these compounds between the compartments of the ecosphere.

In the course of these exchanges, organometal(loid)s are taken up by humans, distributed in tissues, and eliminated *via* breath, urine, or faeces. For example, a variety of organometallic compounds with elements such as germanium, arsenic, selenium, tin, antimony, and mercury have been detected in human urine following consumption of seafood (Kresimon et al. 2001). The occurrence of organometal(loid)s in human tissues may also result from bacterial activities in the intestinal tract (Kresimon et al. 2001).

Genotoxicity of organometallic species

The potential of organometallic compounds for adversely affecting human health is well documented. This applies in particular to the neurotoxic and teratogenic effects of organomercurials. Examples are the poisoning of the Minamata Bay population with methylmercury (MeHg) (Harada 1978), the lethal epidemic in the Iraq where people ingested MeHg-contaminated corn-products (Bakir et al. 1973), or, more recently, the death of the American chemist Karen E. Wetterhahn following accidental dermal exposure to dimethylmercury (Me_2Hg) (Nierenberg et al. 1998). There are only few data published on genotoxic and carcinogenic effects of organometal(loid) compounds and on the mechanisms of their action at the cellular level. Increased interest in these topics came from the recent finding that methylated trivalent metabolites might contribute to the carcinogenicity of arsenic (Styblo et al. 2002).

In the following chapters, studies on the genotoxic activity of organomercurials, organoarsenic compounds, and organotin compounds are briefly reviewed.

Genotoxic effects of organomercury compounds

In a Brazilian population exposed to MeHg in drinking water, a positive relationship between Hg contamination of hair and impairment of lymphocyte proliferation *in vitro* as well as cytogenetic damage were observed at levels as low as 50 µg Hg/g hair where indications of mercury poisoning were seen (Amorim et al. 2000).

Several studies were performed on the genotoxic effects of organomercury compounds in cultured mammalian cells. MeHg and Me_2Hg produced DNA damage in a number of different tests. These compounds induced micronuclei (MN), structural and numerical chromosomal aberrations (CA), and sister chromatid exchanges (SCE) in a dose-dependent manner (for review see De Flora et al. 1994). MeHg was 6-fold more active in inducing CA than Me_2Hg (Betti et al. 1992). Results of more recent studies are presented in Tab. 1. Ehrenstein et al. (2002) observed significantly elevated frequencies of CA and SCE at a MeHg concentration of $1x10^{-6}$ M in the culture medium which led to an intracellular MeHg concentration of $1.99x10^{-16}$ M/cell volume. Ogura et al. (1996) analyzed the induction of MN, CA, and 8-hydroxy-2'-deoxyguanosine (8-OHdG) in human peripheral lymphocytes in culture by MeHgCl. CA increased in lymphocytes exposed to MeHgCl at a concentration of $2 x 10^{-6}$ M and to $HgCl_2$ at a concentration of 10^{-5} M. The increase in the incidence of micronucleated lymphocytes was significant at doses of $2x10^{-5}$ M $HgCl_2$ and $5x10^{-6}$ M MeHgCl when compared to the control. MeHgCl was approximately 4-fold more potent than $HgCl_2$. The 8-OHdG levels were significantly higher in lymphocytes exposed to $1x10^{-5}$ M $HgCl_2$ or to $5x10^{-6}$ M MeHgCl compared to nonexposed cells. The authors conclude that the effects observed in peripheral lymphocytes exposed to mercury compounds could be due to either disturbances of spindle mechanisms or to elevated levels of 8-OHdG caused by reactive oxygen species.

Likewise in human lymphocytes, MeHgCl and Me_2Hg induced DNA fragmentation, as analysed by microgel electrophoresis (Betti et al. 1992; 1993a,b). MeHgCl inhibited DNA synthesis in mouse foetal astrocytes at a 10-fold lower dose than $HgCl_2$ (Choi and Kim 1984). MeHgCl also altered gene expression in rat glioma cells assessed by changes in the amounts and net phosphorylation profiles of specific proteins (Ramanujan and Prasad 1979). Moreover, these compounds interfere with DNA repair (Yamada et al. 1993).

In Chinese hamster V79 cells, MeHgOH induced C-mitoses, concomitantly with a decrease of nonprotein SH groups (mainly GSH), and increased the level of GSH peroxidase (Önfelt 1983).

Phenylmercury acetate caused a dose-dependent elevation of SCE and of endoreduplicated mitoses in cultured human lymphocytes (Lee et al. 1997).

Further studies revealed that organomercury compounds lead to reproductive dysfunctions (Dufresne and Cyr 1999) and are renal carcinogens in mice (De Flora et al. 1994).

Table 1. Recent studies on the genotoxic activity of organomercury compounds

	MeHgCl	Me_2Hg	PhHgAc
MN[1)2)]	+	n.d.	n.d.
CA[1)3)4)]	+	+	n.d.
SCE[4)]	+	n.d.	+
Endoreduplications[5)]	+	n.d.	+
8-OHdG[1)]	+	n.d.	n.d.

n.d.: not determined, +: positive, -: negative

MeHgCl: methylmercury chloride, Me_2Hg: dimethylmercury; PhHgAc: phenylmercury acetate; MN: micronuclei, CA: chromosomal aberrations, SCE: sister chromatid exchanges, 8-OHdG: 8-hydroxy-2'deoxyguanosine. 1) Ogura et al. 1996; 2) Migliore et al. 1999; 3) Amorim et al. 2000; 4) Ehrenstein et al. 2002; 5) Lee et al. 1997

In the majority of comparative studies, ionizable organomercury compounds proved to be much more active in short-term tests than either non-ionizable organomercury compounds (Me_2Hg) or inorganic mercury salts ($HgCl_2$). It has been proposed that there is a common mechanism of genotoxicity, but different bioavailability of inorganic and organic mercury compounds (De Flora et al. 1994).

Genotoxic effects of organoarsenic compounds

Although it is known that humans and many other species convert inorganic arsenic to monomethylated and dimethylated metabolites, relatively little attention has been payed to the biological effects of these methylated products. Observations made during the last few years suggest that methylation of inorganic arsenic is a toxification rather than a detoxification pathway. Accumulating evidence indicates that, in particular, trivalent methylated arsenic metabolites such as monomethylarsonous acid ($MeAs(OH)_2$) and dimethylarsinous acid (Me_2AsOH) interact with cellular targets such as proteins and DNA (Kitchin 2001). As compared to inorganic arsenic compounds in the trivalent oxidation state, methylated trivalent arsenic species are more cytotoxic and genotoxic, and are more potent enzyme inhibitors (Thomas et al. 2001). According to Mass et al. (2001), Me_2AsOH is about 400 times more genotoxic than $MeAs(OH)_2$, which in turn is about 80 times more genotoxic than arsenite.

Arsenic is methylated via glutathione conjugation in two steps: (1) to monomethylarsonic acid ($MeAsO(OH)_2$) and (2) to dimethylarsinic acid (Me_2AsOOH) (Fig. 1). Methylation requires the metabolic reduction of arsenate to arsenite with ($MeAs(OH)_2$) as an intermediary metabolite. Both methylated and dimethylated arsenicals that contain arsenic in the trivalent oxidation state were identified as intermediates in the metabolic pathway (Fig. 1). These compounds have been detected in human cells cultured in the presence of inorganic arsenic and in urine of individuals chronically exposed to inorganic arsenic (for review see Thomas et al. 2001).

Inorganic arsenic is enzymatically methylated by a As-methyltransferase using S-adenosyl-methionine (SAM) as a donor of methyl groups. The fact that DNA methyltransferases require the same methyl donor suggests an important role for methylation in arsenic carcinogenicity. Studies with rat liver epithelial cells conducted by Zhao et al. (1997) revealed that global DNA hypomethylation occurres concurrently with arsenic-induced malignant transformation under conditions of suppressed levels of SAM. This effect was dependent on dose and duration of exposure. It was also found in this animal model that changes of gene expression and malignant transformation occur after long-term exposure to arsenic (Zhao et al. 1997).

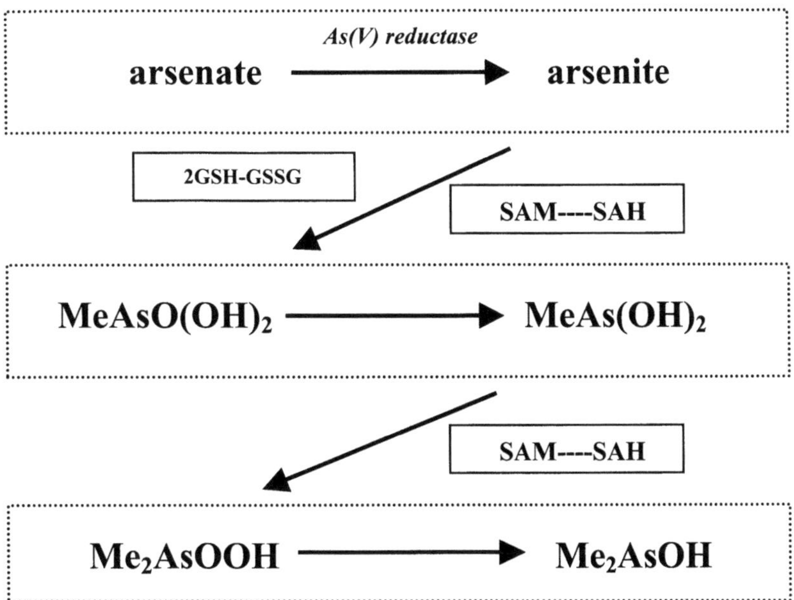

Fig.1. Metabolism of arsenic

GSSG, glutathione (oxidized); GSH, glutathione (reduced); SAM, S-adenosylmethionine; SAH, S-adenosylhomocysteine; $MeAsO(OH)_2$, monomethylarsonic acid; $MeAs(OH)_2$, monomethylarsonous acid; Me_2AsOOH, dimethylarsinic acid; Me_2AsOH, dimethylarsinous acid

Methylation of arsenic occurs mainly in the liver, a suspected target organ for arsenic carcinogenesis. In some animal species (such as guinea pigs, chimpanzees and marmosets), rates of arsenic methylation are extremely low. Therefore, these species were described as "nonmethylating" (Kitchin 2001). Rats produce urinary trimethylated arsenic species [trimethylarsine oxide (Me_3AsO)] following administration of Me_2AsOOH (Kitchin 2001).

1. Mutagenicity of inorganic and organic arsenicals (colony forming assay):	
	iAs(III)>iAs(V)> $MeAsO(OH)_2$ > Me_2AsOOH (Moore et al., 1997)
2. Genotoxicity of arsenic compounds (SCE, CA, MN, Comet-assay):	
	iAs(III)>iAs(V) > $MeAsO(OH)_2$ > Me_2AsOOH (Moore et al., 1997) iAs(III)>iAs(V)> Me_2AsOOH > $MeAsO(OH)_2$>Me_3AsO (Oya-Ohta et al., 1996) iAs(III)>iAs(V)> Me_2AsOOH > $MeAsO(OH)_2$ > Me_3AsO > tetramethylarsoniumiodide > arsenobetaine = arsenocholine (Kaise et al., 1998) Me_2AsOH > $MeAs(OH)_2$ >> iAs(V) ≈ iAs(III)> $MeAsO(OH)_2$ > Me_2AsOOH (Mass et al., 2001)
3. Cytotoxicity (LDH release, $[K^+]_i$, cell growth inhibition):	
	$MeAs(OH)_2$ > iAs(III)>iAs(V) > $MeAsO(OH)_2$ = Me_2AsOOH (Petrick et al., 2000) iAs(III) > iAs(V) > Me_2AsOOH > $MeAsO(OH)_2$ >tetramethyl-arsoniumiodide > Me_3AsO = arsenobetaine = arsenocholine (Kaise et al., 1998)
4. Viability test (cell proliferation):	
	iAs(III) > MMAO(III) > Me_2AsOH –GS > Me_2AsOOH > $MeAsO(OH)_2$ > iAs(V) (Vega et al., 2001)

Fig. 2. Toxic potency of inorganic and organic arsenic compounds in different test systems

iAs(III), arsenite; iAs(V), arsenate; $MeAsO(OH)_2$, monomethylarsonic acid; Me_2AsOOH, dimethylarsinic acid; Me_3AsO, trimethylarsine oxide; MMAO(III), monomethylarsine oxide; Me_2AsOH–GS, complex of Me_2AsOH with glutathione; LDH, lactate dehydrogenase; $[K^+]_i$, intracellular potassium

Although much progress has recently been made to understand the possible modes of carcinogenic action of arsenic, a scientific consensus has not been reached yet. Nine different possible modes of arsenic carcinogenesis are discussed: (1) induction of chromosomal abnormalities, (2) oxidative stress, (3) altered DNA repair, (4) altered DNA methylation patterns, (5) altered growth factors, (6) enhanced cell proliferation, (7) enhanced promotion/progression, (8) enhanced gene amplification, (9) suppression of p53 gene expression (Kitchin, 2001).

In genotoxicity studies, Me_2AsOOH in millimolar doses was found to induce MN, CA, aneuploidy and chromosome pulverisations (Oya-Ohta et al. 1996). Me_2AsOOH produced apurinic/apyrimidinic sites, protein cross-links, and DNA base damage as a consequence of oxidative stress (Kitchin 2001). Me_2AsOOH adversely interacts with DNA repair (Kawaguchi et al. 1996). In rats, this compound was found to be carcinogenic (Huff et al. 2000). Administration of Me_2AsOOH to p53(+/-) mice and K6/ODC transgenic mice (mice which overexpress a truncated ornithine decarboxylase protein in hair follicle keratinocytes) resulted in weak carcinogenic, cocarcinogenic, or promotional activity in skin or bladder (for review see Kitchin 2001). Me_2AsOOH was found to be a promotor of urinary bladder, kidney, liver and thyroid gland carcinogenesis (Yamamoto et al. 1997).

Tab. 2 summarizes the cytogenetic studies with organoarsenic compounds. The relative toxic potency of inorganic and organic arsenic compounds in different test systems is presented in Fig. 2.

Genotoxic effects of organotin compounds

Studies on the genotoxicity of organotin compounds are rare compared to the numerous reports on their toxicity. Hadjispyrou et al. (2001) investigated the toxic effects of the three organotin compounds trimethyltin chloride (Me_3SnCl), dimethyltin dichloride (Me_2SnCl_2), and dibutyltin diacetate (Bu_2SnAc_2) on *Artemia franciscana*. Me_3SnCl showed the highest toxicity in terms of LC_{50} values (concentration at which 50 % of the animals die) compared with Me_2SnCl_2 or Bu_2SnAc_2.

Hamasaki et al. (1993) studied the mutagenicity of 14 organotin compounds reported to be environmental pollutants. Butyltin oxide (BuSnO), butyltin trichloride ($BuSnCl_3$), dibutyltin dichloride (Bu_2SnCl_2), tributyltin chloride (Bu_3SnCl), tributyltin hydroxide, Bu_3SnOH and Me_2SnCl_2 were mutagenic in *Salmonella typhimurium TA100*. Bu_3SnCl_2 showed the highest mutagenicity. Bu_2SnCl_2 was mutagenic in *S. typhimurium TA98*. BuSnO, $BuSnCl_3$, and Bu_2SnCl_2 were positive in the SOS chromotest with *Escherichia coli PQ37*. Bu_2SnCl_2, Bu_3SnCl, Bu_3SnOH, Me_2SnCl_2 and Me_3SnCl were genotoxic in the rec-assay (Hamasaki et al. 1992). Tab. 3 summarizes the genotoxic effects of different organotin compounds.

In general, the toxicity of organotin compounds in microorganisms increases with number and chain-length of the organic groups bound to the tin atom. Tetraorganotin and inorganic tin have only low toxicity. Because of their lipophilicity, organotins are regarded as membrane active. There is evidence that the site of action of organotins may be at the cytoplasmatic membrane and at the intracellular level. Consequently, it is not known whether cell surface adsorption, accumulation within the cell, or both are responsible for toxicity (White et al. 1999).

The clastogenicity of Me_3SnCl was evaluated in human peripheral lymphocytes using the MN test. Lymphocytes were exposed to 0.1 μg/ml and 0.2 μg/ml of Me_3SnCl for 48 h and a significant increase in MN frequency was observed with both doses (Ghosh et al. 1990).

Exposure of human lymphocytes in culture to organotin compounds resulted in an increased frequency of aneuploid, especially of hyperdiploid cells, probably by affecting the spindle mechanism (Jensen et al. 1991).

Ghosh et al. (1992) analyzed CA, SCE and cell cycle kinetics in human lymphocytes exposed to Me_3SnCl. At a concentration of 0.4 μg/ml, the replicative index was reduced by approximately 50 % and the frequencies of CA and SCE were increased.

The ability of three tributyltins and three triphenyltins to induce CA in Chinese hamster ovary cells (CHO K1) was studied by Sasaki et al. (1993). None of the organotins showed any clastogenic activity, however, post-treatment with organotins increased the frequencies of chromatid aberrations induced by the clastogens mitomycin C, cisplatin, 4-nitroquinoline-1-oxide, methyl methane sulfonate, and actinomycin D.

Some organotin compounds are embryotoxic and teratogenic in mice (Boyer 1989)

Table 2. Genotoxicity of organoarsenic compounds

Tests	Me_4AsI	Me_3AsO	Me_2AsOOH	Me_2AsI	Me_2As-GS	$MeAsO(OH)_2$	$MeAs(OH)_2$	MeAsO	$MeAsI_2$	Arseno-sugar	Arseno-choline	Arseno-betaine
Apoptosis[1,2]	n.d.	n.d.	+	n.d.	n.d.	n.d.	n.d.	n.d.	n.d.	n.d.	n.d.	n.d.
AP[3]	n.d.	n.d.	+	n.d.	n.d.	n.d.	n.d.	n.d.	n.d.	n.d.	n.d.	n.d.
CA[4,5,6]	+	n.d.	-[4]/+[5]/+[6]	n.d.	n.d.	+	n.d.	n.d.	n.d.	+	+	+
Carcinog.[7]	n.d.	n.d.	+	n.d.	n.d.	n.d.	n.d.	n.d.	n.d.	n.d.	n.d.	n.d.
CGI[6]	-	n.d.	+	n.d.	n.d.	+	n.d.	n.d.	n.d.	n.d.	-	-
COMET[8,9]	n.d.	n.d.	(-/+)[8]/(+/-[9])	+	n.d.	+[8]/-[9]	n.d.	+	n.d.	n.d.	n.d.	n.d.
CP[10]	n.d.	n.d.	-	n.d.	+	-	n.d.	+	n.d.	n.d.	n.d.	n.d.
CV[10]	n.d.	n.d.	+	n.d.	+	+	n.d.	+	n.d.	n.d.	n.d.	n.d.
DNA-nick[9]	n.d.	n.d.	-	+	n.d.	-	n.d.	+	n.d.	n.d.	n.d.	n.d.
DNA-PC[3]	n.d.	n.d.	+	n.d.	n.d.	n.d.	n.d.	n.d.	n.d.	n.d.	n.d.	n.d.
HC[7]	n.d.	n.d.	+	n.d.	n.d.	n.d.	n.d.	n.d.	n.d.	n.d.	n.d.	n.d.
Inh.TR[11]	n.d.	n.d.	n.d.	-	n.d.	n.d.	n.d.	+	n.d.	n.d.	n.d.	n.d.
K^+[12]	n.d.	n.d.	-	n.d.	n.d.	-	+	n.d.	+	n.d.	n.d.	n.d.
LDH[12]	n.d.	n.d.	-	n.d.	n.d.	-	+	n.d.	+	n.d.	n.d.	n.d.
MI[8]	n.d.	n.d.	-	n.d.	n.d.	-	n.d.	n.d.	n.d.	n.d.	n.d.	n.d.
MN[4]	n.d.	n.d.	-	n.d.	n.d.	+	n.d.	n.d.	n.d.	n.d.	n.d.	n.d.
MTS[1]	n.d.	n.d.	+	n.d.	n.d.	-	n.d.	n.d.	n.d.	n.d.	n.d.	n.d.
MTT[13]	n.d.	n.d.	-	+	n.d.	-	n.d.	+	n.d.	n.d.	n.d.	n.d.
MUT[4]	n.d.	n.d.	+	n.d.	n.d.	+	n.d.	n.d.	n.d.	n.d.	n.d.	n.d.
8-oxodG[14]	n.d.	n.d.	+	n.d.	n.d.	n.d.	n.d.	n.d.	n.d.	n.d.	n.d.	n.d.
RI[8]	n.d.	n.d.	-	n.d.	n.d.	-	n.d.	n.d.	n.d.	n.d.	n.d.	n.d.
SCE[6,15]	-	-	-	n.d.	n.d.	n.d.	n.d.	n.d.	n.d.	n.d.	-	-
SSB[3,16]	n.d.	n.d.	+	n.d.	n.d.	n.d.	n.d.	n.d.	n.d.	n.d.	n.d.	n.d.
XTT[12]	n.d.	n.d.	-	n.d.	n.d.	-	+	n.d.	+	n.d.	n.d.	n.d.

n.d., not determined; +, positive; -, negative; Me_4AsI, tetramethylarsonium oxide; Me_3AsO, trimethylarsine oxide; Me_2AsOOH, dimethylarsinic acid; Me_2AsI, iododimethylarsine; Me_2As-GS, dimethylarsinous acid with glutathione; $MeAsO(OH)_2$, monomethylarsonic acid; $MeAs(OH)_2$, monomethylarsonous acid; MeAsO, monomethylarsine oxide; $MeAsI_2$, diiodomethylarsine. AP, apurinic/apyrimidinic sites; CA, chromosomal aberrations; CGI, cell growth inhibition; COMET, COMET-assay; CP, cell proliferation; CV, cell viability (neutral red assay); DNA-nick, DNA nicking assay; DNA-PC, DNA protein cross-links; HC, clumpsing of heterochromatin; Inh.TR, inhibition of thioredoxin reductase; K^+, intracellular potassium; LDH, LDH activity test (cytotoxicity); MI, mitotic index; MN, micronucleus test; MTS-assay, cytolethality test; MTT, thiazolyl blue assay; MUT, mutagenicity test; 8-oxodG, test for oxidative damage (8-oxo-2'deoxyguanosine); RI, replication index; SCE, sister chromatide exchanges; SSB, single strand breaks; XTT, XTT-assay for mitochondrial damage. 1) Sakurai et al. 2002; 2) Namgung and Xia, 2001; 3) Yamanaka et al.

1995; 4) Moore et al., 1997; 5) Oya-Ohta et al., 1996; 6) Kaise et al., 1998; 7) Wei et al., 2002; 8) Sordo et al. 2001; 9) Mass et al., 2001; 10) Vega et al., 2001; 11) Lin et al., 2001; 12) Petrick et al. 2000; 13) Styblo et al. 2000; 14) Yamanaka et al. 2001; 15) Rasmussen and Menzel 1997; 16) Yamanaka and Okada 1994.

Table 3. Genotoxicity of organotin compounds

	AT[1)]	ATP[2)]	CA[3)]	IMFT[4)]	MN[5)6)]	REC[7)]	SCE[3)]	SOS[7)]
Me_4Sn	n.d.	n.d.	n.d.	-	n.d.	-	n.d.	-
Me_3SnX	+	n.d.	n.d.	n.d.	n.d.	n.d.	n.d.	n.d.
Me_3SnCl	n.d.	n.d.	+	-	+	+	+	-
Me_2SnX_2	-	n.d.	n.d.	n.d.	n.d.	n.d.	n.d.	n.d.
Me_2SnCl_2	n.d.	n.d.	n.d.	+	n.d.	+	n.d.	-
$MeSnCl_3$	n.d.	n.d.	n.d.	-	n.d.	-	n.d.	-
Bu_4Sn	n.d.	n.d.	n.d.	-	n.d.	-	n.d.	-
Bu_3SnX	+	-	n.d.	-	n.d.	n.d.	n.d.	n.d.
Bu_3SnCl	n.d.	n.d.	n.d.	+	n.d.	+	n.d.	-
Bu_3SnOH	n.d.	n.d.	n.d.	+	n.d.	+	n.d.	-
Bu_2SnX_2	-	-	n.d.	-	n.d.	n.d.	n.d.	n.d.
Bu_2SnCl_2	n.d.	n.d.	n.d.	+	n.d.	+	n.d.	+
$BuSnCl_3$	n.d.	n.d.	n.d.	+	n.d.	-	n.d.	+
BuSnO	n.d.	n.d.	n.d.	+	n.d.	-	n.d.	+
Ph_4Sn	n.d.	n.d.	n.d.	-	n.d.	-	n.d.	-
Ph_3SnX	+	n.d.	n.d.	-	n.d.	n.d.	n.d.	n.d.
Ph_3SnCl	n.d.	n.d.	n.d.	-	n.d.	-	n.d.	-
Ph_2SnX_2	+	n.d.	n.d.	-	n.d.	n.d.	n.d.	n.d.
Ph_2SnCl_2	n.d.	n.d.	n.d.	-	n.d.	-	n.d.	-
$PhSnCl_3$	n.d.	n.d.	n.d.	-	n.d.	-	n.d.	-

+, positive; -, negative; n.d., not determined. Me_4Sn, tetramethyltin; Me_3SnX, trimethyltin; Me_3SnCl, trimethyltin chloride; Me_2SnCl_2, dimethyltin dichloride; $MeSnCl_3$, methyltin trichloride; Bu_4Sn, tetrabutyltin; Bu_3SnX, tributyltin, Bu_3SnCl, tributyltin chloride; Bu_3SnOH, tributyltin hydroxide; Bu_2SnX_2, dibutyltin; Bu_2SnCl_2, dibutyltin dichloride; $BuSnCl_3$, butyltin trichloride; BuSnO, butyltin oxide; TetraPhT, tetraphenyltin; Ph_3SnX, triphenyltin; Ph_3SnCl, triphenyltin chloride; Ph_2X_2, diphenyltin; Ph_2SnCl_2, diphenyltin dichloride; $PhSnCl_3$, phenyltin trichloride. Note: X = unknown side groups. AT, aneuploidy test; ATP, ATP-levels are affected; CA, chromosomal aberrations; IMFT, induced mutation frequency test in *Salmonella typhymurium* (modification of the conventional *Salmonella* assay); MN, micronucleus assay; REC, rec-assay in *Bacillus subtilis*; SCE, sister chromatid exchanges; SOS, SOS chromotest with *Escherichia coli PQ37* (quantitative bacterial colorimetric assay for genotoxins). 1) Jensen KG et al. (1991); 2) Wahlen MM et al. (2002); 3) Ghosh BB et al. (1992); 4) Hamasaki T et al. (1993); 5) Ghosh BB et al. (1990); 6) Yamada H (1993); 7) Hamasaki T et al. (1992).

In vitro genotoxicity of different organometal(loid) compounds in CHO cells

In ongoing studies, we analyzed the genotoxic effects of organic derivatives of metal(loid)s such as tin and arsenic, and organoantimony compounds, the genotoxicity of which have not yet been studied in detail.

In our initial studies, we analysed four organotin compounds (monomethyl trichloride ($MeSnCl_3$), dimethyltin dichloride (Me_2SnCl_2), trimethyltin chloride Me_3SnCl, tetramethyltin (Me_4Sn), two pentavalent organoarsenic compounds (monomethylarsonic acid ($MeAsO(OH)_2$), trimethylarsine oxide (Me_3AsO), trivalent organoarsenic compounds, and one antimony compound (trimethylantimony dichloride (Me_3SbCl_2)). To test the genotoxic effects of these compounds, MN, CA, and SCE were analyzed in CHO cells.

Micronuclei (MN)

In 1985, Fenech and Morley developed the cytokinesis-blocked method (with cytochalasin B) which allows to identify cells that have been divided once in culture (binucleated cells). In cells with CA or failure of chromosome segregation MN are formed in addition to the main nuclei. MN require one cell division to be expressed. Cytochalasin B blocks the cell division but not the nuclear division and leads to binucleated cells in which MN can be easily seen.

Chromosomal Aberrations (CA) and Sister Chromatid Exchanges (SCE)

CA are structural changes in metaphase chromosomes and are the microscopically visible part of a wide spectrum of DNA changes (Obe et al. 2002).
The SCE test indicates exchanges of sister chromatids in metaphase chromosomes following exposure to genotoxic compounds. SCE are visible in chromosomes differentially substituted with bromodeoxyuridine (Perry and Wolff 1974; Bruckmann et al. 1999).

Summary

Organometal(oid)s are considered to be much more toxic than the inorganic species, but the genotoxic potential has only been investigated in a few compounds such as organomercury. For the majority of organometallic species, no or only few data are available on genotoxic, mutagenic, and carcinogenic effects. Recent findings that trivalent organoarsenic compounds, known products of arsenic metabolism, exhibit genotoxic effects indicates that the previous view on the toxic potential of organometal(loid) compounds has to be reconsidered. Therefore, studies on genotoxicity of organotin, organoarsenical, and organoantimony compounds using the MN assay, the CA test, and the SCE test in CHO cells are currently under way in our laboratory.

Literature

Amorim MI, Mergler D, Bahia MO, Dubeau H, Miranda D, Lebel J, Burbano RR, Lucotte M (2000) Cytogenetic damage related to low levels of methyl mercury contamination in the Brazilian Amazon. An Acad Bras Cienc 72:497-507

Bakir F, Damluji SF, Amin-Zaki L, Murtadha M, Khalidi A, al-Rawi NY, Tikriti S, Dahahir HI, Clarkson TW, Smith JC, Doherty RA (1973) Methylmercury poisoning in Iraq. Science 181:230-241

Betti C, Barale R, Pool-Zobel BL (1993a) Comparative studies on cytotoxic and genotoxic effects of two organic mercury compounds in lymphocytes and gastric mucosa cells of Sprague-Dawley rats. Environ Mol Mutagen 22:172-180

Betti C, Davini T, Barale R (1992) Genotoxic activity of methyl mercury chloride and dimethyl mercury in human lymphocytes. Mutat Res 281:255-260

Betti C, Davini T, He J, Barale R (1993b) Liquid holding effects on methylmercury genotoxicity in human lymphocytes. Mutat Res 301:267-273

Boyer IJ (1989) Toxicity of dibutyltin, tributyltin and other organotin compounds to humans and to experimental animals. Toxicology 55: 253-298

Chau YK, Wong PTS (1978) Occurrence of biological methylation of elements in the environment. In: Brinckman FE, Bellama JM (eds) Organometals and organometalloids - Occurence and fate in the environment. ACS Symp Ser, Washington DC, 39-53pp

Bruckmann E, Wojcik A, Obe G (1999) X-irradiation of G1 CHO cells induces SCE which are both true and false in BrdU-substituted cells but only false in biotin-dUTP-substituted cells. Chromosome Res 7:277-288

Choi BH, Kim RC (1984) The comparative effects of methylmercuric chloride and mercuric chloride upon DNA synthesis in mouse fetal astrocytes *in vitro*. Exp Mol Pathol 41:371-376

Craig PJ, Glockling F (1988) The biological alkylation of heavy elements. The Royal Society of Chemistry, London, 298pp

De Flora S, Bennicelli C, Bagnasco M (1994) Genotoxicity of mercury compounds. A review. Mutat Res 317:57-79

Dufresne J, Cyr DG (1999) Effects of short-term methylmercury exposure on metallothionein mRNA levels in the testis and epididymis of the rat. J Androl 20:769-778

Ehrenstein C, Shu P, Wickenheiser EB, Hirner AV, Dolfen M, Emons H, Obe G (2002) Methyl mercury uptake and associations with the induction of chromosomal aberrations in Chinese hamster ovary (CHO) cells. Chem Biol Interact 141:259-264

Fenech M, Morley A (1985) Solutions to the kinetic problem in the micronucleus assay. Cytobios 43:233-246

Ghosh BB, Talukder G, Sharma A (1990) Frequency of micronuclei induced in peripheral lymphocytes by trimethyltin chloride. Mutat Res 245:33-39

Ghosh BB, Talukder G, Sharma A (1992) Cytotoxicity of tin on human peripheral lymphocytes *in vitro*. Mutat Res 282: 61-67

Hamasaki T, Sato T, Nagase H, Kito H (1992) The genotoxicity of organotin compounds in SOS chromotest and rec-assay. Mutat Res 280:195-203

Hamasaki T, Sato T, Nagase H, Kito H (1993) The mutagenicity of organotin compounds as environmental pollutants. Mutat Res 300:265-271

Hadjispyrou S, Kungolos A, Anagnostopoulos A (2001) Toxicity, bioaccumulation, and interactive effects of organotin, cadmium, and chromium on Artemia franciscana. Ecotoxicol Environ Saf 49:179-186

Harada M (1978) Congenital Minamata disease: intrauterine methylmercury poisoning. Teratology 18:285-288

Huff J, Chan P, Nyska A (2000) Is the human carcinogen arsenic carcinogenic to laboratory animals? Toxicol Sci 55:17-23

Jensen KG, Andersen O, Ronne M (1991) Organotin compounds induce aneuploidy in human peripheral lymphocytes *in vitro*. Mutat Res 246:109-112

Kaise T, Ochi T, Oya-Otha Y, Hanaoka K, Sakurai T, Saitoh T, Matsubara C (1998) Cytotoxicological aspects of organic arsenic compounds contained in marine products using the mammalian cell culture technique. Appl Organometal Chem 12:137-143

Kawaguchi K, Oku N, Rin K,Yamanaka K, Okada S (1996) Dimethylarsenics reveal DNA damage induced by superoxide anion radicals. Biol Pharmacol Bull 19:551-553

Kitchin KT (2001) Recent advances in carcinogenesis: modes of action, animal model systems, and methylated arsenic metabolites. Toxicol Appl Pharmacol 172:249-261

Kresimon J, Grueter UM, Hirner AV (2001) HG/LT-GC/ICP-MS coupling for identification of metal(loid) species in human urine after fish consumption. Fresenius J Anal Chem 371:586-590

Lee CH, Lin RH, Liu SH, Lin-Shiau SY (1997) Distinct genotoxicity of phenylmercury acetate in human lymphocytes as compared with other mercury compounds. Mutat Res 392:269-276

Lin S, Del Razo LM, Styblo M, Wang C, Cullen WR, Thomas DJ (2001) Arsenicals inhibit thioredoxin reductase in cultured rat hepatocytes. Chem Res Toxicol, 14:305-311

Mass MJ, Tennant A, Roop BC, Cullen WR, Styblo M, Thomas DJ, Kligerman AD (2001) Methylated trivalent arsenic species are genotoxic. Chem Res Toxicol 14:355-361

Migliore L, Cocchi L, Nesti C, Sabbioni E (1999) Micronuclei assay and FISH analysis in human lymphocytes treated with six metal salts. Environ Mol Mutagen 34:279-284

Moore MM, Harrington-Brock K, Doerr LC (1997) Relative genotoxic potency of arsenic and its methylated metabolites. Mutat Res 38:279-290

Namgung U, Xia Z (2001) Arsenic induces apoptosis in rat cerebellar neurons via activation of JNK3 and p38 MAP kinases. Toxicol Appl Pharmacol 174:130-138

Nierenberg DW, Nordgren RE, Chang MB , Siegler RW, Blayney MB, Hochberg F, Toribara TY, Cernichiari E, Clarkson T (1998) Delayed cerebellar disease and death after accidental exposure to dimethylmercury. N Engl J Med 338:1672-1676

Obe G, Pfeiffer P, Savage JR, Johannes C, Goedecke W, Jeppesen P, Natarajan AT, Martinez-Lopez W, Folle GA, Drets ME (2002) Chromosomal aberrations: formation, identification and distribution. Mutat Res 504:17-36.

Ogura H, Takeuchi T, Morimoto K (1996) A comparison of the 8-hydroxydeoxyguanosine, chromosomal aberrations and micronucleus techniques for the assessment of the genotoxicity of mercury compounds in human blood lymphocytes. Mutat Res 340:175-182

Önfelt A (1983) Spindle disturbances in mammalian cells. I. Changes in the quantity of free sulfhydryl groups in relation to survival and C-mitosis in V79 Chinese hamster cells after treatment with colcemid, diamide, carbaryl and methyl mercury. Chem Biol Interact 46:201-217

Oya-Ohta Y, Kaise T, Ochi T (1996) Induction of chromosomal aberrations in cultured human fibroblasts by inorganic and organic arsenic compounds and the different roles of glutathione in such induction. Mutat Res 357:123-129

Perry P, Wolff S (1974) New Giemsa method for the differential staining of sister chromatids. Nature 251:156-158

Petrick JS, Ayala-Fierro F, Cullen WR, Carter DE, Vasken Aposhian H (2000) Monomethylarsonous acid (MMA(III)) is more toxic than arsenite in Chang human hepatocytes. Toxicol Appl Pharmacol 163:203-207

Ramanujam M, Prasad KN (1979) Alterations in gene expression after chronic treatment of glioma cells in culture with methylmercuric chloride. Biochem Pharmacol 28:2979-2984

Rasmussen RE, Menzel DB (1997) Variation in arsenic-induced sister chromatid exchange in human lymphocytes and lymphoblastoid cell lines. Mutat Res 386:299-306

Sakurai T, Qu W, Sakurai MH, Waalkes MP (2002) A major human arsenic metabolite, dimethylarsinic acid, requires reduced glutathione to induce apoptosis. Chem Res Toxicol 15:629-637

Sasaki YF, Yamada H, Sugiyama C, Kinae N (1993) Increasing effect of tri-n-butyltins and triphenyltins on the frequency of chemically induced chromosomal aberrations in cultured Chinese hamster cells. Mutat Res 300:5-14

Sordo M, Herrera LA, Ostrosky-Wegman P, Rojas E (2001) Cytotoxic and genotoxic effects of As, MMA, and DMA on leukocytes and stimulated human lymphocytes. Teratog Carcinog Mutagen 21:249-260

Styblo M, Del Razo LM, Vega L, Germolec DR, LeCluyse EL, Hamilton GA, Reed W, Wang C, Cullen WR, Thomas DJ (2000) Comparative toxicity of trivalent and pentavalent inorganic and methylated arsenicals in rat and human cells. Arch Toxicol 74:289-299

Styblo M, Drobna Z, Jaspers I, Lin S, Thomas DJ (2002) The role of biomethylation in toxicity and carcinogenicity of arsenic: a research update. Environ Health Perspect 110: 767-771

Thomas DJ, Styblo M, Lin S (2001) The cellular metabolism and systemic toxicity of arsenic. Toxicol Appl Pharmacol 176:127-144

Vega L, Styblo M, Patterson R, Cullen W, Wang C, Germolec D (2001) Differential effects of trivalent and pentavalent arsenicals on cell proliferation and cytokine secretion in normal human epidermal keratinocytes. Toxicol Appl Pharmacol 172:225-232

Wei M, Wanibuchi H, Morimura K, Iwai S, Yoshida K, Endo G, Nakae D, Fukushima S (2002) Carcinogenicity of dimethylarsinic acid in male F344 rats and genetic alterations in induced urinary bladder tumors. Carcinogenesis 23:1387-1397

Whalen MM, Green SA, Loganathan BG (2002) Butyltin exposure induces irreversible inhibition of the cytotoxic function on human natural killer cells *in vitro*. Environ Res 88:19-29

White JS, Tobin JM, Cooney JJ (1999) Organotin compounds and their interactions with microorganisms. Can J Microbiol 45:541-554

Yamada H, Miyahara T, Kozuka H, Matsuhashi T, Sasaki YF (1993) Potentiating effects of organomercuries on clastogen-induced chromosomal aberrations in cultured Chinese hamster cells. Mutat Res 290:281-291

Yamamoto S, Wanibuchi H, Hori T, Yano Y, Matsui-Yuasa I, Otani S, Chen H, Yoshida K, Kuroda K, Endo G, Fukushima S (1997) Possible carcinogenic potential of dimethylarsinic acid as assessed in rat in vivo models: a review. Mut Res 386:353-361

Yamanaka K, Hayashi H, Kato K, Hasegawa A, Okada S (1995) Involvement of preferential formation of apurinic/apyrimidinic sites in dimethylarsenic-induced DNA strand breaks and DNA-protein crosslinks in cultured alveolar epithelial cells. Biochem Biophys Res Commun 207:244-249

Yamanaka K, Okada S (1994) Induction of lung-specific DNA damage by metabolically methylated arsenics via the production of free radicals. Environ Health Perspect 102:37-40

Yamanaka K, Takabayashi F, Mizoi M, An Y, Hasegawa A, Okada S (2001) Oral exposure of dimethylarsinic acid, a main metabolite of inorganic arsenics, in mice leads to an increase in 8-oxo-2'-deoxyguanosine level, specifically in the target organs for arsenic carcinogenesis. Biochem Biophys Res Commun 287:66-70

Zhao CQ, Young MR, Diwan BA, Coogan TP, Waalkes MP (1997) Association of arsenic-induced malignant transformation with DNA hypomethylation and aberrant gene expression. Proc Natl Acad Sci USA 94:10907-10912

Yamanaka K, Hayashi H, Kato K, Hasegawa A, Okada S (1995) [illegible]

[illegible]

[illegible]

[illegible]

Chapter 12

Current aspects on the genotoxicity of arsenite and its methylated metabolites: Oxidative stress and interactions with the cellular response to DNA damage

A. Hartwig, T. Schwerdtle, I. Walter

Introduction

Increased cancer incidences associated with arsenic exposure have long been recognized. Significant levels of arsenic exist at a variety of workplaces including copper, zinc and lead smelters or glass manufacturing, as well as during the production and use of pesticides and herbicides (IARC 1980). Although the commercial use of arsenicals has been largely reduced over the last decades, there is, nevertheless, persistent concern about high levels of arsenic in drinking water in some regions of Argentina, Canada, India, Japan, Taiwan, and Thailand due to natural sources (National Research Council 2001). While exposure to arsenic compounds by inhalation increases the risk of lung cancer (IARC 1980), oral ingestion of arsenic has been associated with increased incidence of skin, lung, kidney, bladder, and liver tumours (Chiou et al. 1995; National Research Council 2001). However, the reason for the carcinogenicity of arsenic compounds is not fully understood. Inorganic arsenic compounds are not significantly mutagenic in bacterial test systems or in mammalian cells in culture, but they are clastogenic, causing mainly chromatid type chromosomal aberrations, sister-chromatid exchanges and micronuclei (for recent reviews see Gebel 2001, Basu et al. 2001).

With respect to arsenite, one proposed mechanism consists in the generation of reactive oxygen species, even though in frequently applied DNA strand break assays DNA damage was mainly observed at high, cytotoxic concentrations. At lower concentrations arsenite has been shown to interfere with different types of DNA repair systems, but molecular targets have not yet been identified.

Furthermore, the role of biomethylation in arsenic-induced genotoxicity and carcinogenicity is still a matter of debate. In humans, like in many mammalian species, inorganic arsenic is almost quantitatively reduced from pentavalent to trivalent arsenic in plasma and subsequently methylated to the trivalent and pentavalent methylated metabolites in the liver (summarized in Pott et al. 2001; see figure 1). Human urinary excretion profiles exhibit about 10 – 20 % inorganic arsenic, 10 – 20 % monomethylarsonic acid ($MeAsO(OH)_2$) and 60 – 80 % dimethylarsinic acid Me_2AsOOH (Vahter 1999). In addition to the pentavalent metabolites, monomethylarsonous ($MeAs(OH)_2$) and dimethylarsinous acid (Me_2AsOH) have been identified as intermediates in the metabolic pathway in cultured human cells treated with inorganic arsenic (summarized in Thomas et al. 2001) and in urine of people chronically exposed to inorganic arsenic *via* drinking water in West Bengal, India (Mandal et al. 2001).

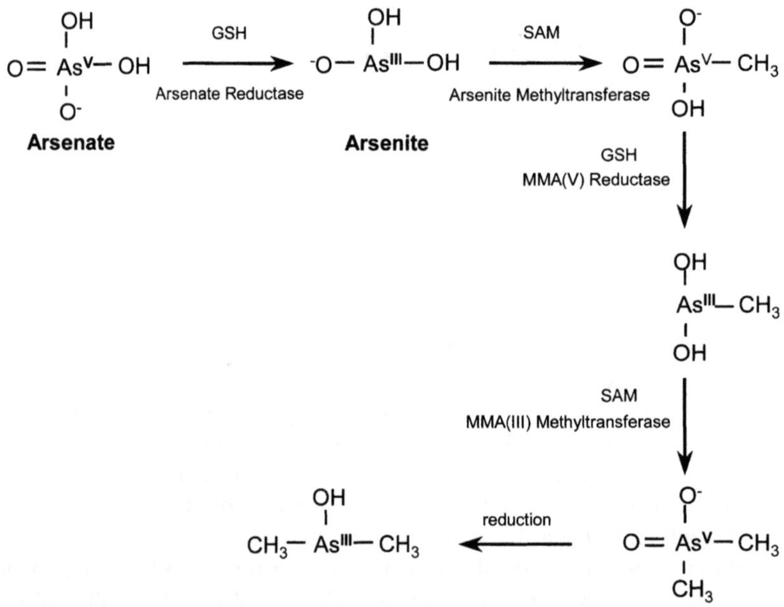

Fig. 1. Proposed metabolism of inorganic arsenic; for details see Pott et al. (2001). SAM: S-adenosylmethionine.

Based on rapid excretion, weak cytotoxicity and the restriction of genotoxicity to high, in case of most cellular systems millimolar concentrations, until recently the *in vivo* biomethylation to Me_2AsOOH and $MeAsO(OH)_2$has been considered as one major detoxification process. Only one study showed increased DNA migration in single cell gel electrophoresis (Comet Assay) in lymphocytes treated with micromolar concentrations of $McAsO(OH)_2$ and Me_2AsOOH, indicating DNA strand break formation (Sordo et al. 2001). Trivalent methylated metabolites are cytotoxic in cultured mammalian cells in the low micromolar range, but only few data are available with respect to their genotoxicity (Thomas et al. 2001; Wang et al. 2002).

In recent studies, we aimed to elucidate the role of oxidative DNA damage in arsenite-induced genotoxicity, to identify particularly sensitive molecular targets related to DNA repair inhibition and to understand potential contributions of the trivalent and pentavalent methylated metabolites at low concentrations relevant for environmental exposure conditions.

Oxidative DNA damage

There are several lines of evidence suggesting that arsenite promotes the formation of reactive oxygen species, particularly superoxide radical anions and hydrogen peroxide, which may lead to genotoxicity. Thus, reactive oxygen species scavengers such as superoxide dismutase, catalase, glutathione peroxidase and DMSO counteracted the formation of deletion mutations in human chromosome 11 in a human-hamster hybrid cell line (Hei et al. 1998), of micronuclei and of sister-chromatid exchanges (Nordenson and Beckman 1991; Wang and Huang 1994; Wang et al. 1997). Even though DNA strand break induction was mostly restricted to comparatively high concentrations, recent studies demonstrated that the inclusion of proteinase K or the bacterial formamidopyrimidin-DNA glycosylase (Fpg) greatly increased the sensitivity of the Comet Assay, indicating the formation of DNA-protein crosslinks and the generation of oxidative DNA base damage (Lynn et al. 2000; Li et al. 2001; Wang et al. 2001). Much less is known about the genotoxic potential of the trivalent and pentavalent methylated metabolites. By applying the Comet Assay, Mass et al. (2001) demonstrated that Me_2AsOH and $MeAs(OH)_2$ were more potent in generating DNA strand breaks in human lymphocytes as compared to arsenite. Furthermore there is evidence that Me_2AsOH and at very high concentrations (30 mM) $MeAs(OH)_2$ are able to nick isolated DNA without enzymatic or chemical activation (Mass et al. 2001; Ahmad et al. 2002).

In our experiments, we applied the Alkaline Unwinding technique in combination with Fpg (Hartwig et al. 1996) to quantify the induction of DNA strand breaks and oxidative DNA base modifications by arsenite and its trivalent and pentavalent metabolites $MeAs(OH)_2$, Me_2AsOH, $MeAsO(OH)_2$ and Me_2AsOOH in cultured human cells. Alkaline Unwinding is a sensitive procedure to detect DNA strand breaks by partial unwinding of double-stranded DNA on moderate alkaline conditions, thus increasing the fraction of single-stranded DNA. The oxidative DNA base modifications 7,8-dihydro-8-oxoguanine (8-oxoguanine), 2,6-diamino-4-hydroxy-5-formamidopyrimidine (Fapy-Gua), 4,6-diamino-5-formamidopyrimi-dine (Fapy-Ade) and to a smaller extent 7,8-dihydro-8-oxoadenine (8-oxoadenine) as well as apurinic/apyrimidinic sites (AP-sites) are recognized by the glycosylase and/or AP endonuclease activity of Fpg, which converts them into DNA strand breaks (Tchou et al. 1991; Boiteux et al. 1992); therefore these lesions can be quantified by Alkaline Unwinding as well. To elucidate the role of cellular reactions involved in DNA damage induction by the arsenicals, comparative studies with isolated supercoiled DNA derived from bacteriophage PM2 were performed.

With respect to arsenite, only few DNA strand breaks but pronounced frequencies of oxidative DNA base modifications recognized by Fpg were visible at concentrations as low as 10 nM after short-term (0.5 – 3 h) incubation of human HeLa S3 or at 100 nM in human A549 cells. This was surprising, since even though previous studies reported an increase in DNA damage when including Fpg or proteinase K in the Comet Assay (e.g., Li et al. 2001; Wang et al. 2001), the effective concentrations were much lower in our experiments. Thus, it appears that DNA strand break assays in the absence of Fpg or proteinase K in general underestimate the genotoxic potential of arsenite. However, less DNA damage was visible after 18 h incubation, indicating repair of the lesions at low concentrations of arsenite.

Concerning the trivalent and pentavalent metabolites, only $MeAs(OH)_2$ and Me_2AsOOH generated DNA strand breaks in a concentration-dependent manner. However, both trivalent and both pentavalent methylated arsenic compounds induced pronounced levels of Fpg-sensitive sites in the nanomolar or micromolar non-cytotoxic concentration range, respectively. Thus, again, the inclusion of Fpg greatly facilitates the detection of DNA damage also in case of the metabolites. In contrast to arsenite, these lesions were present after short-term and long-term incubations, indicating their continuous generation and/or their persistence due to diminished repair. Interestingly, only Me_2AsOH induced DNA strand breaks in isolated PM2 DNA starting at concentrations of 10 μM, but no Fpg-sensitive sites. All other compounds including arsenite were inactive at concentrations up to 10 mM, strongly suggesting the need of cellular reactions in the generation of DNA reactive species (Schwerdtle et al. 2003; summarized in Table 1).

Table 1. Induction of oxidative DNA damage by arsenite and its methylated metabolites in HeLa S3 cells and isolated DNA.

	Arsenite	**$MeAs(OH)_2$**	**Me_2AsOH**	**$MeAsO(OH)_2$**	**Me_2AsOOH**
HeLa S3, 3 h	+++ > 1 nM	++ ≥ 0.1 μM	++ ≥ 0.1 μM	+++ ≥ 100 μM	+++ ≥ 100 μM
HeLa S3, 18 h	+ ≥ 0.1 μM	++ ≥ 0.1 μM	++ ≥ 0.1 μM	+++ ≥ 10 μM	+++ ≥ 10 μM
Isolated DNA	- ≤ 10 mM	- ≤ 10 mM	+++ ≥ 10 μM	- ≤ 10 mM	- ≤ 10 mM

+++: strong effect, ++: intermediate effect, +: small effect but still significant, -: no significant effect.

The underlying mechanisms for the induction of oxidative DNA damage by the different arsenic compounds remain unclear. With respect to arsenite, HeLa S3 cells have only very limited methylation capability (Styblo et al. 1999; Peel et al. 1991); therefore, the participation of the methylated metabolites in DNA damage induction at low nanomolar concentrations after short-term incubation is highly unlikely. One possible explanation for the very fast induction of oxidative DNA damage at very low concentrations by arsenite may be the formation of reactive oxygen species. Thus, recent studies have provided evidence that arsenite can induce DNA damage by promoting the formation of reactive oxygen species, particularly superoxide radical anions and hydrogen peroxide (Lynn et al. 2000; Li et al. 2001; Wang et al. 2001). Concerning the methylated metabolites, our data as well as some reports by other authors suggest that no common mechanism for DNA damage induction applies for all investigated compounds. As described above, only Me_2AsOH induced DNA strand breaks in isolated DNA in the absence of cellular reactions. With respect to Me_2AsOOH, electron spin resonance studies provided evidence that in addition to the superoxide radical anion a dimethylarsenic peroxyl radical was formed by the reaction of molecular oxygen with dimethylarsine, a product in the further metabolic processing of Me_2AsOOH, which may form DNA adducts and subsequently AP sites and DNA protein crosslinks (Yamanaka et al. 1990; Tezuka et al. 1993). Thus, at least some Fpg-sensitive sites in our study may be AP sites. As another potential mechanism for the generation of ROS and oxidative DNA damage, Ahmad et al. (2000, 2002) reported the release of iron from horse spleen and human liver ferritin in the presence of arsenite, arsenate and the methylated metabolites with strongest effects exerted by Me_2AsOH. Furthermore, the addition of human liver ferritin increased the DNA damage on isolated pBR322 DNA by Me_2AsOH. Nevertheless, whether this mechanism is relevant in intact cells has to be further elucidated. In cellular systems, in addition to a direct increase in ROS, arsenite and especially some of the methylated metabolites may induce oxidative DNA damage indirectly by inhibition of important detoxifying enzymes. Thus, both trivalent methylated metabolites $MeAs(OH)_2$ and Me_2AsOH are more potent inhibitors of isolated glutathion reductase as compared to arsenite, which may be due to the interaction of trivalent arsenic with critical thiol groups and may alter the cellular redox status (summarized in Thomas et al. 2001).

DNA repair systems and interactions by arsenite

Maintenance of genetic information and thus the correct sequence of nucleotides in DNA is essential for replication, gene expression and protein synthesis; DNA lesions at critical sites like oncogenes or tumour suppressor genes may lead to cell cycle arrest, programmed cell death, mutagenesis, genomic instability and cancer. In addition to replication errors, DNA is not only damaged by environmental mutagens including UV-light and polycyclic aromatic hydrocarbons, but also permanently by reactive oxygen species generated endogenously due to oxygen metabolism (Lindahl 1993). To minimize adverse consequences, a complex network of different repair systems has evolved to maintain genomic integrity. Thus, replication errors are repaired by the mismatch repair pathway and double strand breaks are repaired by recombination processes. The major pathway eliminating DNA base damage and helix distortions is the excision repair pathway, subdivided into nucleotide excision repair (NER) and base excision repair (BER) (de Boer et al. 2000). During the last years, there has been accumulating evidence that both BER and NER are inhibited at low, non-cytotoxic concentrations of As(III). Regarding NER, arsenite impaired the incision step at low micromolar and the ligation at 10-fold higher concentrations after induction of UVC-induced DNA damage, affecting both global genome repair and transcription-coupled repair (Hartwig et al. 1997). In addition, arsenite was shown to inhibit the ligation step during the removal of N-methyl-N-nitrosourea induced DNA damage by BER (for recent review see Hartwig et al. 2002). To elucidate whether the observed repair inhibitions also apply to other environmentally relevant DNA damaging agents, we investigated the effects of arsenite and its metabolites on the removal of benzo[*a*]pyrene (B[*a*]P)-induced DNA damage. B[*a*]P belongs to the class of polycyclic aromatic hydrocarbons (PAH) generated by incomplete combustion of organic matter and is therefore not only present at many work places but also in the ambient air as well as in cooked foods. Its carcinogenic activity is attributed to the formation of DNA adducts, resulting from electrophilic attack predominantly at guanine residues by metabolically activated intermediates formed from the parent hydrocarbon. Routes of metabolic activation include the formation of radical cations *via* P450 and/or peroxidases and the formation of *o*-quinones *via* dihydrodiol dehydrogenases. For carcinogenicity, the probably most relevant metabolic pathway is connected to the action of cytochromes P450 1A1 and 1B1 and epoxide hydrolase, yielding *syn*- and *anti*-B[*a*]P-7,8-diol 9,10-epoxides (BPDE) which form stable adducts at the N^2 position of guanine. When replicated prior to repair, these adducts can lead to mutations and cancer (Drouin and Loechler 1993). Thus, the induction and removal of the latter adducts generated by the active metabolite (+)-*anti*-BPDE in the absence and presence of arsenic compounds was quantified by a very sensitive HPLC/fluorescence method (Alexandrov et al. 1992; Rojas et al. 1994; Schwerdtle et al. 2002). This study was conducted in A549 human lung cancer cells which retained important characteristics of pneumocytes type II with respect to metabolism and phagocytotic capability (Foster et al. 1998). In the absence of arsenite, about 45 % of BPDE-DNA adducts were repaired within 8 h. In the presence of arsenite, there was a significant increase of adduct formation. Additionally, the repair capacity towards the stable lesions was decreased in a concentration-dependent manner reaching about 25 % of the control at 75 μM, which was still only slightly cytotoxic in this cell line as determined by colony forming ability. Regarding the metabolites, similar effects on adduct formation and repair were observed in case of $MeAs(OH)_2$, but at much lower concentrations. Thus, an increased lesion generation by 60 % and only 40 % of repair compared to the control were detected at 5 μM. Repair inhibition was also observed with 5 μM Me_2AsOH, but no effect on adduct generation was evident. Nevertheless, the cytotoxicity of the trivalent metabolites was higher than that of arsenite. Finally, significant but less marked repair inhibition was mediated by 250 and 500 μM of Me_2AsOOH or $MeAsO(OH)_2$. Altogether, the results demonstrate that arsenite as well as the methylated metabolites interfere with cellular repair systems; lowest inhibitory concentrations were found for the trivalent metabolites (Schwerdtle, Walter and Hartwig, submitted for publication).

Poly(ADP-ribosyl)ation as sensitive intracellular target for arsenite

In spite of diminished DNA repair capacities observed in the low micromolar concentration range of arsenite, no single DNA repair protein has so far been identified that is particularly sensitive to arsenic compounds. For example, *in vitro* experiments on nuclear extracts required 10 mM arsenite to detect an inhibition of DNA ligase (Li and Rossman 1989). Similarly, no other isolated DNA repair enzyme has been found to be inhibited at relevant concentrations of arsenite (Hu et al. 1998).

Thus, even though arsenite has been shown to inhibit more than 200 enzymes, the issue of the inhibitory concentration has to be addressed, since only those effects would appear to be biologically relevant that are observed at rather low, noncytotoxic concentrations. One group of potential molecular targets for toxic metal compounds are the so-called zinc finger proteins. They represent a family of proteins where zinc is complexed through four invariant cysteine and/or histidine residues forming a zinc finger domain, which is mostly involved in DNA binding, but also in protein-protein interactions (Mackay and Crossley 1998). Besides transcription factors, several DNA repair enzymes belong to this family, including the Fpg protein involved in the repair of certain types of oxidative DNA base damage (O'Connor et al. 1993) and the xeroderma pigmentosum group A protein (XPA) essential for the assembly of the DNA damage recognition/incision complex during NER in mammalian cells (Miyamoto et al. 1992). Nevertheless, neither the isolated Fpg protein nor XPA were inhibited at concentrations up to 1 mM arsenite (Asmuss et al. 2000).

One of the immediate cellular responses to DNA damage induced by ionizing radiation, alkylating agents and oxidants is the poly(ADP-ribosyl)ation of nuclear proteins. This reaction is catalysed predominantly by poly(ADP-ribose) polymerase-1 (PARP-1) and to a lesser extent by additional, recently identified and less well characterized members of the "PARP family" (for recent reviews see de Murcia and Shall 2000; Bürkle 2001). Upon stimulation by DNA single or double strand breaks, PARP-1 catalyses the sequential covalent attachment of multiple ADP-ribosyl moieties from NAD^+ to glutamate, aspartate, and lysine residues of itself ("automodification") and of other nuclear proteins including histones H1, H2B and topoisomerases (Le Rhun et al. 1998), thus forming protein-conjugated poly(ADP-ribose). This posttranslational modification leads to the dissociation of the proteins from DNA, which thus should become more accessible for repair enzymes (de Murcia and Shall 2000), while PARP-1 automodification could lead to localized chromatin relaxation and/or damage signalling and inactivation of the catalytic function of the enzyme (Le Rhun et al. 1998; Althaus 1992; Pleschke et al. 2000). Binding of PARP-1 to DNA strand breaks occurs *via* two zinc finger motifs in its aminoterminal domain where zinc is complexed to three cysteine residues and one histidine (Gradwohl et al. 1990; Ikejima et al. 1990). The effect of arsenic on the extent of poly(ADP-ribosyl)ation has been investigated previously in two studies yielding controversial results. Yager and Wiencke (1997) observed a decreased amount of poly(ADP-ribose) in human T-cell lymphoma-derived Molt-3 cells at arsenite concentrations above 5 μM. In contrast, an increase of poly(ADP-ribosyl)ation reaction at higher concentrations was reported in CHO-K1 cells (Lynn et al. 1998). In our laboratory we investigated the effects of arsenite on poly(ADP-ribosyl)ation stimulated by H_2O_2 in intact HeLa S3 cells by *in situ* immunofluorescence using an anti-poly(ADP-ribose) monoclonal antibody.

Our experiments demonstrate a markedly reduced level of poly(ADP-ribosyl)ation becoming apparent at the extremely low and non-cytotoxic concentration of 10 nM arsenite and reaching about 40 % residual activity at 0.5 μM arsenite. Also, there was an increase in H_2O_2-induced DNA strand break formation by arsenite in agreement with the assumed role of poly(ADP-ribosyl)ation in DNA strand break repair (Hartwig et al. 2003). The inhibition of poly(ADP-ribosyl)ation provides the first molecular target linking the impairment of DNA repair processes with a distinct enzymatic reaction at very low concentrations of arsenite. Although the reaction was not completely suppressed, the relevance of even partial inhibition of poly(ADP-ribosyl)ation has been demonstrated previously applying the PARP inhibitor 3-aminobenzamide at concentrations ranging from 0.1-2 mM which led to comparable levels of PARP inhibition as arsenite and a potentiation of MNNG-induced DNA amplification (Bürkle et al.1987), indicating decreased genomic stability. Nevertheless, the reasons for decreased levels of poly(ADP-ribosyl)ation are still unclear. One potential mechanism consists in direct interaction of arsenite with PARP-1. As stated above, binding of PARP-1 to DNA strand breaks is mediated *via* two zinc finger motifs, and due to the high affinity of arsenite to vicinal SH groups (Knowles and Benson 1983), it may disturb the zinc finger structures, thus preventing PARP-1-DNA interactions. Further studies with isolated PARP-1 are required to test this hypothesis. However, at present, other mechanisms leading to reduced levels of PARP protein cannot be excluded, such as interference with gene expression and/or PARP protein synthesis/degradation.

Conclusions and perspectives

In summary, arsenite as well as the trivalent and pentavalent methylated metabolites decrease genetic stability directly by inducing mainly oxidative DNA base modifications, and indirectly by interfering with the repair of DNA lesions as demonstrated for the removal of BPDE-induced DNA adducts (see Fig. 2). Our results show that very low concentrations of arsenite and the methylated metabolites induce high levels of oxidative DNA damage in cultured human cells. With respect to arsenite, the inclusion of the bacterial Fpg protein to quantify oxidative DNA base modifications in addition to DNA strand breaks has revealed DNA damage at very low, environmentally relevant concentrations. Among the lesions recognized by Fpg, 8-oxoguanine has attracted special attention, since it is mutagenic by causing G to T transversions and thus may play an important role in carcinogenesis (Wood et al., 1990; Floyd, 1990). Interestingly, oxidative DNA damage even more persistent as compared to arsenite was induced also by nanomolar concentrations of the trivalent and by micromolar concentrations of the pentavalent methylated metabolites, suggesting that at least $MeAs(OH)_2$ and Me_2AsOH may contribute *in vivo* to the genotoxic/carcinogenic potential of arsenite.

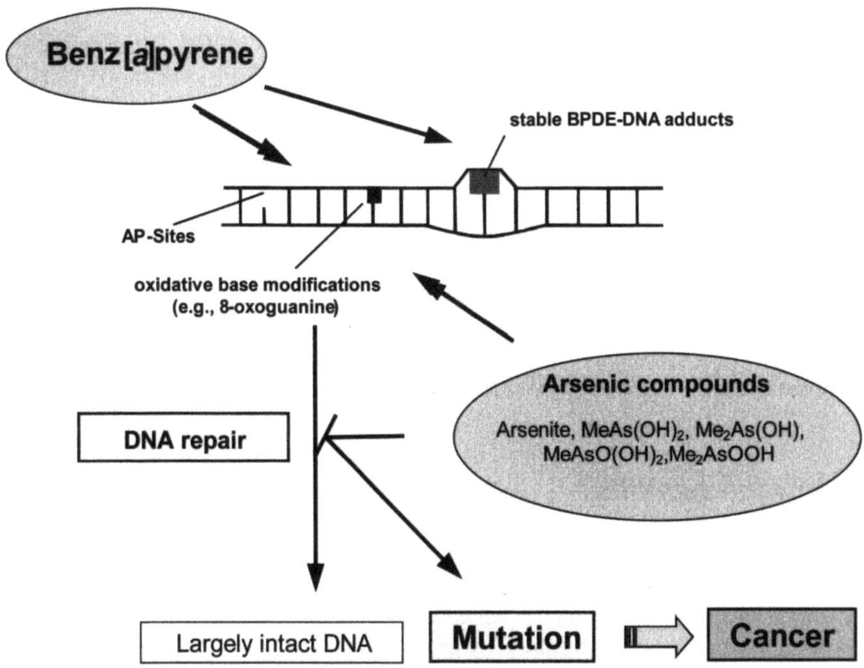

Fig 2. Proposed factors contributing to carcinogenicity induced by inorganic arsenic and its methylated metabolites.

With respect to DNA repair inhibition, recently intracellular targets have been identified which are particularly sensitive towards carcinogenic metal ions. In case of arsenite the poly(ADP-ribosyl)ation appears to be one critical step suppressed at extremely low concentrations, which may explain at least interactions with BER. What are the toxicological implications of the observed repair inhibitions? During the last years, there has been accumulating evidence that the DNA is continuously damaged not only by environmental agents like UV radiation, but also by endogenous processes, leading to several thousand lesions per cell and day (Lindahl 1993). The efficient repair of these lesions is thus an important prerequisite to maintain DNA integrity. If repair is not functioning, cells may accumulate DNA damage, leading to increased probabilities of mutations in DNA stability genes involved in DNA repair and cell cycle control (mutator phenotype) and thus to tumour formation (Loeb 1994). This model is supported by high tumour incidences of patients suffering from rare nucleotide excision repair deficiency syndromes like xeroderma pigmentosum (de Boer and Hoeijmakers 2000) or defects in mismatch repair involved in the correction of DNA replication errors, which have been associated with increased susceptibility to heriditary nonpolyposis colon cancer (HNPCC) (Peltomäki 2001).

Concerning the methylated metabolites of arsenite, in cultured mammalian cells they induce DNA damage and interfere with DNA repair. In case of $MeAs(OH)_2$ and Me_2AsOH even lower concentrations as compared to arsenite inhibited NER, while higher, but still noncytotoxic concentrations of $MeAsO(OH)_2$ or Me_2AsOOH were required. Nevertheless, it has to be taken into account that the experiments were conducted in cell cultures, and uptake of $MeAsO(OH)_2$ and Me_2AsOOH are probably very limited. In contrast, *in vivo* these metabolites are generated in cells, and therefore may damage DNA and other critical targets at much lower concentrations. Thus, to further explore the role of the metabolites the identification and quantification of intracellular species is urgently needed. However, the results suggest that biomethylation of inorganic arsenic may be involved in inorganic arsenic-induced genetic instability.

Acknowledgements

The Fpg protein was a kind gift of Dr. Serge Boiteux, Commissariat a l'Energie Atomique, Fontenay aux Roses, France. $MeAs(OH)_2$ and Me_2AsOH were kindly provided by Prof. W. R. Cullen, Vancouver, Canada. The work conducted in our laboratory was supported by the EU, grant no. QLK4-1999-01142, by the Deutsche Forschungsgemeinschaft grant no. Ha 2372/1-3 and by BWPLUS, grant no. BW BG 99012.

Literature

Ahmad S, Kitchin KT, Cullen WR (2000) Arsenic species that cause release of iron from ferritin and generation of activated oxygen. Arch Biochem Biophys 382:195-202

Ahmad S, Kitchin KT, Cullen WR (2002) Plasmid DNA damage caused by methylated arsenicals, ascorbic acid and human liver ferritin. Toxicol Lett 133:47-57

Alexandrov K, Rojas M, Geneste O, Castegnaro M, Camus AM, Petruzzelli S, Giuntini C, Bartsch H (1992) An improved fluorometric assay for dosimetry of benzo[*a*]pyrene diol-epoxide-DNA adducts in smokers' lung: comparisons with total bulky adducts and aryl hydrocarbon hydroxylase activity. Cancer Res 52:6248-6253

Althaus FR (1992) Poly ADP-ribosylation: a histone shuttle mechanism in DNA excision repair. J Cell Sci 102:663-670

Asmuss M, Mullenders LHF, Eker A, Hartwig A (2000) Differential effects of toxic metal compounds on the activities of Fpg and XPA, two zinc finger proteins involved in DNA repair. Carcinogenesis 21:2097-2104

Basu A, Mahata J, Gupta S, Giri AK (2001) Genetic toxicology of a paradoxical human carcinogen, arsenic: a review. Mutat Res 488:171-194

Boiteux S, Gajewski E, Laval J, Dizdaroglu M (1992) Substrate specificity of the Escherichia coli Fpg protein (formamidopyrimidine-DNA glycosylase): excision of purine lesions in DNA produced by ionizing radiation or photosensitization. Biochemistry 31:106-110

Bürkle A (2001) Physiology and pathophysiology of poly(ADP-ribosyl)ation. Bioessays 23:795-806

Bürkle A, Meyer T, Hilz H, zur Hausen H. (1987) Enhancement of N-methyl-N'-nitro-N-nitrosoguanidine-induced DNA amplification in a simian virus 40-transformed Chinese hamster cell line by 3-aminobenzamide. Cancer Res 47:3632-3636

Chiou HY, Hsueh YM, Liaw KF, Horng SF, Chiang MH, Pu YS, Lin JS, Huang CH, Chen CJ (1995) Incidence of internal cancers and ingested inorganic arsenic: a seven- year follow-up study in Taiwan. Cancer Res 55:1296-1300

de Boer J, Hoeijmakers JH (2000) Nucleotide excision repair and human syndromes. Carcinogenesis 21:453-460

de Murcia G, Shall S (2000) From DNA damage and stress signalling to cell death. Oxford University Press, Oxford

Drouin EE, Loechler EL (1993) AP sites are not significantly involved in mutagenesis by the (+)-anti diol epoxide of benzo[a]pyrene: the complexity of its mutagenic specificity is likely to arise from adduct conformational polymorphism. Biochemistry 32:6555-62

Floyd RA (1990) The role of 8-hydroxyguanine in carcinogenesis. Carcinogenesis 11:1447-1450.

Foster KA, Oster CG, Mayer MM, Avery ML, Audus KL (1998) Characterization of the A549 cell line as a type II pulmonary epithelial cell model for drug metabolism. Exp Cell Res 243:359-366

Gebel TW (2001) Genotoxicity of arsenical compounds. Int J Hyg Environ Health 203:249-262

Gradwohl G, Menissier de Murcia JM, Molinete M, Simonin F, Koken M, Hoeijmakers JH, de Murcia G (1990) The second zinc-finger domain of poly(ADP-ribose) polymerase determines specificity for single-stranded breaks in DNA. Proc Natl Acad Sci USA 87:2990-2994

Hartwig A, Dally H, Schlepegrell R (1996) Sensitive analysis of oxidative DNA damage in mammalian cells: use of the bacterial Fpg protein in combination with alkaline unwinding. Toxicol Lett 88:85-90

Hartwig A, Gröblinghoff UD, Beyersmann D, Natarajan AT, Filon R, Mullenders LHF (1997) Interaction of arsenic(III) with nucleotide excision repair in UV-irradiated human fibroblasts. Carcinogenesis 18:399-405

Hartwig A, Asmuss M, Ehleben I, Herzer U, Kostelac D, Pelzer A, Schwerdtle T, Bürkle A (2002) Interference by toxic metal ions with DNA repair processes and cell cycle control: molecular mechanisms. Environ Health Perspect 110 Suppl 5:797-799

Hartwig A, Asmuss M, Pelzer A, Bürkle, A (2003) Very low concentrations of arsenite suppress poly(ADP-ribosyl)ation in mammalian cells. Int J Cancer in press

Hei TK, Liu SX, Waldren C (1998) Mutagenicity of arsenic in mammalian cells: role of reactive oxygen species. Proc Natl Acad Sci U S A 95:8103-8107

Hu Y, Su L, Snow ET (1998) Arsenic toxicity is enzyme specific and its affects on ligation are not caused by the direct inhibition of DNA repair enzymes. Mutat Res 408:203-218

IARC (1980) Monographs on the Evaluation of the Carcinogenic Risk of Chemicals to Humans, Vol. 23: Some Metals and Metallic Compounds. IARC, Lyon

Ikejima M, Noguchi S, Yamashita R, Ogura T, Sugimura T, Gill DM, Miwa M (1990) The zinc fingers of human poly (ADP-ribose) polymerase are differentially required for the recognition of DNA breaks and nicks and the consequent enzyme activation. Other structures recognize intact DNA. J Biol Chem 265:21907-21913

Knowles FC, Benson AA (1983) The biochemistry of arsenic. Trends Biochem. Sci. (Pers. Ed.) 8:178-180

Le Rhun Y, Kirkland JB, Shah GM (1998) Cellular responses to DNA damage in the absence of poly(ADP-ribose) polymerase. Biochem Biophys Res Commun 245:1-10

Li D, Morimoto K, Takeshita T, Lu Y (2001) Formamidopyrimidine-DNA glycosylase enhances arsenic-induced DNA strand breaks in PHA-stimulated and unstimulated human lymphocytes. Environ Health Perspect 109:523-526

Li JH, Rossman TG (1989) Inhibition of DNA ligase activity by arsenite: a possible mechanism of its comutagenesis. Mol Toxicol 2:1-9

Lindahl T (1993) Instability and decay of the primary structure of DNA. Nature 362:709-15

Loeb LA (1994) Microsatellite instability: marker of mutator phenotype in cancer. Cancer Res 54:5059-5063

Lynn S, Shiung JN, Gurr JR, Jan KY (1998) Arsenite stimulates poly(ADP-ribosylation) by generation of nitric oxide. Free Radical Biol Med 24:442-449

Lynn S, Gurr JR, Lai HT, Jan KY (2000) NADH oxidase activation is involved in arsenite-induced oxidative DNA damage in human vascular smooth muscle cells. Circ Res 86:514-519

Mackay JP, Crossley M (1998) Zinc fingers are sticking together. Trends Biochem Sci 23:1-4

Mandal BK, Ogra Y, Suzuki KT (2001) Identification of dimethylarsinous and monomethylarsonous acids in human urine of the arsenic-affected areas in West Bengal, India. Chem Res Toxicol 14:371-378

Mass MJ, Tennant A, Roop BC, Cullen WR, Styblo M, Thomas DJ, Kligerman AD (2001) Methylated trivalent arsenic species are genotoxic. Chem Res Toxicol 14:355-361

Miyamoto I, Miura N, Niwa H, Miyazaki J, Tanaka K (1992) Mutational analysis of the structure and function of the xeroderma pigmentosum group A complementing protein. Identification of essential domains for nuclear localization and DNA excision repair. J Biol Chem 267:12182-12187

National Research Council (2001) Arsenic in Drinking Water. 2001 Update. National Academy Press, Washington, DC

Nordenson I, Beckman L (1991) Is the genotoxic effect of arsenic mediated by oxygen free radicals? Hum Hered 41:71-73

O'Connor TR, Graves RJ, de Murcia G, Castaing B, Laval J (1993) Fpg protein of Escherichia coli is a zinc finger protein whose cysteine residues have a structural and/or functional role. J Biol Chem 268:9063-9070

Peel AE, Brice A., Marzin D, Erb F (1991) Cellular uptake and biotransformation of arsenic V in transformed human cell lines HeLa S3 and Hep G2. Toxicol in Vitro 5:165-168

Peltomäki P (2001) Deficient DNA mismatch repair: a common etiolic factor for colon cancer. Hum Mol Gen 10:735-740

Pleschke JM, Kleczkowska HE, Strohm M, Althaus FR (2000) Poly(ADP-ribose) binds to specific domains in DNA damage checkpoint proteins. J Biol Chem 275:40974-40980

Pott WA, Benjamin SA, Yang RS (2001) Pharmacokinetics, metabolism, and carcinogenicity of arsenic. Rev Environ Contam Toxicol 169:165-214

Rojas M, Alexandrov K, van Schooten FJ, Hillebrand M, Kriek E, Bartsch H (1994) Validation of a new fluorometric assay for benzo[*a*]pyrene diolepoxide- DNA adducts in human white blood cells: comparisons with ^{32}P- postlabeling and ELISA. Carcinogenesis 15:557-560

Schwerdtle T, Seidel A, Hartwig A (2002) Effect of soluble and particulate nickel compounds on the formation and repair of stable benzo[*a*]pyrene DNA adducts in human lung cells. Carcinogenesis 23:47-53

Schwerdtle T, Walter I, Mackiw I, Hartwig A (2003) Induction of oxidative DNA damage by arsenite and its trivalent and pentavalent methylated metabolites in cultured human cells and isolated DNA. Carcinogenesis 24:967-974

Sordo M, Herrera LA, Ostrosky-Wegman P, Rojas E (2001) Cytotoxic and genotoxic effects of As, MMA, and DMA on leukocytes and stimulated human lymphocytes. Teratog Carcinog Mutagen 21:249-260

Styblo M, Del Razo LM, LeCluyse EL, Hamilton GA, Wang C, Cullen WR, Thomas DJ (1999) Metabolism of arsenic in primary cultures of human and rat hepatocytes. Chem Res Toxicol 12:560-565

Tchou J, Kasai H, Shibutani S, Chung MH, Laval J, Grollman AP, Nishimura S (1991) 8-oxoguanine (8-hydroxyguanine) DNA glycosylase and its substrate specificity. Proc Natl Acad Sci U S A 88:4690-4694

Tezuka M, Hanioka K, Yamanaka K, Okada S (1993) Gene damage induced in human alveolar type II (L-132) cells by exposure to dimethylarsinic acid. Biochem Biophys Res Commun 191:1178-1183

Thomas DJ, Styblo M, Lin S (2001) The cellular metabolism and systemic toxicity of arsenic. Toxicol Appl Pharmacol 176:127-144

Vahter M (1999) Methylation of inorganic arsenic in different mammalian species and population groups. Sci Prog 82:69-88

Wang TS, Huang H (1994) Active oxygen species are involved in the induction of micronuclei by arsenite in XRS-5 cells. Mutagenesis 9:253-257

Wang TS, Shu YF, Liu YC, Jan KY, Huang H (1997). Glutathione peroxidase and catalase modulate the genotoxicity of arsenite. Toxicol 121:229-237

Wang TS, Hsu TY, Chung CH, Wang AS, Bau DT, Jan KY (2001) Arsenite induces oxidative DNA adducts and DNA-protein cross-links in mammalian cells. Free Radic Biol Med 31:321-330

Wang TS, Chung CH, Wang AS, Bau DT, Samikkannu T, Jan KY, Cheng YM, Lee TC (2002) Endonuclease III, formamidopyrimidine-DNA glycosylase, and proteinase K additively enhance arsenic-induced DNA strand breaks in human cells. Chem Res Toxicol 15:1254-1258

Wood ML, Dizdaroglu M, Gajewski E, Essigman JM (1990) Mechanistic studies of ionizing radiation and oxidative mutagenesis: genetic effects of a single 8-hydroxyguanine (7-hydro-8-oxoguanine) residue inserted at a unique site in a viral genome. Biochemistry 29:7024-7032

Yager JW, Wiencke JK (1997) Inhibition of poly(ADP-ribose) polymerase by arsenite. Mutat Res 386:345-351

Yamanaka K, Hoshino M, Okamoto M, Sawamura R, Hasegawa A, Okada S (1990) Induction of DNA damage by dimethylarsine, a metabolite of inorganic arsenics, is for the major part likely due to its peroxyl radical. Biochem Biophys Res Commun 168:58-64

Chapter 13

Cytogenetic investigations in employees from waste industries

Helga Fender, Giesela Wolf

Introduction

In 1960 Moorhead developed the method of chromosomal analysis in peripheral lymphocytes. It was used for the first time in 1962 by Bender and Gooch for detecting biological effects of ionising radiation. In 1964 Pollini reported chromosomal alterations in bone marrow cells and peripheral lymphocytes of a patient suffering from benzene haemopathia. During the following years more and more cytogenetic investigations of groups of people exposed to chemical and physical noxae were done to look for possible mutagenic and/or genotoxic effects of the substances in question. In addition to many other occupational and environmental noxae the biologic effect of metals and their organic compounds has been investigated with cytogenetic methods, for example arsenic, lead, cadmium mercury and nickel (for review see Ashby 1985 and Au 1991).

Exposure to gaseous and particulate releases of waste tips may bear a potential health risk to workers and possibly to residents of the surroundings of waste disposal sites. Sometimes clusters of diseases have been described in such residents and links have been assumed between exposure to the releases and diseases (Lotz et al. 1991; Geschwind et al. 1992). Chemical analyses of gaseous releases of waste disposal sites revealed more than 400 compounds, most of them arising by degradation inside of the tip. The main components are methane and carbon dioxide but substances with known or suspected carcinogenic potential are found too, for example benzene, epichlorohydrine and vinylchloride. Cytogenetic investigations of workers exposed to these substances revealed in most studies increases of the frequencies of chromosomal aberrations (Pollini 1964; de Jong 1988, Yardley-Jones 1990; Fučić 1996; Hüttner et al. 1992). The occurrence of volatile organometalloid species including tin, antimony, arsenic and lead in landfill gases was described by Hirner et al. (1994) and Feldmann and Hirner (1995). Hirner (2003) stated that organometalloid emissions from solid waste are at least two magnitudes higher than the biogenic background, "and require toxicological evaluations in respect to the health of people working or living near these sites."

Since 1990, reports were published concerning possible links between chromosomal aberrations and cancer risk (Brøgger et al. 1990; Hagmar et al. 1994; Bonassi et al. 1995; Hagmar et al. 1998). Therefore results of cytogenetic studies gain more and more significance for risk assessment in exposed populations.

In connection with investigations of the state of health of employees of waste disposal sites in North Rhine-Westphalia and of a copper recycling plant in Ilsenburg/Harz, cytogenetic investigations had been carried out in employees and control persons. Owing to processing copper scrap mainly from electronic industries in the copper recycling plant the workers are exposed to metal compounds but also to dioxins and furans.

Materials and methods

Subjects

Exposed groups:

A. Employees of two regular waste disposal sites which have been operated for about 10 years

Site 1: investigation 1991
39 employees, 24 control persons

Site 2: investigation 1994
43 employees, 47 control persons

Duration of employment in tip workers: 0.5 to 13 years (5.1 ± 3.5 years)

Control group: staff members of local district administrations

Because there were no significant differences concerning age and sex distribution, smoking status and duration of employment as well as in cytogenetic results between groups investigated in 1991 and 1994, they were combined. Table 1 summarizes the main characteristics of study groups.

B. Copper recycling plant: investigation 1995

12 employees, 12 controls

Control group: staff members from public services in Berlin

Table 1. Population Characteristics of Groups Investigated – Waste Disposal Sites

		Employees	Controls
N	Whole Group	82	71
	Men	75	65
	Women	7	6
Age (Years)	Mean ± SD	36.6 ± 10.1	39.6 ± 9.1
	Median	37	38
	Min - Max	18 - 59	21 - 59
Smoking Status	Smokers	48	33
	Nonsmokers	34	38
Duration of Employment (Years)	Mean ± SD	5.1 ± 3.5	No Data
	Median	4.0	
	Min - max	0.5 - 13	

Table 2. Population Characteristics of Groups Investigated – Copper Recycling Plant

		Copper Recycling	Controls
N		12	12
Age		52.3 ± 10.8 32 - 69	50.3 ± 7.6 34 - 58
Smoking status		Smokers: 7 Ex-Smokers: 2 Non-Smokers: 3	Smokers: 4 Ex-Smokers: 4 Non-Smokers: 4
Polychlorinated Dioxins and Furans in Blood*			
I – TEQ (pg/g blood lipid)	Mean ± SD Min - Max	127.4 ± 65.7 48.6 – 299.1	No Data
EQ-BGA (pg/g blood lipid)	Mean ± SD Min - Max	65.2 ± 30.7 25.3 – 141.9	No Data
Lead (μg/dl blood)*	Mean ± SD Min - Max	16.0 ± 5.3 8.2 – 26.6	No Data
Arsenic (hair) (μg/g)*	Mean ± SD Min - Max	0.1 ± 0.1 0.05 – 0.3	No Data

* For methods see Kersten and Bräunlich (2002)

Cytogenetic methods

Per person 4.5 ml blood were drawn by venipuncture with heparinized SARSTEDT monovets, coded and sent to Berlin by train. Whole blood cultures (RPMI 1640 medium with 10% fetal calf serum, PHA and 8μg/ml BrdUrd) were initiated within 24 hours after venipuncture. Colchicine (0.33 μg/ml) was added 45 hours after culture initiation. The cultures were harvested after 48 hours for chromosome analysis as well as for SCE analysis[1]. Slides were stained by a modified FGP method (Goto et al. 1978; Perry and Wolff 1974). For chromosome analysis 1000 M1[2] cells per person were scored for structural aberrations as achromatic lesion, chromatid break, chromatid exchange, atypical monocentric chromosome, excess acentric fragment, dicentric chromosome (dic). Polycentric

[1] In most cases we find sufficient M2 cells for SCE analysis in 48 h cultures

[2] M1, M2: Metaphases of first resp. second cell division in vitro

chromosomes are recorded as (n - 1) dic (n: number of centromeres); rings are recorded as dic. Sister chromatid exchange (SCE) analysis was performed by scoring of 50 M2 cells per person. After decoding results were compared by non-parametric U-test (Wilcoxon, Mann & Whitney).

Results

Chromosomal aberrations are more frequent in waste disposal site workers than in control persons. The differences turned out to be significant in nearly all types of aberrations. Table 3 summarizes the results for cells with dicentric chromosomes, cells with chromatid aberrations, cells with chromosome aberrations, and percentage of aberrant cells (without gaps). Remarkable in waste workers is the frequent occurrence of multi-aberrant cells. Twenty three cells with multiple chromosome aberrations, most of them fulfilling the criteria of rogue cells, i. e. several double minutes in addition to fragments, dicentric, polycentric and/or atypical chromosomes, were found in the employees' group, whereas only one occurred in the control group. One of these cells is shown in Figure 1 to demonstrate the typical feature of rogue cells. Additional to these cells with chromosome aberrations several cells with multiple chromatid breaks occurred in waste workers (10) but also in controls (13).

On the other hand, there was no difference in the frequencies of sister chromatid exchanges (SCE) between the two groups.

Duration of employment did not influence the rate of chromosome aberrations significantly.

An effect of smoking was apparent only in the control group with a clear but not significant tendency to higher frequencies of chromosome aberrations and a significant difference in frequencies of SCE (6.7 ± 1.5 in smokers, 5.7 ± 0.9 in non-smokers, $p < 0.05$).

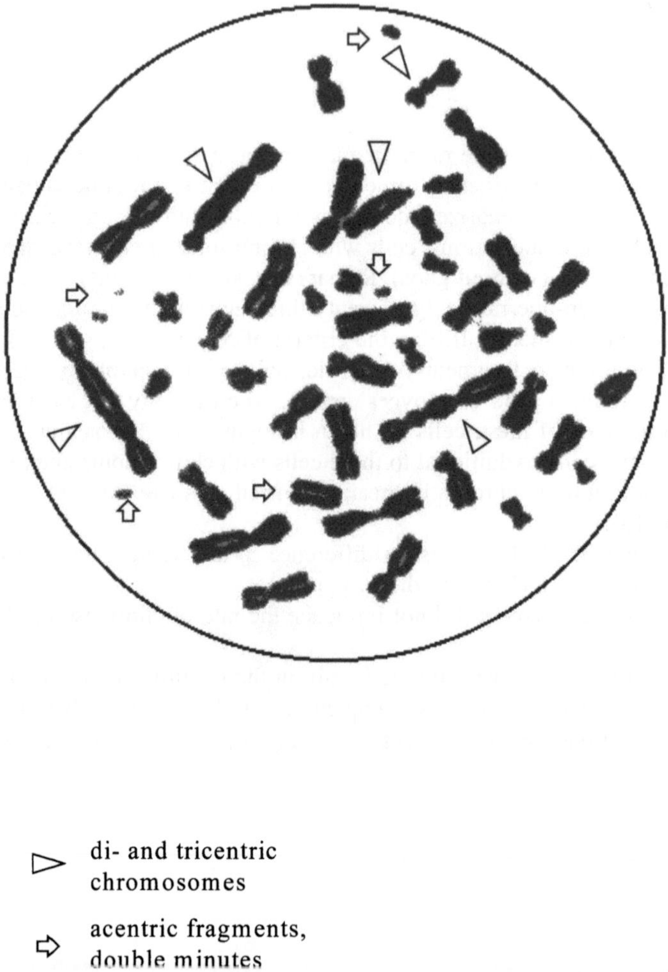

Fig. 1. Multi-aberrant cell from a waste disposal site employee

Table 3. Combined Results of the Waste Disposal Site Studies

		Employees	Control	
Persons		82	69*	
Cells Scored		82.000	69.000	
Cells with Chromatid Aberrations	Mean	7.3	5.0	
	95 % CI Poisson	6.7 – 7.9	4.5 – 5.6	$p < 0.01$
	Min - Max	0 - 20	0 - 18	
Cells with Chromosome Aberrations	Mean	6.3	4.4	
	95 % CI Poisson	5.8 – 6.9	3.9 – 4.9	$p < 0.01$
	Min - Max	1 - 20	0 - 21	
Cells with Dicentric Chromosomes	Mean	2.3	1.4	
	95 % CI Poisson	1.95 – 2.6	1.1 – 1.7	$p < 0.01$
	Min - Max	0 - 8	0 - 5	
Percentage of Cells with Aberrations	Mean ± SD	1.4 ± 0.7	0.9 ±0.6	
	Min - Max	0.2 – 3.6	0.1 – 3.1	$p < 0.01$
Multi-aberrant Cells (rogue cells)	N	33	14	
		(23)	(1)	
Sister Chromatid Exchanges		6.1 ± 1.4	6.2 ± 1.4	n.s.

* Two persons were excluded from statistical analysis of chromosomal aberrations because of insufficient sample size (< 1000 scorable metaphases), n.s. = not significant

Table 4. Results of the Copper Smelter Study

		Copper recycling	Control (Berlin)	
Persons		**12**	**12**	
Cells Scored		**12.000**	**12.000**	
Cells with Chromatid Aberrations	Mean	9.3	6.1	
	95 % CI Poisson	7.6 – 11.1	4.8 – 7.7	n.s.
	Min - Max	1 - 28	2 - 11	
Cells with Chromosome Aberrations	Mean	5.2	2.6	
	95 % CI Poisson	4.0 – 6.6	1.8 – 3.8	n.s.
	min - max	1 - 16	0 - 9	
Cells with Dicentric Chromosomes	Mean	1.7	0.4	
	95 % CI Poisson	1.0– 2.6	0.1 – 1.0	$p < 0.01$
	Min - Max	0 - 4	0 - 3	
Percentage of Cells with Aberrations	Mean ± SD	1.4 ± 1.1	0.9 ±0.4	
	Min - Max	0.3 – 4.1	0.3 – 1.8	n.s.
Multi-aberrant Cells	N	1	0	
SCE	Mean ± SD	8.3 ± 1.0	6.4 ± 1.0	$p < 0.01$
	Min - Max	7.0 – 10.4	4.5 – 7.4	

n.s. = not significant

The multiple exposure to heavy metals and dioxins and furans in the copper smelter workers was mainly caused by processing of copper scrap containing chlorinated plastic materials. Lead and dioxin/furan in blood values exceed considerably the normal background (lead: 16.0 ± 5.3 µg/dl blood vs. 5.5 µg/dl blood; dioxins/furans: I-TEQ 127.4 ± 65.7 pg/g vs. 40.8 pg/g).

Compared to controls, frequencies of chromosomal aberrations are enhanced in copper smelter workers, but only in cells with dicentric chromosomes this turned out to be significant. There was no significant correlation of aberration frequencies with blood contents of lead or dioxin. In contrast to the waste disposal site workers, we found significant more SCE in smelter workers than in control persons. Because of the small group size, dependency of frequencies of chromosomal aberrations on smoking status was not investigated.

Discussion

Numerous cytogenetic studies in persons exposed to chemicals have been done in the last 40 years. Especially in investigations of occupationally exposed groups often significant increased frequencies of chromosomal aberrations had been observed. For some metals and/or metalloid compounds, benzene, vinylchloride, acrylnitrite, ethylenoxide and other substances a genotoxic potential has been demonstrated by cytogenetic methods (Anwar 1991; Al-Hakkak et al. 1986; Bauchinger et al. 1976; Deknudt et al. 1977; Forni 1996; Fučić et al. 1996; Hüttner et al. 1992; de Jong et al. 1988; Yardley –Jones et al. 1990).

In some cases of assumed health effects by tip releases in residents of surroundings of waste disposal sites, e.g. in Love Canal (USA) (Picciano 1980; Heath et al. 1984) or Mellery (Belgium) (Lakhanisky et al. 1993; Klemans et al. 1995) cytogenetic investigations had been used to get information about biologic effects. Whereas the Love Canal chromosome studies caused controversial discussion (Gage 1980; Picciano 1980; Shaw 1980), a significant increase in mean SCE levels and number of HFC (high frequency cells) was observed in inhabitants of Mellery compared with a control collective.

Similar findings in 12 tip workers and 7 control persons were published by Gonsebatt et al. (1995). They too found significant differences in chromosome aberrations only. SCE analysis alone may not always reveal the effect, therefore we propose to apply chromosome analysis and SCE analysis in human population monitoring. Furthermore cytogenetic methods have to be standardized and evaluated since standardization and evaluation are the basis for utilization of the full potential of cytogenetic methods as a tool of risk assessment in human population monitoring.

The results presented here demonstrate an enhanced level of chromosomal damage in persons with occupational exposure to tip releases or metal compounds and dioxins and furanes. Especially remarkable is the occurrence of multi-aberrant cells observed in waste disposal site workers. Up to now the cause and signifi-

cance of these cells is unknown and remains open to speculation (Neel et al. 1992).
Obviously long-lasting exposure is able to cause biologic effects detectable by cytogenetic methods. In consideration of epidemiologic investigations of Scandinavian and Italian research teams (Bonassi et al. 1995; Brøgger et al. 1990; Hagmar et al. 1994; Hagmar et al. 1998) revealing a connection between frequency of chromosome aberrations and cancer risk, the biologic effect found requires attention. Further research is needed urgently.

Acknowledgements

The projects were supported by RWTH Aachen, Institute of Hygiene and Occupational Medicine (waste disposal site studies) and Bundesanstalt für Arbeitsschutz und Arbeitsmedizin (copper recycling plant study).
Especially we acknowledge the engagement and cooperation of Prof. Dr. med. H.-J. Einbrodt and Dr. med. A. Engler.

Literature

Al-Hakkak ZS, Hamamy HA, Murad AMB, Hussain AF (1986) Chromosome aberrations in workers at a storage battery plant in Iraq. Mutat Res 171:53-60

Anwar WA(1991) Cytogenetic monitoring of human populations at risk in Egypt: Role of cytogenetic data in cancer risk assessment. Environ Health Perspect 96:91-95

Ashby J, Richardson CR (1985) Tabulation and assessment of 113 human surveillance cytogenetic studies conducted between 1965 and 1984. Mutat Res 154:111-133

Au WW (1991) Monitoring human populations for effects of radiation and chemical exposures using cytogenetic techniques. Occupational Medicine: State of Art Rev 4:597-611

Bauchinger M, Schmid E, Einbrodt HJ, Dresp J (1976) Chromosome aberrations in lymphocytes after occupational exposure to lead and cadmium. Mutat Res 40:57-62

Bender, MA; Gooch PC (1962) Types and rates of X-ray-induced chromosome aberrations in human blood irradiated in vitro. Proc Natl Acad Sci (USA) 48:522-532

Bonassi S, Abbondandolo A, Camurri L, Pra LD, de Ferrari M, Degrassi F, Forni A, Lamberti, L, Lando C, Padovani P, Sbrana I, Vecchio D, Puntoni R (1995) Are chromosome aberrations in circulating lymphocytes predictive of future cancer onset in humans? Preliminary results of italian cohort study. Cancer Genet Cytogenet 79:133-135

Brøgger A, Hagmar L, Hansten I, Heim S, Högsted B, Knudsen L, Lambert B, Linnainmaa K, Mitelman F, Nordenson I, Reuterwall Ch, Salomaa S, Skerfving S, Sorsa M (1990) An internordic prospective study on cytogenetic endpoints and cancer risk. Cancer Genet Cytogenet 45:85-92

Deknudt Gh, Manuel Y, Gerber GB (1977) Chromosomal aberrations in workers professionally exposed to lead. J Toxicol Environ Health 3:885-891

Feldmann J and Hirner AV (1995) Occurrence of volatile metalloid species in landfill and sewage gases. Intern J Environ Anal Chem 60:339-359

Forni A (1996) Benzene-induced chromosome aberrations: a follow-up study. Environ Health Perspect 104:1309-1312

Fučić A, Barković D, Garaj-Vrhohac V, Kubelka D, Ivanic B, Dabo T, Mijić A (1996) A nine-year follow up study of a population occupationally exposed to vinyl chloride monomer. Mutat Res 361:49-53

Gage S (1980) Love Canal chromosome study. Science 209:752-754

Gonsebatt ME, Salazar AM, Montero R, Barriga FD, Yanez L, Gomez H, Ostrosky-Wegman P (1995) Genotoxic monitoring of workers at a hazardous waste disposal site in Mexico. Environ Health Perspect 103 Suppl. 1:111-113

Geschwind SA, Stolwijk, JAJ, Bracken M, Fitzgerald E, Stark A, Olsen C, Melius J (1992) Risk of congenital malformations associated with proximity to hazardous waste sites. Am J Epidemiol 135:1197-1207

Goto K, Maeda S, Kano Y, Sugiyama T (1978) Factors involved in differential Giemsa-staining of sister chromatids. Chromosoma 66:351-359

Hagmar L, Brogger A, Hansteen I-L, Heim S, Högstedt B, Knudsen L, Lambert B, Linnainmaa K, Mitelman F, Nordenson I, Reuterwall Ch, Salomaa S, Skerfving S, Sorsa M (1994) Cancer risk in humans predicted by increased levels of chromosomal aberrations in lymphocytes: Nordic Study Group on the health risk of chromosome damage. Cancer Research 54:2919-2922

Hagmar L, Bonassi S, Stömberg U, Mikoczy Z, Lando C, Hansteen I-L, Montagud AH, Knudsen L, Norppa H, Reuterwall C, Tinnerberg H et al. (1998) Cancer predictive value of cytogenetic markers used in occupational health surveillance programs. Recent Results in Cancer Research 154:177-184

Heath CW jr, Nadel MR, Zack MM, Chen ATL, Bender MA, Preston RJ (1984) Cytogenetic findings in persons living near the Love Canal JAMA 251 Nr.11:1437-1440

Hirner AV, Feldmann J, Goguel R, Rapsomanikis S, Fischer R, Andreae MO (1994) Volatile metal and metalloid species in gases from municipal waste deposits. Appl Organomet Chem 8:65-69

Hirner AV (2003) Volatile metal(loid) species associated with waste materials: chemical and toxicological aspects. In: Cai Y, Braids OC (eds) Biogeochemistry of Environmentally Important Trace Elements. ACS Symp Ser 835 pp 141-150

de Jong G, Van Sittert NJ, Natarajan AT (1988) Cytogenetic monitoring of industrial populations exposed to genotoxic chemicals and of control populations. Mutat Res 204:451-464

Kersten N and Bräunlich A (2002) Fb 9 60 Gesundheitsrisiken durch Schwermetalle und Dioxine/Furane in Sekundärkupferhütten. Schriftenreihe der Bundesanstalt für Arbeitsschutz und Arbeitsmedizin, Dortmund Berlin 2002

Klemans W,Vleminckx C, Schriewer L, Joris I, Lijsen N, Maes A, Ottogali M, Pays C, Planard C, Rigaux G, Ros Y, Vande Riviere M, Vandenvelde J, Verschaeve L, Deplaen P, Lakhanisky Th (1995) Cytogenetic biomonitoring of a population of children allegedly exposed to environmental pollutants - Phase 2: Results of a three-year longitudinal study. Mutat Res 342:147-156

Lakhanisky T, Bazzoni D, Jadot P, Laurent C, Ottogali M, Pays C, Planard C, Ros Y, Vleminckx C (1993) Cytogenetic monitoring of a village population potentially exposed to a low level of environmental pollutants: Phase 1: SCE analysis. Mutat Res 319:317-323

Lotz I, Brand H, Greiser E (1991) Erhöhung der Leukämieinzidenz in der Nähe einer ehemaligen Sondermülldeponie. Öff Gesundh Wes 53:579-580

Moorhead PS, Nowell WJ, Mellmann WJ, Battips DM, Hungerford DA (1960) Chromosome preparations of leukocytes cultured from human peripheral blood. Exp Cell Res 20:613-616

Neel JV, Awa AA, Kodama Y, Nakano M, Mabuchi K (1992) "Rogue" lymphocytes among Ukrainians not exposed to radioactive fall-out from the Chernobyl accident: The possible role of this phenomenon in oncogenesis, teratogenesis, and mutagenesis. Proc Natl Acad Sci USA 89:6973-6977

Perry P, Wolff S (1974) New Giemsa method for differential staining of sister chromatids. Nature 261:156-158

Picciano D (1980) Love canal chromosome study. Science 209:754-756

Picciano D (1980) A pilot cytogenetic study of the residents living near Love Canal, a hazardous waste site. Mammalian Chromosome Newsletters 21:86-99

Pollini G, Colombi R (1964) Il danno cromosomico dei linfociti nell'empoatia benzenica. Medicina del Lavoro 55:641-654

Shaw MW (1980) Love canal chromosome study. Science 209:751-752

Yardley-Jones A, Anderson D, Lovell DP, Jenkinson PC (1990) Analysis of chromosomal aberrations in workers exposed to low level benzene. Brit J Ind Med 47:48-51

Chapter 14

Neurotoxicity of metals

G. Stoltenburg-Didinger

Introduction

The upsurge of interest in recent years in science, industry, and government in the effects of toxic chemicals on the nervous system has created a new discipline of neurotoxicology.

For a morphologist, the overview of where neurotoxins act in the nervous system, the type of damage which ensues, and the relative and special vulnerabilities of different nervous system components, both during development and adult life is prevalent. This attempt includes the review of experimental studies designed to determine mechanisms of toxic damage to the neuron, axon, synapse, myelinating cell and astroglia. These data are correlated with existing data for established neurotoxic substances which have received attention either because of their environmental significance or for their potential as causes of nervous system dysfunction.

The collaboration FOR 415 between chemists, analysts, microbiologists and physicians focuses on "Organometal(loid) compounds in the environment". The significance of metals in causing adverse neurotoxic effects is already well known for peripheral neuropathies (Behse and Carlsen 1978; Phillippe and Gothard 1903). For the central nervous system, the neuroscience research community paid little attention to the neurometabolism of metal ions. Fundamental to an appriciation of the interface between neuroscience and metallobiology is an awareness that the brain is a specialized organ that concentrates metal ions (Bush 2000). Neurotoxicity has to address the molecular, pathological, and functional responses of astroglia, microglia, and oligodendroglia to neurotoxicant exposure. Neurons are the signaling cells of the nervous system, and as such are responsible for the perception of sensory stimuli and the coordination of cellular and organismal responses to stimuli from the environment. Neuronal function and nervous tissue structure requires the participation of neuroglia, or glia, including astroglia (which participate in neurotransmitter metabolism, induce and maintain the blood-brain barrier and responses to stress and injury), oligodendroglia (which myelinate axons in the central nervous system) and ependymal cells (which border the ventricles). Microglia (which mediate inflammatory responses in the central nervous system) are not of neuroectodermal origin. The roles of each of these cells as target cells and participants in various aspects of neurotoxicity has to be considered.

For the purposes of this contribution, the following descriptions will confine to neurotoxic effects to the metal ions of lead (Pb) and mercury (Hg).

Cell-specific compartmentation of metals in the brain is an important consideration for neurotoxicity induced by metals, because the concentrations of several metals in the brain tend to be in the same order of magnitude as magnesium, i. e. 0.1-0.5 mM (Lovell et al. 1998). Astroglia are hypothesized to selectively accumulate lead and methylmercury (MeHg), thus constituting brain depots for these well-known exogenous neurotoxicants. Astroglia possess several characteristics that would allow them to serve as depots for metals in the brain. First, cytoplasmic processes of astroglia known as footplates or endfeet surround all vascular surfaces of the brain, as well as the pia (Vaguera-Orte et al. 1981). The endfeet form permeable morphologic layers. In addition, the cell bodies of astroglia are juxtaposed between neuronal cell bodies and the capillary endothelium that forms the blood-brain barrier. Thus, astroglia are the first cells of the brain parenchyma to encounter metals crossing the blood-brain barrier (Tiffany-Castiglioni et al. 1989). A second property of astroglia that suggests metal-handling ability is high cytosolic levels of metallothioneins I and II, in contrast to low levels in neurons. Metallothionein III (MT-III) is localized by some studies to hippocampal neurons, though principal localization to astroglia has been reported in normal rat brain (Carrasco et al. 1999). Evidence also exists that reactive astroglia express MT-III during tissue reconstruction after damage to the immature rodent brain (Acarin et al. 1999). Metallthioneins are cysteine-rich proteins with high affinities for divalent metal ions. These proteins may allow astroglia to chelate free metals in the brain as a neuroprotective action (Aschner et al. 1998). Third, astroglia have high cytosolic levels of glutathione, a tripeptide (glycine-cystein-glutamate) that confers most of the redox potential available in the cytoplasm. Levels in mature astroglia are about 2 mM, which exceeds neuronal levels. Fourth, key metal dependent enzymes in the brain are localized to astroglia, such as glutamine synthetase (Norenberg and Martinez-Hernandez 1979). Fifth, astroglia have putative metal transport or carrier proteins. A schematic model for metal detoxification shows that cells handle metals by metalloregulation *via* proteins such as metallothioneins, incorporation into metal-requiring enzymes, extrusion from the cell and/or mineralization in organelles such as lysosomes, mitochondria and nucleus (Tiffany-Castiglioni and Qian 2001).

Mercury

Mercury and its compounds may enter the animal or human body *via* inhalation (lung), ingestion (gastrointestinal tract) or inunction (skin). Hg can exist in physiological conditions as Hg°, Hg^{+} or Hg^{2+}. Because microorganisms metabolize inorganic mercury to methylmercuric compounds ($MeHg^{+}$), MeHg is a significant environmental contaminant that accumulates in the food chain. The relative level of accumulation of the various forms of Hg in the nervous system is $MeHg^{+}$>Hg°>Hg^{2+} (Schiønning 2000).

Cellular localization of Hg in adult MeHg exposures has been studied in rats and monkeys. In rats treated with MeHg, Hg is predominantly accumulated in neurons, with less accumulation in glia and ependymal cells (Schiønning and Møller-Madsen 1992). In contrast, in monkeys (Macaca) treated with MeHg, localization of Hg is predominantly astroglial and microglial (Charleston et al. 1995, 1996). Most of the experiments examining Hg distribution in the brain of rats and mice have been performed after a short-term survival period following cessation of exposure, even if exposure times have been lengthy. Thus, long-term redistribution of inorganic Hg has not been tracked systematically in rodents. This is an important issue because the autometallographic method detects inorganic Hg but not MeHg, and MeHg is apparently demethylated slowly in brain. Charleston et al. (1996) provided evidence in the above mentioned study with macaques that MeHg is converted to inorganic Hg over time in the thalamus of MeHg-exposed animals. Measurements of inorganic Hg, MeHg, and total Hg were carried out by cold vapor atomic absorption spectroscopy in the contralateral thalamus of animals examined for Hg localization by autometallography. Two observations were striking: First, animals showed a doubling of inorganic Hg content in the thalamus between 12 and 18 months of exposure (50μg Hg as MeHg/kg body weight per day per observation), but a loss of MeHg content. Second, animals exposed for 12 months, followed by 6 months without exposure (clearance group), showed an eightfold reduction in total Hg, compared to animals treated for 12 months and then killed for examination with no clearance period. The Hg loss in the clearance group could be attributed to MeHg, as levels of inorganic Hg were similar between the two groups. Cell counts demonstrated the loss of astrocytes at 6 months, followed by an increase at 18 months in the numbers of both astroglia and microglia in MeHg treated animals. The evidence from these studies, coupled with an earlier study in rats by Hargreaves et al. (1985), supports a glial buffering hypothesis for MeHg exposure, whereby astroglia initially accumulate Hg and may subsequently either die or undergo astrogliosis and proliferation. Dead astrocytes are phagocytized by microglia, which then accumulate the Hg. Much later, neurons may accumulate lesser amounts of Hg. Most rodent data do not fit this model, although the idea that transitory Hg accumulation by astroglia precedes neuronal distribution was first presented in rats treated with clinically neurotoxic levels of MeHg (Hargreaves et al. 1985). The glial buffer hypothesis proposed by Hargreaves et al. (1985) and Charleston et al. (1995) is based on the fact that Hg can readily enter cells (here: astroglia) by diffusion as MeHg or Hg° (Schiønning 2000) or *via* an unknown transporter as Hg^{2+}. MeHg may be demethylated to Hg^{2+} in the cytoplasm, alternatively Hg° may be oxidized to Hg^{2+} by catalase. The only reported mechanism for Hg^{2+}storage in the cell is mineralization in lysosomes, which has been observed in neurons by several investigators (see Schiønning 2000, for review).

The first cases of fatal poisoning by organic mercurials were reported over a century ago. Edwards (1865) described the clinical and pathological features of two laboratory technicians who were fatally exposed to some organic mercury compounds. The classical pathological description of MeHg poisoning is Hunter and Russell's report on one of their patients who died as a result of accidental exposure to MeHg (1954). MeHg intoxication occurs after massive environmental contamination. Inorganic mercury is discharged as a waste product, usually from paper or chloralkali manufacture, into rivers or the sea. Inorganic Hg is metabolized by microorganisms to MeHg compounds. Dangerous concentrations built up in water-living animals like fish.

Methylmercury was not recognized as an environmental health hazard until the massive outbreak of hundreds of human cases in Minamata Bay and the Nigata District of Japan in the 1950s. Residents of these areas were exposed through the consumption of mercury-contaminated fish. Since then MeHg poisoning has been referred to as Minamata disease (Takeuchi et al. 1968).

Contaminated foodstuffs have also been a source of human intoxication by mercury compounds. McHg is used as a fungicide. Treated wheat seed is harmless if sown. Humans have eaten treated seed with disastrous consequences. In Iraq (1971) about 10000 developed toxic effects. The follow-up of the Iraq cases showed high developmental neurotoxicity (Marsh et al. 1980). Although the level of exposure at which damage occurred was similar in mothers and fetuses, the degree of damage to fetuses was more severe and diffuse. An increased incidence of mental retardation and spasticity was found in a group exposed *in utero* and reexamined at age 5, the mothers had only mild paresthesias during pregnancy (Marsh et al. 1980).

The pollutant MeHg is as a well established neurotoxicant of considerable public health concern, because it is found in seafood and freshwater fish throughout the world (WHO 1990). Given the potential threat that MeHg poses to the optimal development of cognitive functioning, clinicians and regulatory agencies are concerned about the amount of MeHg a pregnant woman can safely ingest. A cohort of over 1000 singleton births from the Faroer Islands was generated and then neuropsychological testing was conducted at school age. The MeHg exposure originated from fish consumption, mainly whale meat, that is shared in the communities where the whales are killed. Mercury-related cognitive deficits were most pronounced in the domains of language, attention, and memory. The adverse effects on brain function were detected at exposure levels currently considered safe (Hg concentration in maternal scalp hair 10 µg/g or 50 nmol/g (Grandjean et al. 1997).

Behavioral alterations in animals also have been described at dose levels below those associated with overt symptoms of neurotoxicity (Chang 1977). The neuropathological investigation of the somatosensory cortex of rats prenatally exposed to low doses of MeHg (0.025, 0.05, 0.5, and 5.0 mg/kg/day on days 6 to 9 after conception, behavioral testing of the offspring from the second to the seventh month, histopathological investigation at the age of twelve months) revealed dendritic spine dysgenesis (Stoltenburg-Didinger and Markwort 1990). The spine abnormalities in the experimental animals consisted of a reduction of stubby and mushroom-shaped soines and a predominance of long and tortuous spines. Dendritic spine dysgenesis implies defective development and might be the pathological feature of the impaired behavior and learning of the prenatally MeHg-exposed animals. The structural abnormality underlines the persistence of the pathology in developmental neurotoxicity.

Lead

The presence of lead in the environment arises from both natural and anthropogenic sources. The environmental significance reflects both its utility and relative abundance, being comparatively more abundant geologically than most of the other toxic heavy elements. Lead in the various forms has long been known and utilized by man, its extensive mining and processing dating back to pre-Christian times. Current world production exceeds the amount of any other toxic heavy metal.

Given the ubiquitous distribution of lead, due in large measure to a long history of use, it is not surprising that this element is present in food, water, air, soil, dustfall, and other materials that are in use in developed countries. Lead has no known beneficial biological or health effects and thus, one must view lead healthwise purely as a deleterious agent.

Man is exposed to lead chiefly *via* inhalation or ingestion and the relative absorption of lead into the body is dependent on both the chemical and physical character of lead intake. Irrespective of the particular form of lead, it is the ionic species that is absorbed. In addition, specific host factors such as age and nutritional status appear to play an important role in the absorption of lead; e. g. children consistently show higher blood-lead values than adults in identical environmental settings and the distinction is most pronounced with children under 5 years of age. This presumably is a reflection not only of enhanced rate of lead absorption but also distinctive behavior of young children, such as mouthing-activity and ingestion of lead-contaminated non-foodstuffs. Lead toxicity in childhood was thought to be without residual effects in nonlethal cases; a misconception corrected by the studies of Byers (1943) who asserted that lead not only killed cells but interfered with normal neuronal development. While the inorganic compounds of lead are recognized neurotoxicants in human and animal models, alkyl derivates of lead are equipotent on a dosage basis and equally neurotoxic, comparable to observations on lead toxicity in bacteria and fungi (Verity 1990).

The characteristic acute, predominantly cerebellar encephalopathy associated with neonatal high lead exposure contrasts to the subtle, axo-dendritic disorganization shown to be associated with low-level neonatal inorganic Pb^{2+} exposure. There is a preferential involvement of the hippocampus in both low-level inorganic Pb^{2+} and organolead exposure, and the clinical syndromes of irritability, hyperactivity, aggression, and seizures are common features of disturbed hippocampal function.

Neurotransmitter system abnormalities have been described with inorganic Pb^{2+}, but recent attention has focused on the abnormalities in glutamate, dopamine and gamma-aminobutyric acid (GABA) uptake, efflux and metabolism. Abnormalities of GABA and glutamate metabolism are also found with the organolead species. While the pathogenesis is still unclear, the interactive role of Pb^{2+} on mitochondrial energy metabolism, Ca^{2+} uptake, intracellular Ca^{2+} homeostasis, and the influx or efflux of neurotransmitters is considered. Low dose inorganic Pb^{2+}

and organolead effects on mitochondrial membranes including inhibition of intracellular ATP biosynthesis could underlie the neurotoxicity.

The neurotoxicity of Pb has been the subject of intensive investigation. The existing paradigm is that Pb enters the brain from blood by crossing the blood-brain barrier and encounters astroglia, which are interposed between blood-brain barrier and neurons and participate in sequestration of Pb. This paradigm is supported by the observations of Thomas et al. (1973) that, 72h after 1-day-old rat pups are injected with ^{210}Pb, Pb is localized autoradiographically to capillary endothelium and astroglial footplates in the cerebellum. Many other *in vivo* experiments support the astroglial deposition of Pb. Holtzman et al. (1984) showed that in weanling rats dosed daily with high amounts of Pb, a Pb-sequestering property is exhibited by astroglia in mature, but not immature brain tissue. For example, 18-day-old animals were given a dose of Pb that caused encephalopathy in the cerebellum (hemorrhage, edema, and neuronal necrosis), but not in the more mature cerebrum in the same animals. The subcellular distribution of Pb in cerebral tissue was found to be in astroglial cytoplasm, lysosomes and nuclei, but not in neurons. In contrast, Pb was distributed throughout the organelles of both astroglia and neurons in the cerebellum. From these observations arose the hypothesis that mature astroglia in situ have the capacity to take up Pb into non-toxic subcellular sites, potentially protecting neurons. This fact was later apostrophized as "lead sink hypothesis" (Tiffany-Castigliony et al. 1989). This hypothesis states that mature astroglia accumulate Pb in the brain.

Pb exists as a divalent cation under physiologic conditions and is redox inactive. Metallothionein is not induced by Pb in cultured murine astroglia (Kramer et al. 1996), and its ability to interact with Pb in the brain has not been established. However, other proteins bind to lead. Pb also interacts with several other Ca^{2+}-regulated enzymes, including protein kinase C (Johnston and Goldstein 1998) and calmodulin (Noack et al. 1996; Kern et al. 2000).

Many of the biological aberrations produced by lead appear related to the ability of this heavy metal to mimic and sometimes inhibit the action of calcium (Bressler and Goldstein 1991). Lead modulates neurotransmitter release by altering calcium metabolism - either by competing with calcium for entry into the cell or by increasing intracellular calcium levels. The biphasic response on transmitter release to lead with stimulation of the basal rate and inhibition of the depolarization-induced fraction may have special relevance to the immature nervous system. The continuous release of subthreshold amounts of neurotransmitter into the synaptic cleft is thought to have a trophic influence on maintaining the efficiency of a synaptic connection and the survival of the postsynaptic cell. Depolarization-induced release of presynaptic neurotransmitter, on the other hand, is responsible for producing the signal summation that excites or inhibits the postsynaptic cell. The trophic and functional events that these two types of activity produced in neuronal networks are particularly important during development. Since the pattern of neuronal activity seems to be the pruning process, exposure to lead early in life may have a lasting adverse effect on synaptic anatomy and brain function. Neuronal development and plasticity are known to be dependent on N-methyl-D-aspartate (NMDA) receptors. Activation of NMDA receptors promotes neurite outgrowth in cultured hippocampal neurons (Brewer and Cotman 1989). Such changes may underlie the effect of low level lead exposure upon the learning skills and behavior of young children in the absence of overt pathological damage.

The hippocampal region deserves particular attention when studying the effects of lead on the central nervous system. It accumulates lead to a higher degree than other divisions of the brain (Fjerdingstad et al. 1974; Collins et al. 1982). Behavioral alterations following lead exposure have been related to hippocampal dysfunction. Earlier studies have demonstrated that lead affects the morphology (Alfano et al. 1982, Slomianka et al. 1989) and the connective plasticity of the hippocampal region (Alkondon et al. 1990).

The inhibition of NMDA-activated current in cultured fetal hippocampal neurons by Pb^{2+} was investigated in various stages of cell development (Ujihara and Albuquerque 1992). Pb inhibited NMDA currents recorded from young cultured neurons. In the first week of culture, Pb^{2+} showed the most prominent inhibition, which was gradually attenuated in the following weeks. Pb^{2+} is one of the most common heavy metals in the environment which, even at low doses of exposure, can cause learning and behavior deficits (Bellinger et al 1987; Bellinger and Stiles 1993). In rat hippocampal slice preparations, Pb^{2+} impairs the formation of long-term potentiation which is considered to be a model of the learning process.

The astroglial intermediate filament protein, glial fibrillary acidic protein (GFAP), is a generally accepted sensitive indicator for neurotoxic effects in the mature brain. Astrogliosis, often referred to as reactive gliosis or simply gliosis, is detectable by immunohistochemistry of GFAP. Astrogliosis can be quantified by assaying GFAP or its mRNA (Peters et al. 1994). Neurotoxicologists are interested in GFAP because astrocytes undergo hypertrophy when damage occurs and accumulate glial filaments, the main component of which is GFAP. Hypertrophy has been widely documented by measurements of GFAP. Reactive gliosis is characterized by hypertrophy of the astrocyte cell body and lengthening and thickening of the processes.

The signaling mechanisms underlying the astroglial response to chemically induced injury remain largely unknown. Glial activation is rapid. The current dogma is that astrogliosis is a dominant response of the adult central nervous system to all types of injuries. Postnatal exposure to low levels of lead result in the presence of a large number of reactive astrocytes, distinguished from normal astrocytes by their greater size, longer, thicker processes, and increased amount of GFAP (Selvin-Testa et al. 1991; Stoltenburg-Didinger et al. 1996). Astrogliosis is diminished or absent in the developing CNS (Stoltenburg-Didinger et al. 1996).

Astrocytes failed to react to the toxic exposure with an adequate increase of GFAP and GFAP gene transcripts when the intoxication fell into the developmental period of the brain. The majority of astrocytes that have been exposed during their development and differentiation fail to react even if the exposure is continued to adulthood. This suggests an irreversible insult by low level lead exposure during this period of time.

That lead is a developmental toxin is already indicated by studies revealing behavioral impairment in children, including deficits in learning and memory, with blood lead levels as low as 15 μg/dl (Needleman 1990; Goldstein 1992). However, no such alterations related to similarly low blood lead levels are known to occur in adults. On the basis of these studies, the maximally tolerable serum level after lead exposure was reduced to 10 µg/dl by the Centers for Disease Control (1991). Experimental studies have confirmed that the developing nervous system is particularly vulnerable to lead exposure, resulting in irreversible damage.

The adult rats that were exposed to Pb^{2+} during development but removed from the Pb^{2+} exposure at weaning remained impaired in learning the spatial task despite the fact that they no longer possessed elevated levels of blood or brain Pb^{2+} at the time of testing. These data suggest that Pb^{2+}-induced insults occurring early in life have long-lasting cffects. The importance of studying post-natal versus post-weaning exposures is underscored by how differently Pb^{2+} exposure affects developing versus mature neuronal systems (Altman et al. 1993).

Having established that exposure to low levels of Pb^{2+} during development produce cognitive deficits in laboratory animals, the apparent question comes, are we able to identify impairment at the cellular level?

The processing of information and refinement of synaptic connections is thought to occur during development by activity-dependent changes in synaptic strength, such as long-term potentiation (LTP). LTP induced in vivo at the dentate gyrus or CA1 region of the hippocampus has been demonstrated to be impaired by chronic exposure to low levels of Pb^{2+}. Other studies, where the exposure was performed in vivo but LTP was done ex vivo (Altmann et al. 1993) also show significant deficits in the in LTP in the CA1 region in the hippocampus. LTP in the dentate gyrus of the hippocampus is an N-methyl-D-aspartate receptor (NMDAR)-dependent event. The experiment connecting behavioral deficits and impairment of LTP in littermates, has also demonstrated molecular changes in NMDARs of rats exposed to Pb^{2+} (Nihei et al. 2000). This study confirms deficits of behavior and synaptic plasticity and describes significant changes in the gene and protein expression of NMDAR subunits (Nihei et al. 2000). Glutamate is the major excitatory transmitter in the brain and mediates activity dependent processes. Receptor subtypes mediate the actions of glutamate. Distinct genes encode five subunits of NMDA-type ionotropic glutamate receptors. Of these five subunits, presence of NMDAR1 (NR1) is obligatory for functional NMDARs. There are diverse data suggesting that NMDARs in the developing brain are targets for Pb^{2+}-induced neurotoxicity (Nihei and Guilarte 2001). Recent evidence shows that adult rats exposed to 1500 ppm PbAc throughout life possess marked reductions in both gene and protein expression of thc NR1 subunit in the hippocampus (Nihei et al. 2000). The pivotal aspects of this result were that changes in NR1 gene and protein expression detected in littermates exposed to the same amount of Pb^{2+}, also exhibited deficits of spatial learning and LTP. The changes of NR1 subunits in the developing hippocampus from Pb^{2+}-exposed rats have shown the critical nature of the NMDARs in cognitive function and synaptic plasticity and for the mandatory presence in vivo of the NR1 subunit. The reduction of NR1 expression in the hippocampus that is associated with impaired learning and memory, deficits of LTP and exposure to Pb^{2+} , equally support the critical role of the NR1 subunit. These correlations are persuasive in implicating a molecular mechanism for Pb^{2+} -neurotoxicity at the focal point of synaptic plasticity (Nihei and Guilarte 2001).

The cell that controls glutamate concentrations on the synaptic cleft is the astrocyte. The numerous sites of interaction between Pb and metal-handling machinery of astroglia offer many mechanisms for metabolic damage to the cell by Pb. Among them are alterations of enzyme activities (such as stimulation of PKC or inhibition of glutamin synthetase), interference with Ca^{2+} or secondary oxidative stress by intracellular Cu accumulation (Tiffany-Castigliony and Qian 2001). These interactions between a xenobiotic metal and the mechanisms by which cells handle essential metals stress the importance of examining metal interactions when studying the mechanisms of actions of neurotoxic metals.

The preponderance of available evidence supports the concept that astroglia are the major site for the deposition of several neurotoxic metals in the brain. In the case of lead, astroglia appear to be the principal cellular site of accumulation. However, information on the accumulation of toxic metals, particularly its selective accumulation, is incomplete. Among the areas of greatest need for further research are the complete chronological fate of metals accumulated by Astroglia; mechanisms of metal uptake, storage and release; comparisons of differences in metal handling between mature and immature brains and interactions of nonphysiologic toxic metals with essential metals. These investigations could contribute to disease etiologies associated with metals.

Literature

Alfano DP, Petit TL (1982) Neonatal lead exposure alters the dendritic development of hippocampal dentate granule cells. Exp Neurol 75:275-288

Altmann L, Weinsberg F, Sveinsson K, Lilienthal H, Winneke G (1993) Impairment of long-term potentiation and learning following chronic lead exposure. Toxicol Lett 66:105-112

Ascarin l, Carrasco J, Gonzalez B, Hidalgo J, Castellano B (1999) Expression of growth inhibitory factor (metallothionein-III) mRNA and protein following excitotoxic immature brain injury. J Neuropathol Exp Neurol 58:389-397

Aschner M, Conklin DR, Yao CP, Allen JW, Tan KH (1998) Induction of astrocyte metallothioneins (MTs) by zinc confers resistance against the acute cytotoxic effects of methylmercury on cell swelling, Na^+ uptake, and K^+ release. Brain Res 813:254-261

Bellinger D, Levitan A, Watermaux C, Needleman H, Rabinowitz M (1987) Longitudinal analyses of prenatal and postnatal lead exposure and early cognitive development. N Eng J Med 316:1037-1043

Bellinger D, Stiles KM (1993) Epidemiologic approaches to assessing the developmental toxicity of lead. Neurotoxicology 14:151-160

Behse F, Carlsen F (1978) Histology and ultrastructure of alterations in neuropathy. Muscle and Nerve 1:368-374

Bush AI (1989) Metals and neuroscience. Curr Opinion Chem Biol 2004:184-191

Brewer GJ, Cotman GW NMDA receptor regulation in neuronal morphology in cultured hippocampal neurons. Neurosci Lett 99:268-273

Byers RK (1959) Lead poisoning. Review of the literature and report on 45 cases. Pediatrics 23:585

Carrasco J, Giralt M, Molinero A, Penkowa M, Moos T, Hidalgo J (1999) Metallothionein (MT)-III: generation of polyclonal antibodies, comparison with MT-I + II in the freeze lesioned rat brain and in a bioassay with astrocytes, and analysis of Alzheimer's disease brains. J Neurotrauma 16:1115-1129

Centers for Disease Control. Preventing lead poisoning in young children. Atlanta 1991

Chang LW (1977) Neurotoxic effects of mercury - a review. Environ Res 14:329-337

Charleston JS, Body RL, Mottet NK, Vahter ME, Burbacher TM (1995) Autometallographic determination of inorganic mercury distribution in the cortex of the calcarine sulcus of the monkey *Macaca fascicularis* following long-term subclinical exposure to methylmercury and mercuric chloride. Toxicol appl Pharmacol 132:325-333

Charleston JS, Body RL, Bolender RP, Mottet NK, Vahter ME, Burbacher TM (1996) Changes in the number of astrocytes and microglia in the thalamus of the monkey *Macaca fascicularis* following long-term subclinical methylmercury exposure. Neurotoxicology 17:127-138

Collins MF, Hrdina PD, Whittle E, Singhal RL (1982) Lead in blood and brain regions of rats chronically exposed to low doses of the metal. Toxicol Appl Pharmacol 65:341-322

Edwards GN (1865) Two cases of poisoning by mercuric methide. St. Bartholomew's Hospital Reports 18652:141-150

Fjerdinstad EJ, Danscher G, Fjerdingstad E (1974) Hippocampus: selective concentration of lead in the normal rat brain. Brain Res 80:350-356

Goldstein GW (1992) Neurological concepts of lead poisoning in children. Pediatr Ann 21:384-388

Grandjean P, Weihe P, White RF, Debes F, Araki S, Yokoyama K, Murata K, Sørensen N, Dahl R, Jørgensen PJ (1997) Cognitive deficit in 7-year-old children with prenatal exposure to methylmercury. Neurotoxicol Teratol 19:417-428

Hargreaves RJ, Foster JR, Pelling D, Moorhouse SR, Gangolli SD, Rowland JR (1985) Changes in the distribution of of histochemically localized mercury in the CNS and in tissue-levels of organic and inorganic mercury during the development of intoxication in methylmercury treated rats. Neuropathol Appl Neurobiol 11:383-401

Holtzman D, DeVries C, Nguyen H, Olson J, Bensch K (1984) Maturation and resistance to lead encephalopathy: cellular and subcellular mechanisms. Neurotoxicology 5:97-124

Hunter D, Russell DS (1954) Focal cerebellar atrophy in human subject due to organic mercury compounds. J Neurol Neurosurg Psychiatry 17:235-241

Johnston MV, Goldstein GW (1998) Selective vulnerability of the developing brain to lead. Curr Opinion Neurol 11:689-693

Kern M, Wisniewski M, Cabell L, Audesirk G (2000) Inorganic lead and calcium interact positively in activation of calmodulin. Neurotoxicology 21:353-364

Kramer KK, Liu J, Choudhuri S, Klaassen CD (1996) Induction of metollo-thionein mRNA and protein in murine astrocyte cultures. Toxicol Appl Pharmacol 136:94-100

Marsh DO, Myers GJ, Clarkson TW, Amin-Zaki L, Tigriti S, Majeed M (1980) Fetal methylmercury poisoning: Clinical and pathologic features. Ann Neurol 7:348-353

Needleman HL (1990) The future challenge of lead toxicity. Environ Health Perspect 89:85-89

Nihei MK, Desmond NL, McGlothan JL, Kuhlmann AC, Guilarte TR (2000) NMDA receptor subunit changes are associated with Pb^{2+} -induced deficits of LTP and spatial learning. Neuroscience 99:233-242

Nihei MK, Guilarte TR (2001) Molecular changes in glutamatergic synapses induced by Pb2+: Association with deficits of LTP and spatial learning. Neurotoxicology 22:635-643

Noack S, Lilienthal H, Winneke G, Stoltenburg-Didinger G (1996) Immunohistochemical localization of neuronal and glial calcium-binding proteins in hippocampus of chronically low level lead exposed Rhesus monkeys. Neurotoxicology 17:679-684

Norenberg MD, Martinez-Hernandez A (1979) Fine structural localization of glutamine synthetase in astrocytes of rat brain. Brain Res 21:199-202

Peters B, Stoltenburg-Didinger G, Hummel M, Herbst H, Altmann L, Wiegand H (1994) Effects of chronic low level lead exposure on the expression of GFAP and vimentin mRNA in the rat brain hippocampus analysed by in situ hibridization. Neurotoxicology 15:685-693

Phillippe M, Gothard M (1903) Contribution à l'étude de l'origine centrale de la paralyse saturnine. Rev Neurol 11:117

Schiønning JD (2000) Experimental neurotoxicity of mercury autometallographic and stereologic studies on rat spinal root ganglion and spinal cord. APMIS 108:5-32

Schiønning JD, Møller-Madsen B. (1992) allographic detection of mercury in rat spinal cord after treatment with organic mercury. Virchows Arch B: Cell Pathol 61:307-313

Selvin-Testa A, Lopez-Costa JJ, Nessi-de-Avinon AC, Pessi-Saavedra J (1991) Astroglial reactions in rat hippocampus during chronic lead exposure. Glia 4:384-392

Slomianka L, Rungby J, West MJ, Danscher G, Andersen AH (1989) Dose-dependent bimodal effect of low level lead exposure on the developing hippocampal region of the rat: a volumetric study. Neurotoxicology 10:177-190

Stoltenburg-Didinger G, Markwort S (1990) Prenatal methylmercury exposure results in dendritic spine dysgenesis. Neurotoxicol Teratol 12:573-576

Stoltenburg-Didinger G, Pünder I, Peters B, Marcinkowski M, Herbst H, Winneke G, Wiegand H (1996) Glial fibrillary acidic protein and RNA expression in adult rat hippocampus following low-level lead exposure during development. Histochem Cell Biol 105:431-442

Takeuchi T, Matusmoto H, Sasaki M, Kambara T, Shiraishi Y, Hirata Y, Nobuhiro M, Ito H. (1968) Pathology of Minamata disease. Kumamoto Medical Journal 34:521

Thomas JA, Dallenbeck FD, Thomas M (1973). The distribution of radioactive lead (^{210}Pb) in the cerebellum of developing rats. J Pathol 109:45-50

Tiffany-Castiglioni E, Sierra EM, Wu JN, Rowles TK. (1989) Lead toxicity in neuroglia. Neurotoxicology 10:383-410

Tiffany-Castiglioni E, Qian Y. (2001) Astroglia as metal depots: Molecular mechanism for metal accumulation, storage and release. Neurotoxicology 22:577-592

Ujihara H, Albuquerque EX. (1992) Developmental change in the inhibition by lead of NMDA-activated currents of cultured hippocampal neurons. J Pharmacol Exp Ther 263:868-875

Vaguera-Orte J, Cervós-Navarro J, Martin-Giron F, Beccera-Ratia J (1981) Fine structure of the perivascular-limiting membrane. In: Cervós-Navarro J, Fritschka E (eds) Cerebral Microcirculation and Metabolism. Raven Press, New York, pp129-138

Verity MA (1990) Comparative observations on inorganic and organic lead neurotoxicity. Environ Health Perspect 89:43-48

WHO Environmental Health Criteria 101 (1990): Methylmercury. Geneva: World Health Organization

Chapter 15

Actions of metals on membrane channels, calcium homeostasis and synaptic plasticity

D. Büsselberg

Relevance of metals

This report gives a short overview of the actions of different metals on membrane currents, calcium homeostasis and higher neuronal functions like the long-term potentiation. The results focus on experiments conducted over the last decade. Due to the importance of lead, the actions of this metal are emphasized in this report, while the actions of other metals such as mercury, zinc and aluminium as well as the organic metal compounds triethyllead and methylmercury are included for purposes of comparison. The range of metal functions in biological systems is very diverse. While some metals are essential for cell growth and development others, like lead or mercury, are purely toxic. The toxicity of zinc and aluminium is a matter of discussion.

Lead

Lead in the environment

For several thousand years humans have used lead and its compounds. In the last century B.C. large amounts of lead were used in constructing aqueducts of the Roman Empire, and during the regime of the emperor Titus (A.D. 39-81) more than 40000 people were employed for mining of lead. Lead production was increased in Europe after 1320, when black gunpowder became widely used and more lead was produced to develop and build more powerful weapons. After 1420 lead was used in printing and in the 17th century for roofing and even more lead was used with the invention of the rechargeable battery in 1859. The natural emission of lead rising from volcanoes and wood fires is relatively small (2×10^6 kg/year to 18.6×10^6 kg/year; Settle and Peterson 1980; Pacyna 1986) compared to man-made emission of lead (449×10^6 kg/year; Nriagu 1979). In the 1990s about 300000 t of lead were produced per year, but lead production has decreased slightly in Germany over the last decade.

Dust and water are still polluted in several regions of Germany, mostly due to the earlier use of lead-based paints and lead plumbing. The maximum allowable value for lead in German drinking water is 0.04 mg/l (Trinkwasserverordnung 1990). The European standard recently has been reduced to 0.01 mg/l in accordance with the recommendation of the World Health Organization (WHO), which should become German law soon (Mehlhorn 2000). A recent study of lead in drinking water has shown that more than 5% of the samples of drinking water in Berlin (Zietz et al. 2001a) and about 3.1% of the samples from Lower Saxony (Zietz et al. 2001b) exceeded the WHO recommended limit.

Lead in biological systems

Lead has no biological function and it is highly toxic. Nonetheless, lead and some lead compounds (e.g. PbS) have good antiseptic properties and were used more than 600 years ago (Nriagu 1992). Paracelsus (1493-1541) wrote that "salts of lead to the witches brew of plants and animal matter as treatment of virtually all ailments" (cited in Wedeen 1984). In 1766 Goulard described more than 100 therapeutic "uses" of lead and lead compounds.

Although lead has been banned in paints and in gasoline, it is accumulated in biological systems in higher concentrations than in ancient times. In man the lead amount in bone substance is 10 to 100 times higher than in the bones of prehistoric man. Adults absorb 8-10% of the lead present in their food through the gastrointestinal tract; and children absorb up to 50% of the lead intake. In the latter population, this might result in a daily lead intake of up to 170 μg (Michaelson 1980). Until the late sixties, blood-lead concentrations up to 80 μg/100 ml (60 μg/100 ml for children) were allowed by law. In 1987 the WHO fixed 20 μg/100 ml as the maximal allowable blood-lead concentration. Today the maximal blood lead concentration recommended by the WHO is 10 μg/100 ml.

The NHANES II study (Second National Health and Nutrition Examination Survey, published by the National Centre for Health Statistics 1984) has shown that 15% of the population in the USA has blood-lead levels above 10 μg/100 ml. Blood lead levels in Germany have decreased during the last 25 years since, among other factors, leaded gasoline has been taken from the market. The average blood lead concentration of the German population is 3.1 μg/100 ml (Umweltbundesamt 2002; data from 1998).

That lead intake changes behaviour is known for more than 2000 years, since the Greek physician Dioscorides (70 B.C – A.D. 20) wrote, "Lead makes the mind give way" (cited in Needleman 1988). But it took until 1897 until Turner published the first clinical report about lead neurotoxicity in children.

The studies of Needleman et al. have demonstrated that blood-lead concentrations as low as 10 μg/100 ml have clear negative effects on development, memory and behaviour (Needleman 1979; Needelman et al. 1979; Needelman and Bellinger 1991; Pocock et al. 1994). Furthermore a blood-lead concentration of 25 μg/100 ml shortens pregnancy and reduces birth weight (Padich 1985; Bellinger et al. 1987; Dietrich et al.; 1986, Bellinger et al. 1984a,b, 1986).

It has been shown that lead exert several effects at the cellular level. Low blood lead concentrations (20 μg/100 ml) block the sodium-potassium pump (Hernberg et al. 1967) and decouple oxidative phosphorylation (Holtzman and Shen Hsu

1976). In addition, 10-15μg/100ml lead in blood is sufficient to reduce haemoglobin synthesis (Piomelli et al. 1982). Additional deleterious effects of lead include reduced production of 1,25-Dihydroxycalciferol (Rosen et al. 1980) and suppressed cholinergic function (Carroll et al. 1977), including reduced presynaptic release of acetylcholine (Suszkiw et al. 1984) and suppression of nicotinic acetylcholine receptor-activated presynaptic current (Ootgiesen 1990a, b). Furthermore, lead depresses protein kinase C (PKC) activity (Markovac and Goldstein 1988; Murakami et al. 1993).

This list of lead actions is by far not complete, but clearly demonstrates that lead acts at different sites and through different mechanisms. Learning and behavioural deficits, as documented by Needleman and others, point to actions within the nervous system. Fjerdingstad et al. (1974) demonstrated that lead accumulates in the hippocampus, a structure of great importance for learning and memory processes. Especially high concentrations of lead were found in the Fascia dentata and the Hilus (Danscher et al. 1976).

Trimethyllead

Organic lead derivates have been used for decades as a gasoline additive to prevent engines from knocking. While the most common gasoline additives are tetraalkyl derivates (usually ethyl or methyl), these are degraded relatively rapidly in the body to the trialkyl derivates, which are more toxic and more persistent (Grandjean and Nielsen 1979). Triethyllead remains for a long time in nervous tissue (Schwartzwelder 1986). It enhances dopaminergic neuronal processes (Walsh et al. 1986), depresses the formation of microtubuli (Zimmermann et al. 1985) and depresses Na^+/K^+-ATPase (Haeffner et al. 1984). Symptoms of intoxication with tetraethyllead are similar to those resulting from Pb^{2+} intake such as loss of memory and behavioural deficits (Avery et al. 1974).

Aluminium

Aluminium is the third most common element on earth and found in soil, water and air. It is chemically very active and forms numerous compounds, dependent on the elements available and the pH. It enters the body by food intake and especially by the use of antacids. Most of the aluminium intake is in a relatively insoluble form and is not reabsorbed in the intestinal tract. Little is known about the metabolism of aluminium (Alfrey 1989). Aluminium has been demonstrated in all tissues from the day of birth, sometimes in relatively high concentrations (lung: 56.9 ± 63.0 μg/g dry weight; skin: 5.9 ± 3.1 μg/g, grey matter: 2.2 ± 1.3 μg/g, heart: 1.1 ± 0.7 μg/g) (Alfrey et al. 1980; Cooke and Gould 1991).

Not a single process is known that depends on aluminium, but neurotoxicity has been reported (Murray et al. 1991; Strong and Garruto 1991a,b). A rise of blood aluminium levels occurring in renal dialysis recipients has been recognized as a possible cause of dementia (Yuan et al. 1989; Arieff et al. 1979; Sideman and Mannor 1982). Aluminium has also been linked to Alzheimer's disease (Birchall and Chappell 1988a,b, 1989; Trapp et al. 1978; Jacobs et al. 1989; Jansson 2001), although this is a controversial issue. While some researchers think that aluminium is a marker for the syndrome (Shoe and Wyatt 1983); others postulate that it is the trigger (Garruto et al. 1991; Crapper et al. 1973, 1975; McLachlan et al. 1989; Perl and Brody 1980; Clauberg and Joshi 1993). Still others do not believe any relationship exists between aluminium and Alzheimer´s disease (McDermott et al. 1979).

Zinc

Zinc is an essential element for all mammals and has several functions that affect cell metabolism (Constantinidis 1991; Halas et al. 1986; Hesse 1979). It is added to most of the vitamin-mineral food supplements and is also the active compound in several pharmaceutical products.

Research concerning the actions of zinc in the nervous system is diverse and findings in some instances appear to be contradictory. For example, there are reports that even small amounts of zinc are toxic (Choi et el. 1988; Yokoyama et al. 1986), however it has been shown that relatively high concentrations of zinc are located close to cell membranes (Bettger and O´Dell 1981). Especially high concentrations are found in mossy fibres of the hippocampus (Fjerdingstad et al. 1974). Zinc has modulatory effects on the release of several neurotransmitters (Nishimura 1988; Forsythe et al. 1988), modulates the neurotoxicity of excitatory amino acids (Koh and Choi 1988; Weiss et al. 1993) and blocks N-methyl-D-aspartate-activated channels in mammalian neurones (Peters et al. 1987; Christine and Choi 1990). It also alters the activation and inactivation kinetics of potassium- and sodium currents (Gilly and Armstrong 1982a,b) and reversibly blocks calcium entry in synaptosomes (Nachsen 1984).

Mercury

Mercury is used in amalgam tooth fillings (Leistevuo et al. 2001), and is added as an antiseptic to widely-used vaccines (Clarkson 2002).

Mercury is known to accumulate in the nervous system (Arvidson 1992; Möller Madson 1992). In humans mercury intoxication results in impairment of coordination, reduced cognition, tremor production and abnormal reflexes (Albers et al. 1988). Several mechanisms have been described: Interaction with GABA by activation of chloride channels (Arakawa et al. 1991), increased release of neurotransmitters (Manalis and Cooper 1975), reduced activity of brain Na^+/K^+-ATPase (Magour 1987), reduced phosphorylation (Kuznetov et al. 1987; Kuznetov and Richter 1987), modulation of m-RNA metabolism (Carty and Malone 1979), reduced calcium currents in synaptosomes (Atchison et al. 1986; Hewett and Atchison 1992) and the release of calcium from intracellular stores (Atchison and Hare 1994).

Methylmercury

A major incident in the early 1950s in the Japanese Minamata Bay directed attention towards methylmercury. The dumping of large quantities of mercury into the ocean led to bioaccumulation of mercury compounds in the marine food chain and resulted in severe brain damage to persons who had consumed contaminated fish or seafood (Tsubaki and Krukuyama 1977). Methylmercury has become a major health hazard because inorganic mercury is methylated by aquatic microorganisms (Jensen and Jernelov 1969). The main source for humans is coming from contaminated fish (Clarkson 2002).

Methylmercury produces behavioural changes (Atchison et al. 1986; Hewett and Atchison 1990) as well as learning deficits in rats after prenatal exposure (Eccles and Annau 1982a,b; Musch et al. 1978), and visual deficits in monkeys (Rice and Gilbert 1990). It is distributed homogeneously throughout body tissues (Clarkson 1987). Orally administered, it accumulates in parts of the brain and the spinal cord (Möller Madson 1991). At a physiological pH (7.2) and chloride concentration (~ 120 mM) of the extracellular fluid, organic mercury is present as a neutral complex that easily penetrates lipid bilayers (permeability > 10^{-2} cm/s; Gutknecht 1981) and readily crosses the blood-brain barrier.

Methylmercury increases spontaneous neurotransmitter release (Atchison and Narahashi 1982). In various preparations it also increases the intracellular free calcium concentration (Hare et al. 1993; Sarafian 1993). Fluorimetric experiments demonstrated, that inorganic as well as organic forms of mercury depolarise the membrane of synaptosomes (Hare and Atchison 1992).

Interference of metals with cell functions

The results of experiments to be described in the following chapter were obtained over the last decade. The dissociated forms of the inorganic metals are believed to be the active forms at neuronal membranes, therefore effects of Pb^{2+}, Zn^{2+}, Al^{3+}, Hg^{2+} were investigated in our research. In reality the relations in physiological solutions – especially for aluminium – are much more complex. The situation depends on the concentration of other ions, the buffer system and the pH. The complex chemistry of aluminium is described by Martin (1986). In our experiments we have not distinguished between the different forms of aluminium or other metals.

In the following chapters actions of these metals on:

- voltage-gated, and ligand-gated membrane channels,
- calcium release activated channels,
- gap-junctions,
- calcium homeostasis,
- cell-cell communication,
- long term potentiation (LTP)

will be described and compared to several organic forms of these metals (trimethyllead, methylmercury).

Metals on Membrane Channels

Cell-cell communication is based on electric signals, action potentials and chemical processes. Water-filled membrane-spanning proteins form pores (channels) which play critical roles. These channels can directly transfer electrical signals from cell to cell (gap junctions) or use the concentration gradient of ions to move ions, depending on the concentration gradient, across the cell membrane whenever the channels are open. While these channels are relatively specific for single ions, they differ in the mechanisms by which they are activated. They are distinguishable by the following activation factors:

- voltage change,
- receptor-/or ligand binding,
- calcium release from internal stores.

Metals interfere with calcium entry

It has been described that a number of divalent and trivalent cations, including Co^{2+}, Cd^{2+}, La^{3+}, Mn^{2+}, Ni^{2+} and Zn^{2+} interfere with calcium entry into neurons (Hagiwara and Byerly 1981a,b).

Ca^{2+} may enter neurons through at least three types of processes:

- There are several types of voltage dependent calcium channels in the membranes of neurons and other cells, and these channels are opened during depolarising events such as action potentials (Hagiwara and Byerly 1981).
- Calcium can also enter neurons through ion channels opened by neurotransmitter action. Some of these transmitter activated channels are specific for calcium, as demonstrated in Aplysia (Pellmar and Carpenter 1980), while others as that activated by one class of excitatory amino acid receptor (the N-methyl-D-aspartate receptor) in mammalian neurons, are large channels where monovalent cations carry much of the current but with a calcium component (Mayer and Westbrook 1985). These channels are blocked by magnesium. When the membrane depolarises critically, it is not surprising that they are also sensitive to blockade by divalent cations such as Co^{2+}, Cd^{2+} and Zn^{2+} (Hori et al. 1987; Pellmar and Carpenter 1980; Westbrook and Mayer 1987).
- Finally, calcium can also cross cell membranes *via* transport and exchange processes.

Voltage activated membrane channels

Voltage-activated channels are most important gates through which calcium enters neurons. In 1987, Tsien et al.(Fox et al. 1987a,b Tsien et al. 1987a) discovered three types of voltage-activated calcium channels in chick dorsal root ganglion neurons: a low voltage-activated, rapidly inactivating current (T); a high voltage-activated, slowly inactivating current (L-type); and a high voltage-activated current with intermediate activating kinetics (N-type). Subsequently, other types of channels, like the high voltage-activated P channel (Mintz et al. 1992a,b) or the Q- and R-type (Neelands at al. 2000) have been described in a variety of neurons.

In our studies (Büsselberg and colleagues), we set out to determine the actions of metals on the inactivating component of the current and the sustained current. All metals tested (Pb^{2+}, Zn^{2+}, Al^{3+}, Hg^{2+}, methylmercury and trimethyllead) blocked transient and sustained components of *voltage gated calcium channel* currents in *Aplysia* abdominal ganglion neurons (Büsselberg et al. 1990 a, b 1991 a, b, 1993 a, Pekel et al. 1993) and in cultured dorsal root ganglion (DRG) cells of rats (Büsselberg et al. 1992, 1993a,b; 1994a,b,c, 1995, 1996, 1998; Evans et al. 1991; Gawrisch et al. 1997; Leonhardt et al. 1996a,b,c; Pekel et al. 1993; Platt et al. 1993; Platt and Büsselberg 1994a,b). The voltage-gated channels were opened by depolarising voltage jumps from a holding potential of –80 mV in DRG neurons and from a holding potential of –40 mV in Aplysia neurons. The metals differed in threshold, reversibility and specificity and on actions on the different voltage activated channels.

Pb^{2+} was most effective in reducing the *voltage-activated calcium channel currents of mammalian DRG neurons.* Threshold concentration was <0.1µM, the threshold for methylmercury was 2.6 µM, for Zn^{2+} action < 5 µM, for Al^{3+} ~ 20 µM, for Hg^{2+} ~ 0.2 µM, for trimethlyllead ~ 0.5 µM. Total blockade (> 80%) was obtained with concentrations > 1 µM Pb^{2+}, ~ 2 µM Hg^{2+}, ~ 10 µM methylmercury, ~ 50 µM trimethyllead and 150-200 µM Zn^{2+} or Al^{3+}. Half of the current was blocked with 0.6 µM Pb^{2+}, 1.1 µM Hg^{2+}, 2.6 µM methylmercury, 1-5 µM trimethyllead, 69 µM Zn^{2+}, and 84 µM Al^{3+}. The Hill slope for Pb^{2+}, Zn^{2+} Hg^{2+} and methylmercury was around 1, whereas for Al^{3+} it was close to 3.

The actions of Al^{3+}, methylmercury and trimethyllead in blocking voltage-gated calcium channels clearly needed an open channel state for their action (use-dependent), whereas this was not the case for either Pb^{2+}, Hg^{2+} or Zn^{2+}.

The blockade of voltage-gated Ca^{2+} channels by the different metals was totally reversible. The best recovery was obtained upon wash after exposure to Pb^{2+} ($\geq$ 60%), some recovery of the blockade was seen with Zn^{2+} ($\geq$ 50%), whereas there was little or no recovery after application of Al^{3+}, methylmercury, trimethyllead or with Hg^{2+}.

With Zn^{2+}, Al^{3+}, methylmercury or Hg^{2+} in the extracellular solution the current-voltage relation of these channel currents was shifted to depolarised voltages. The degree of the shift was a function of concentration, but differed from cell to cell and was most likely due to the charge-screening actions of these metal ions.

The actions of Pb^{2+} and Al^{3+} were found to be relatively specific for voltage-activated calcium channel currents. Concentrations which blocked the voltage-activated calcium channel current by ≥ 80% (1 μM Pb^{2+} or 200 μM Al^{3+}) had little effect (≤ 20%) on voltage-activated potassium or sodium channel currents. The effects of the other metals used in our studies were less specific. In particular, methylmercury had clear effects on voltage-activated potassium currents (IC_{50} = 2.2 μM) as well as on voltage activated sodium channels (IC_{50}=12.3 μM).

The effect of aluminium was highly pH dependent (Platt et al. 1993). At pH 7.3-7.8, the concentration-response curve shifted slightly to higher concentrations, whereas at pH 6.4-6.9 a pronounced shift to lower concentrations was observed. This might be due to changes in the chemical equilibrium of different aluminium species in the extracellular solution.

When two of three cations (Pb^{2+}, Zn^{2+} or Al^{3+}) were applied simultaneously at the concentration of their IC_{50} values, repetitive activation of the voltage-activated calcium channel a 75% ± 9% block of the control current was recorded (Platt and Büsselberg 1994 a). This observation is consistent with an action of two metals at the same site as well as independent actions at different locations of the ion channel.

With increasing concentrations of Hg^{2+} (> 2 μM) or methylmercury (~ 20 μM) a slow membrane current was additionally activated (Pekel et al. 1993; Leonhardt et al. 1996a,b,c). This current was irreversible and might be due to opening of other (non-specific) ion channels by mercury or methylmercury.

We have also investigated the effects of Pb^{2+} on voltage activated calcium channels of *Aplysia californica* (Büsselberg et al. 1990a,b, 1991a,b, 1993a; Pekel et al. 1993). Pb^{2+} was a potent blocker of calcium channel currents at concentrations that did not significantly affect potassium and sodium channel currents. The blockade was concentration-dependent and the percentage of blockade was reduced when the concentration of the charge carrier was elevated. The threshold concentration was about 1 μM, and the Hill coefficient was ~ 1 under all conditions. Pb^{2+} did not significantly change inactivation, but shifted the voltage dependence of activation to hyperpolarized voltages in a dose-dependent manner. The blockade of calcium currents was highly voltage-dependent and increased with depolarisation.

Ligand- or receptor gated membrane channels

A possible site of action for metals is the glutamatergic system of the CNS. The excitatory amino acid glutamate is important to processes such as learning and memory, development and synaptic plasticity, as well as being critically involved in pathological functions such as hypoxia or ischaemia (Weahl and Thomson 1991; Zorumski and Yang 1988). Whereas activation of the α-amino-3-hydroxy-5-methylisoxazole-4-proprionic acid (AMPA) receptor subtype mediates the fast excitatory postsynaptic potential (EPSP), the *N*-methyl-D-aspartate (NMDA) receptor is involved in certain types of neuronal plasticity. Through the NMDA receptor channel complex, calcium enters neurons. Furthermore, the NMDA receptor has modulatory binding sites for Mg^{2+} and Zn^{2+}. A variety of studies have provided evidence for an interaction of the NMDA receptor channel complex with metal cations (Hubbarth et al. 1989; Ascher and Nowak 1988; Uteshev et al. 1993; Ujihara and Albuquerque 1992a,b).

In our experiments, receptor-activated currents in acutely dissociated rat hippocampal neurons were elicited by aspartate and glycine. Lead blocked these NMDA currents in both fast, reversible and slow, irreversible manners (Büsselberg and Platt 1994c; Uteshev et al. 1993). When Pb^{2+} was applied simultaneously with the agonists, the inward currents were rapidly and reversibly reduced in a dose-dependent manner with a minimum effective concentration below 2μM and a total blockade (> 80%) with 100 μM Pb^{2+}. The half maximal concentration was about 45 μM and the Hill coefficient 1.1. Pre-incubation with 50 μM Pb^{2+} resulted in a greater reduction in the response to the agonists with Pb^{2+}. This effect was reversed within 2-5 s after applying a lead free solution. There was a lack of voltage dependence, which suggests that Pb^{2+} does not block the channel, but rather alters the binding of agonists. Prolonged superfusion of neurons with glutamate receptor agonist- and divalent lead-containing external solution resulted in a slow and irreversible decrease of the receptor activated current. No clear threshold concentration was found for the slow and irreversible effect of Pb^{2+}.

Aluminium reduced NMDA, AMPA and glutamate mediated currents by 50% at a concentration of 1.4 μg/ml (Platt and Büsselberg 1994b; Platt et al. 1994). Higher concentrations (≥ 2.7 μg/ml) inhibited the current completely and irreversibly. Examination of the concentration-response relationship for the action of aluminium on NMDA-activated currents revealed a threshold concentration < 0.27 μg/ml.

These data indicate that glutamate receptors are putative sites of action for lead and aluminium. A metal-induced reduction of the channel current will therefore also decrease the calcium rise in the neurons after activation of the receptors.

Calcium release-activated calcium channels (CRAC)

In non-excitable cells, the refilling of intracellular calcium stores (ICS) depends on a voltage-independent current through calcium release-activated channels in the cell membrane (Hoth and Penner 1992). This current is activated by an unknown signal, possibly released from depleted ICS. Excitable cells on the other hand are thought to refill ICS through voltage-dependent calcium channels (Bode and Netter 1996, Razani-Boroujerdi et al. 1994), although some excitable cell types (PC12, GH3) (Razani-Boroujerdi et al. 1994, Villalobos and Garcia-Sancho 1995) and neurons of the developing retina have been shown to express CRAC channels (Sakaki et al. 1997).

Since Pb^{2+} accumulates in bone, from where it is released during bone resorption, thus leading to high local concentrations of lead, we examined the effects of this heavy metal on osteoblast-like (OBL) cells. Five to 12.5 μM Pb^{2+} applied simultaneously with re-added Ca^{2+} reduced the immediate CRAC of OBL cells to 70% or 30% of control value, respectively (Wiemann et al. 1999). An exaggerated influx of Pb^{2+} occurred during CRAC activation led to a 2.7-fold faster increase of the excitation ratio. Inhibitory effects of Pb^{2+} on Ca^{2+} ATPase activity did not contribute to these effects. CRAC channels of OBL cells are blocked as well as permeated by Pb^{+2}.

Gap junctions

Gap junctions are key elements for signal transfer in a variety of different cell types. Gap junctions in bone cells are controlled e.g. by hormones and second messengers (Schirrmacher and Bingmann 1998; Donahue et al. 1995; Xia and Ferrier 1992), whose complex actions might be impaired by heavy metals. Cx43 channels in OBL cells are sensitive to $[Ca^{2+}]_i$ and close with rising $[Ca^{2+}]_i$ (Schirrmacher et al. 1996). Since heavy metals compete with Ca^{2+} for specific binding sites, we investigated whether the coupling might be sensitive to heavy metal cations.

Pb^{2+}, Hg^{2+} and methylmercury were applied extracellularly (e) and intracellularly (i) to OBL cells. The coupling between two neighbouring cells was calculated by dividing the change of membrane potential in the first cell (ΔMP_2) by the change of the membrane potential of the second cell (ΔMP_1) after application of a known current injection (Schirrmacher et al. 1998). $Pb^{2+}_{(e)}$ (5 µM) and $Hg^{2+}_{(e)}$ (5 µM) as well as $Pb^{2+}_{(i)}$ (25 µM) did not change the coupling ($\Delta MP_2/\Delta MP_1$). In contrast, methylmercury (1-10µM) extracellularly applied, as well as $Hg_{(i)}$ ($\geq$ 5 µM) reduced the coupling to 79.5 ± 19.3% and 62.4 ± 15.3%, respectively within 15-20 minutes. The reduction of coupling followed individual time courses, and in no case was a steady state of decoupling reached within 20 minutes.

In conclusion, $Pb^{2+}_{(e)}$, $Pb^{2+}_{(i)}$ and $Hg^{2+}_{(e)}$ do not affect gap junctional coupling per se. Since methylmercury (e) and $Hg^{2+}_{(i)}$ deplete calcium stores, the decrease of the electric coupling after the application of these cations is attributed to an increase of $[Ca^{2+}]_I$, which affects gap junction channels.

Other channels involved in lead toxicity

The neurotoxicity of lead and other divalent and trivalent cations on voltage- or receptor gated membrane channels is widely accepted and has been underlined by research of other scientists (e.g. Audesirk 1987; Shafer 1998). There are several other membrane channels that are affected by lead toxicity. Different types of potassium channels have been shown to be sensitive to lead (Madeja et al. 1997, Dai et al. 2001). Furthermore, ACh receptor- mediated inward currents are suppressed by lead (1 nM - 200 µM) (Zwart et al. 1997) and GABAergic transmission in hippocampal neurons is inhibited by nanomolar concentrations of lead (Braga et al. 1999).

Metals and calcium homeostasis

Neuronal signalling depends in part on intracellular calcium concentrations, on calcium entry through voltage activated calcium channels (VACC) and on calcium elevation in pre-synaptic terminals of neurons. It seems evident therefore that the calcium concentration in neurons, calcium entry and calcium release from internal stores need to be tightly regulated.

We measured calcium entry in cultured dorsal root ganglion neurons by depolarzing the cells with 50 mM potassium and used fura-2 as a calcium indicator (Domann et al. 1997). *Lead* added to the extracellular solution reduced the rise of $[Ca^{2+}]_i$ in a concentration-dependent manner, with a threshold concentration of 0.25 μM. More than 80% of calcium entry was prevented by ~ 5 μM Pb^{2+}. The IC_{50} and the Hill coefficient were 1.3 μM and 1, respectively. This effect was considered to result from block of voltage-activated calcium channels, since applications of NMDA did not result in a rise of intracellular calcium.

In osteoblast-like cells (Wiemann et al. 1999), extracellular application of Pb^{2+} (5 μM) for 20 minutes linearly elevated the fura emission ratio, reflecting transmembrane lead permeation rather than an increase in calcium (note, that lead entry was not seen in DRG neurons). Mercury increased the intracellular calcium concentration from 100 to ~ 200 nmol/l whereas methylmercury (5 μM) released calcium from internal stores, thus increasing calcium up to 2 μmol/l.

Lead and aluminium actions on synaptic plasticity

Long-term-potentiation (LTP) (Bliss and Lomo 1973) is a long lasting elevation of synaptic efficiency and a good electrophysiological model for learning and memory (Teyler and Discenna 1987). In most, but not in all preparations, the generation of LTP was prevented by removing extracellular calcium, by blocking calcium entry into the cells e. g. by blocking the NMDA receptor channel complex during *in vitro* (Harris et al. 1984; Zalutsky and Nicoll 1990) as well as *in vivo* experiments (Morris et al. 1986).

Since NMDA receptors are modulated by zinc or magnesium, Hori et al. (1993) tested the hypothesis that lead or aluminium changes the generation or maintenance of LTP, possibly by actions at the NMDA receptor. Lead in concentrations up to 10 μM had no effect on the synaptic response elicited by stimulation of the lateral olfactory tract in piriform cortex pyramidal neurons of a rat brain slice preparation. On the other hand lead blocked LTP by about 75% at a concentration of 5μM and completely at 10 μM. At these concentrations, Pb^{2+} had no effect on post-tetanic potentiation. Surprisingly, in the concentration range studied, there was no effect of Pb^{2+} on NMDA response. Similar results were reported by Altmann et al.(1991, 1993).

Aluminium at a concentration 0.27 μg/ml did not alter induction and maintenance of LTP in our experiments (Platt et al. 1995), however, 0.68 μg/ml resulted in a smaller increase in the amplitude of the population spike. The duration of LTP was not affected. At a concentration of 2.7 μg/ml aluminium, the population spike was further reduced and LTP declined to baseline within two hours.

Our results do not allow identification of the mechanisms through which LTP is reduced by lead or aluminium. Since post-tetanic potentiation and therefore synaptic transmission is not changed at metal concentrations that block LTP, presynaptic channels such as the VACCs are most likely not crucial for this process. While in some systems postsynaptic channels such as the NMDA receptor-activated types are important, the concentrations of the metals found to block these channels are higher than those required to reduce LTP. A possible candidate mechanism involves PKC. Injection of PKC in pyramidal neurons elicits responses that are consistent with LTP (Hu et al. 1987), while inhibitors of PKC prevent LTP (Reymann et al. 1988). The generation of LTP in the hippocampus is associated with membrane translocalisation of PKC (Akers et al. 1986), which also has been shown by the classical conditioning in the hippocampus (Banks et al. 1988). PKC possibly could be activated by very low threshold concentrations of lead (10^{-13}M; Markovac and Goldstein 1988, Sun et al. 1999) or inhibited by concentrations between 5 μM and 10 μM (Murakami et al. 1993), which are within the range of lead concentrations, which block membrane currents in our experiments. Wehner et al.(1990) demonstrated that PKC activity was reduced in mice with an impaired spatial learning, while others reported that developmental exposure of rats to lead decreased PKC-gamma, but not PKC-alpha, -beta II, and –epsilon (Chen et al. 1999; Nihei et al. 2001). The suppression of PKC could result in a reduction of calcium channel currents (Doerner et al. 1988; Doener and Alger 1992).

The lead concentrations needed to reduce LTP in our experiments are clearly higher than the blood lead concentrations that result in long lasting learning deficits. However, it is very difficult to compare the lead concentrations of our solutions with blood lead levels. We do not know how much of the added lead was in an "active" Pb^{2+} form. Furthermore, blood lead concentrations do not reflect the lead concentrations in the cerebrospinal fluid or extracellular space surrounding nerve cells. In addition, different "forms" of lead might be involved, depending on the pH and the co-existence of other metal ions (Matthews et al. 1993).

Relevance of the data

To obtain the metal burdens of the population, blood concentrations are measured. However, 99.2% of the lead in blood is bound to erythrocytes while 0.8% is present in the serum. Al-Modhefer and colleagues (1991) calculated a relation of complex bound lead and "free" lead of 5200 : 1, taking these values into consideration and assuming a blood lead concentration of 10 μg/100ml (Simons 1993), a concentration of 10^{-13}M in the extracellular fluid is calculated. Their conclusion, that extracellular mechanisms such as a block of membrane channels have not to be considered for the toxicity of the metals, seems to be incorrect. The actual free lead concentration will be much lower since in all experiments the added lead was used to calculate the final lead concentration but in the preparations as well as in the solutions lead will also be complex-bound.

The question, how the variety of actions of the different metal species could be explained cannot finally be answered. But the following possible mechanisms have to be considered:

- Chemical peculiarities of various metal species (especially in the different intra- and extracellular environments).
- Unspecific effects at the cell membrane (e.g. screening of the membrane surface charge "charge screening").
- Specific effects (binding) to proteins at the surface of the cell membrane.
- Specific effects (binding) at the entrance of channel proteins (likely to be one possible mechanisms for divalent metal cations which do not need an open channel state for their action, like Pb^{2+} or Zn^{2+}).
- Specific effects within the channel (a possible mechanism for metals which need an open channel state of action, like Hg^{2+} or MeHg).
- Intracellular changes (for this action the metal has to pass the cell membrane, which is likely to occur for most (uncharged) "organic"-metal compounds and some cations (as it has been demonstrated for lead in osteoblast like cells).

Unfortunately all of these different actions could occur at a different time scale or simultaneously. Furthermore it has to be considered which action occurs first and which effect has the highest "efficiency". Therefore it is extremely difficult to judge the toxicity of specific metal ions or metal components.

General Implications

In the yearly report of the New York State Department of Health (1993) it is stated that more patients were treated for lead toxicity than for alcoholic toxicity. The individual as well as the personal harm due to lead poisoning is immense. Frank (1991) tried to calculate the possible costs of acute lead poisoning in a seven-year-old boy. The child was hospitalised with a blood lead concentration of 70 μg/100 ml for four weeks. Due to the lead the intelligence quotient (IQ) was reduced by 18 points. The cost of hospitalization was \$40.000, the loss in potential salary because of the reduction in IQ was estimated to be \$190.000 and the reduction of life quality was assumed to be \$70.000. This amounts to a total sum of \$300.000 from a single case.

In Western Europe the lead burden has dropped and is no longer considered to be a major problem. But in under-developed countries, especially in the big cities lead concentrations are significantly higher. A report of the World Bank reports blood lead concentrations of 18.5 μg/100 ml in the population of Bangalore, 40 μg/100 ml in Bangkok (1990), 23.3 μg/100 ml in Manila (1994) and 30 μg/100 ml in Cairo.

Literature

Akers RF, Lovinger DM, Colley PA, Linden DJ, Routten-Berg A (1986) Translocation of protein kinase C activity may mediate hippocampal long term potentiation. Science 231:587-589

Albers JW, Kallenbach LR, Fine LJ, Langwolf GD, Wolfe RA, Donogrio PD, Alessi AG, Stolp-Smith KA, Bromberg MB (1988) Neurological abnormalities associated with remote occupational element mercury exposure. Ann Neurol 5:651-659

Alfrey AC, Hegg A, Craswell P (1980) Metabolism and toxicity of aluminum in renal failure. Am J Clin Nutr 33:1509-1516

Alfrey AC (1989) Physiology of aluminum in man. In: Gitelman HJ (ed) Aluminum and Health - A critical review, Dekker, New York, pp 101-124

Alkondon M, Costa AC, Radhakrishnan V, Aronstan RS, Albuquerque EX (1990) Selective blockade of NMDA-activated channel currents may be implicated in learning deficits caused by lead. FEBS Lett 261:124-130

Al-Modhefer AJA, Bradbury MWB, Simons TJB (1991) Observations on the chemical nature of lead in human blood serum. Clin Sci 81:823-829

Altmann L, Sveinsson K, Wiegand H (1991) Long-term potentiation in rat hippocampal slices is impaired following acute lead perfusion. Neurosci Lett 128:109-112

Altmann L, Weinsberg F, Sveinsson K, Lilienthal H, Wiegand H, Winneke G (1993) Impairment of long-term potentiation and learning following chronic lead exposure. Toxicol Lett 66:105-112

Arakawa O, Nakahiro M, Narahashi T (1991) Mercury modulation of GABA-activated chloride channels and non-specific cation channels in rat dorsal root ganglion neurons. Brain Res 551:58-63

Arieff AI, Cooper JD, Armstrong D, Lazarowitz VC (1979) Dementia, renal failure, and brain aluminum. Ann Intern Med 90:741-747

Arvidson B (1992) Accumulation of inorganic mercury in lower motorneurons of mice. Neurotoxicology 13:277-280

Ascher P, Nowak L (1988) The role of divalent cations in the N-methyl-D-aspartate responses of mouse central neurons in culture. J Physiol Lond 399:247-266

Atchison WD, Narahashi T (1982) Methylmercury-induced depression of neuromuscular transmission in the rat. Neurotoxicology 3(3):37-50

Atchison WD, Joshi U, Thornburg JE (1986) Irreversible suppression of calcium entry into nerve terminals by methylmercury. J Pharmacol Exp Ther 238:618-624

Atchison WD, Hare MF (1994) Mechanisms of methylmercury-induced neurotoxicity. FASEB J 8(9):622-629

Audesirk G (1987) Effects of in vitro and in vivo lead exposure on voltage-dependent calcium channels in central neurons of Lymnaea stagnalis. Neurotoxicology. 8:579-592

Avery DD, Cross HA, Schröder T (1974) The effects of tetraethyl lead on behavior in the rat. Pharmacol Biochem Behav 4:473-479

Banks B, Deweer A, Kusian, AM, Rusmussen H, Alkom DL (1988) Classical conditioning induces long-term translocation of protein kinase C in rabbit hippocampal CA1 cells. Proc Natl Acad Sci USA 85:1988-1992

Bellinger D, Leviton A, Needleman HL, Waternaux C, Rabinowitz M (1986) Low-level lead exposure and infant development in the first year. Neurobehav Toxicol Teratol 8:151-161

Bellinger D, Liviton A, Waternaux C, Needleman HL, Rabinowitz M (1987) Longitudinal analyses of prenatal and postnatal lead exposure and early cognitive development. N Engl J Med 316:1037-1043

Bellinger DC, Needleman HL, Leviton A, Waternaux C, Rabinowitz MB, Nichols ML (1984) Early sensory-motor development and prenatal exposure to lead. Neurobehav Toxicol Teratol 6:387-402

Bettger WJ, O'Dell BL (1981) A critical physiological role of zinc in the structure and function of biomembranes. Life Sci 28:1425-1438

Birchall JD, Chappell JS (1988a) The chemistry of aluminum and silicon in relation to Alzheimer's disease. Clin Chem 34:265-267

Birchall JD, Chappell JS (1988b) Aluminum, chemical physiology, and Alzheimer's disease. Lancet 2:1008-1010

Birchall JD, Chappell JS (1989) Aluminium, water chemistry, and Alzheimer's disease [letter]. Lancet 1:953

Bliss TV, Lomo T (1973) Long-lasting potentiation of synaptic transmission in the dentate area of the anaesthetized rabbit following stimulation of the perforant path. J Physiol 232:331-356

Bode HP, Netter KJ (1996) Agonist-releasable intracellular calcium stores and the phenomenon of store-dependent calcium entry. A novel hypothesis based on calcium stores in organelles of the endo- and exocytotic apparatus. Biochem Pharmacol 51:993-1001

Braga MF, Pereira EF, Albuquerque EX (1999) Nanomolar concentrations of lead inhibit glutamatergic and GABAergic transmission in hippocampal neurons. Brain Res 826:22-34

Büsselberg D, Evans ML, Rahamnn H, Carpenter DO (1990a) Zn^{2+} blocks the voltage activated calcium current of Aplysia neurons. Neurosci Lett 117:117-122

Büsselberg D, Evans ML, Rahmann H, Carpenter DO (1990b) Lead inhibits the voltage-activated calcium current of Aplysia neurons. Toxicol Lett 51:51-57

Büsselberg D, Evans, ML, Rahmann H, Carpenter DO (1991a) Effects of inorganic and triethyl lead and inorganic mercury on the voltage activated calcium channel of Aplysia neurons. Neurotoxicology 12:733-744

Büsselberg D, Evans ML, Rahmann H, Carpenter DO (1991b) Lead and zinc block a voltage-activated calcium channel of Aplysia neurons. J Neurophysiol 65:786-795

Büsselberg D, Michael D, Evans ML, Carpenter DO, Haas HL (1992) Zinc (Zn^{2+}) blocks voltage gated calcium channels in cultured rat dorsal root ganglion cells. Brain Res 593:77-81

Büsselberg D, Evans ML, Haas HL, Carpenter DO (1993 a) Blockade of mammalian and invertebrate calcium channels by lead. Neurotoxicology 14:249-258

Büsselberg D, Platt B, Haas HL, Carpenter DO (1993b) Voltage-gated calcium channel currents or rat dorsal root ganglion (DRG) are blocked by Al^{3+}. Brain Res 622:163-168

Büsselberg D, Platt B, Michael D, Haas HL, Carpenter DO (1994a) Mammalian voltage-activated calcium channel currents are blocked by Pb^{2+}, Zn^{2+} and Al^{3+}. J Neurophysiol 71:1491-1497

Büsselberg D, Pekel M, Michael D, Platt B (1994b) Mercury (Hg^{2+}) and zinc (Zn^{2+}): Two divalent cations with different actions at voltage activated calcium channel currents. Cell Molec Neurobiol 14:675-687

Büsselberg D, Platt B (1994c) Pb^{2+} reduces voltage- and NMDA-activated calcium channel currents. Cell Molec Neurobiol 14:711-722

Büsselberg D (1995) Calcium channels as target sites of heavy metals. Toxicol Lett 82/83:255-262

Büsselberg D (1996) Actions of lead on neuronal membrane currents and synaptic plasticity. Metal Ions in Biology and Medicine 4:241-243

Büsselberg D, Schirrmacher K, Domann R, Wiemann M (1998) Lead interferes with calcium entry through membrane pores. Fresenius J Anal Chem 361: 372-376

Carroll PT, Silbergeld EK, Goldberg AM (1977) Alteration of central cholinergic function by chronic lead acetate exposure. Biochem Pharmacol 26:397-402

Carty AJ, Malone SF (1979) The chemistry of mercury in biological system. In: Nriagu JO (ed) The Biogeochemistry of Mercury in the Enviroment, New York, Elsevier Biomedical Press pp 433-480.

Christine CW, Choi DW (1990) Effect of zinc on NMDA receptor-mediated channel currents in cortical neurons. J Neurosci 10:108-116

Clarkson TW (2002) The three modern faces of mercury. Environ Health Perspect 1:11-23

Clauberg M, Joshi JG (1993) Regulation of serine protease activity by aluminum: implications for Alzheimer disease. Proc Natl Acad Sci USA 90:1009-1012

Chen HH, Ma T, Ho IK (1999) Protein kinase C in rat brain is altered by developmental lead exposure. Neurochem Res 24:415-421

Choi DW, Yokoyama M, Koh J (1988) Zinc neurotoxicity in cortical cell culture. Neuroscience 24:67-79

Clarkson TW (1987) Metal toxicity in the central nervous system. Environ Health Perspect 75:59-64

Constantinidis J (1991) The hypothesis of zinc deficiency in the pathogenesis of neurofibrillary tangles. Med Hypotheses 35:319-323

Cooke K, Gould MH (1991) The health effects of aluminium - a review. J R Soc Health 111:163-168

Crapper DR, Krishnan SS, Dalton AJ (1973) Brain aluminum distribution in Alzheimer's disease and experimental neurofibrillary degeneration. Science 180: 511-513

Crapper DR, Krishnan SS, DeBoni U, Tomko GJ (1975) Aluminum: a possible neurotoxic agent in Alzheimer's disease. Trans Am Neurol Assoc 100:154-156

Dai X, Ruan D, Chen j, Wang M, Cai L (2001) The effects of lead on transient outward currents of acutely dissociated rat dorsal root ganglia. Brain Res 904:327-340

Danscher G, Fjerdingstad EL, Fjerdingstad E, Fredens K (1976) Heavy metal content in subdivisions of the rat hippocampus (zinc, lead and copper). Brain Res 112:442-446

Dietrich KN, Krafft KM, Bornschein RL, Hammond PB, Berger O, Succop PA, Bier M (1987) Low-level fetal exposure effect on neurobehavioral development in early infancy. Pediatrics 80:721-730

Doerner D, Pitler TA, Alger BE (1988) Protein kinase C activators block specific calcium and potassium current components in isolated hippocampal neurons. J Neurosci 8:4069-4078

Doerner D, Alger BE (1992) Evidence for hippocampal calcium channel regulation by PKC based on comparison of diacylglycerols and phorbol esters. Brain Res 597:30-40

Domann R, Wunder l, Büsselberg D (1997) Lead reduces calcium entry after depolarisation without passing the cell membrane in cultured DRG neurons: FURA-1 measurements. Cell Molec Neurobiol 17/3:305-314

Donahue HJ, McLeod KJ, Rubin CT, Andersen J, Grine EA, Hertzberg EL, Brink PR (1995) Cell-to-cell communication in osteoblastic networks: cell line-dependent hormonal regulation of gap junction function. J Bone Miner Res10:881-889

Eccles CU, Annau Z (1982a) Prenatal methylmercury exposure: II. Alterations in learning and psychotropic drug sensitivity in adult offspring. Neurobehav Toxicol Teratol 4:377-382

Eccles CU, Annau Z (1982b) Prenatal methylmercury exposure: I. Alterations in neonatal activity. Neurobehav Toxicol Teratol 4:371-376

Evans ML, Büsselberg D, Carpenter DO (1991) Pb^{2+} blocks calcium currents of cultured dorsal root ganglion cells. Neurosci Lett 129:103-106

Fjerdingstad EJ, Danscher G, Fjerdingstad E (1974a) Hippocampus: selective concentration of lead in the normal rat brain. Brain Res 80:350-354

Fjerdingstad E, Danscher G, Fjerdingstad EJ (1974b) Zinc content in hippocampus and whole brain of normal rats. Brain Res 79:338-342

Forsythe ID, Westbrook GL, Mayer ML (1988) Modulation of excitatory synaptic transmission by glycine and zinc in cultures of mouse hippocampal neurons. J Neurosci 8:3733-3741

Fox AP, Nowycky MC, Tsien RW (1987a) Single-channel recordings of three types of calcium channels in chick sensory neurons. J Physiol Lond 394:173-200

Fox AP, Nowycky MC, Tsien RW (1987b) Kinetic and pharmacological properties distinguishing three types of calcium currents in chick sensory neurones. J Physiol Lond 394:149-172

Frank RG (1991) Economic aspects of the ligitimation for harm due to lead poisoning. In: Needleman HL (ed) Human lead exposure., 259-266

Garruto RM, Strong MJ, Yanagihara R (1991) Experimental models of aluminum-induced motor neuron degeneration. Adv Neurol 56:327-340

Gawrisch E, Leonhardt R, Büsselberg D (1997) Voltage-activated calcium channel currents of rat dorsal root ganglion cells are reduced by trimethyllead. Tox Lett 92:117-122

Gilly WF, Aramstrong CM (1982a) Divalent cations and the activation kinetics of potassium channels in squid giant axons. J Gen Physiol 79:965-996

Gilly WF, Aramstrong CM (1982b) Slowing of sodium channel opening kinetics in squid axon by extracellular zinc. J Gen Physiol 79: 935-964

Grandjean P, Nielson T (1979) Organolead compounds: environmental health aspects. Residue Rev 72:97-148

Gutknecht J (1981) Inorganic mercury transport through lipid bilayer membranes. J Membr Biol 61:61-66

Haeffner EW, Zimmermann HP, Hoffmann CJ (1984) Influence of triethyl lead on the activity of enzymes of the ascites tumor cell plasma membrane and its microviscosity. Toxicol Lett 23:183-188

Hagiwara S, Byerly L (1981a) Calcium channel. Ann Rev Neurosci 4:69-125

Hagiwara S, Byerly L (1981b) Membrane biophysics of calcium currents. Fed Proc 40:2220-2225

Halas ES, Hunt CD, Eberhardt MJ (1986) Learning and memory disabilities in young adult rats from mildly zinc deficient dams. Physiol Behav 37:451-458

Hare MF, Atchison WD (1992) Comparative action of methylmercury and divalent inorganic mercury on nerve terminal and intraterminal mitochondrial membrane potentials. J Pharmacol Exp Ther 261:166-172

Hare MF, McGinnis KM, Atchison WD (1993) Methylmercury increases intracellular concentrations of Ca^{2+} and heavy metals in NG108-15 cells. J Pharmacol Exp Ther 266:1626-1635

Harris EW, Ganong AH, Cotman CW (1984) Longterm potentiation in the hippocampus involves activation of N-methyl-D-aspartate receptors. Brain Res 323: 132-137

Hernberg S, Viekko E, Hasan L (1967) Red cell membrane ATPase in workers exposed to inorganic lead. Arch Environ Health 14:319-324

Hesse GW (1979) Chronic zinc deficiency alters neuronal function of hippocampal mossy fibers. Science 205:1005-1007

Hewett SJ, Atchison WD (1992) Effects of charge and lipophilicity on mercurial-induced reduction of $45Ca^{2+}$ uptake in isolated nerve terminals of the rat. Toxicol Appl Pharmacol 113:267-273

Holtzman D, Hsu JS (1976) Early effects of inorganic lead on immature rat brain mitochondrial respiration. Pediatr Res 10:70-75

Hori N, Büsselberg D, Matthews MR, Parsons PJ, Carpenter DO (1993) Lead blocks LTP by an action not at NMDA receptors. Exp Neurol 119:192-197

Hoth M, Penner R (1992) Depletion of intracellular calcium stores activates a calcium current in mast cells. Nature 355:353-356

Hu GY, Hvalby O, Walacers SI, Albert A, Skjeflo P, Andersen P, Greengard P (1987) Protein kinase C injection into hippocampal pyramidal cells elicits features of long term potentiation. Nature 328:426-429

Hubbarth CM, Redpath GT, MacDonald TL, VandenBerg SR (1989) Modulatory effects of aluminum, calcium, lithium, magnesium, and zinc ions on [3H]MK-801 binding in human cerebral cortex. Brain Res 486:170-174

Jacobs RW, Duong T, Jones RE, Trapp GA, Scheibel AB (1989) A reexamination of aluminum in Alzheimer's disease: analysis by energy dispersive X-ray microprobe and flameless atomic absorption spectrophotometry. Can J Neurol Sci 16:498-503

Jansson ET (2001) Aluminum exposure and Alzheimer's disease. J Alzheimers Dis 3(6):541-549

Jensen S, Jernelov A (1969) Biological methylation of mercury in aquatic organisms. Nature 223:753-754

Koh JY, Choi DW (1988) Zinc alters excitatory amino acid neurotoxicity on cortical neurons. J Neurosci 8:2164-2171

Kuznetov DA, Zavijalov NV, Gororkov AV, Silileva TM (1987) Methylmercury induced nonselective blocking of phosphorylation processes as a possible cause of protein synthesis inhibition in vitro and in vivo. Toxicol Lett 36:153-160

Kuznetov DA, Richter V (1987) Modulation of messenger RNA metabolism in experimental methylmercury neurotoxicity. J Neurosci 34:1-17

Leistevuo J, Leistevuo T, Helenius H Pyy L, Osterblad M, Houvinen P, Tenovuo J (2001) Dental amalgam fillings and the amount of organic mercury in human saliva. Caries Res 35:163-166

Leonhardt R, Pekel M, Platt B, Haas HL, Büsselberg D (1996a) Voltage activated calcium channel currents of rat DRG neurons are reduced by mercuric chloride and mehtylmercury. Neurotoxicology 17:85-91

Leonhardt R, Haas HL, Büsselberg D (1996b) Voltage gated calcium, potassium and sodium channel currents of rat DRG neurones are reduced by mehtylmercury. Naunyn Schmiederbergs Arch Pharmacol 354:532-538

Leonhardt R, Pekel M, Platt B, Haas HL, Büsselberg D (1996c) Mercury compounds reduce voltage activated ion channel currents. Metal ions in biology and Medicine 4:204-206

Madeja M, Mußhoff U, Binding N, Witting U, Speckmann EJ (1997) Effects of Pb^{2+} on delayed-rectifier potassium channels in acutely isolated hippocampal neurons. J Neurophysiol 78:2649-2654

Magour S (1987) Studies on the inhibition of brain synaptosomal Na^{+}/K^{+}-ATPase by mercury chloride and methylmercury chloride. Arch Toxicol 9:393-396

Manalis RS, Cooper GP (1975) Evoked transmitter release increased by inorganic mercury at frog neuromuscular junction. Nature 257:256-257

Markovac J, Goldstein GW (1988) Lead activates protein kinase C in immature rat brain microvessels. Toxicol Appl Pharmacol 96:14-23

Martin RB (1986) The chemistry of aluminium as related to biology and medicine. Clin Chem 32:1797-1806

Matthews MR, Parsons PJ, Carpenter DO (1993) Solubility of lead as lead (II) chloride in Hepes-Ringer and artificial seawater solutions. Neurotoxicology 14:283-290

Mayer ML, Westbrook GL (1987) Permeation and block of N-methyl-aspartic acid receptor channels by divalent cations in mouse cultured central neurons. J Physiol Lond 394:501-527

McDermott JR, Smith AI, Iqbal K, Wisniewski HM (1979) Brain aluminum in aging and Alzheimer disease. Neurology 29:809-814

McLachlan DR, Lukiw WJ, Kruck TP (1989) New evidence for an active role of aluminum in Alzheimer's disease. Can J Neurol Sci 16:490-497

Mehlhorn H (2000) Die Novellierung der Trinkwasserverordnung – technische Aspekte. gwf Wasser/Abwasser 141:424-430

Michaelson TA (1980) An appraisal of rodent studies on the behavioral toxicity of lead. In: Singhal RL, Thomas JA (eds) Lead Toxicity, Baltimore: Urban & Schwarzenberg pp 301

Mintz IM, Adams ME, Bean BP (1992a) P-type calcium channels in rat central and peripheral neurons. Neuron 9:85-95

Mintz IM, Venema VJ, Swiderek KM, Lee TD, Bean BP, Adams ME (1992 b) P-type calcium channels blocked by the spider toxin omega-Aga-IVA. Nature 355:827-829

Möller Madson B (1991) Localisation of mercury in the CNS of the rat, III. Oral administration of methylmercuric chloride. Fund Appl Toxicol 16:172-187

Möller Madson B (1992) Localization of mercury in CNS of the rat. V. Inhalation exposure to metallic mercury. Arch Toxicol 66:79-89

Morris RG, Anderson E, Lynch GS, Baudry M (1986) Selective impairment of learning and blockade of long-term potentiation by an N-methyl-D-aspartate receptor antagonist, AP5. Nature 319:774-776

Murakami K, Feng G, Chen SG (1993) Inhibition of brain protein kinase C subtypes by lead. J Pharmacol Experimental Therap 265:757-761

Murray JC, Tanner CM, Sprague SM (1991) Aluminum neurotoxicity: a reevaluation. Clin Neuropharmacol 14:179-185

Musch HR, Bornhausen M, Kriegel H, Greim H (1978) Methylmercury chloride induces learning deficits in prenatally treated rats. Arch Toxicol 40:103-108

Nachshen DA (1984) Selectivity of the Ca binding site in synaptosome Ca channels. Inhibition of Ca influx by multivalent metal cations. J Gen Physiol 83:941-967

National Centre for Health Statistics (1984) Blood lead levels for persons ages 6 months - 74 years: United States, 1976 - 1980. Washington, DC: DHHS Publication, US Government Printing Office

Needleman HL (1979) Lead levels and children's psychologic performance [letter]. N Engl J Med 301:163

Needleman HL, Gunnoe C, Leviton A, Reed R, Peresie H, Maher C, Barrett P (1979) Deficits in psychologic and classroom performance of children with elevated dentine lead levels. N Engl J Med 300:689-695

Needleman HL (1988) The persistent threat of lead: medical and sociological issues. Curr Probl Pediatr 18:697-744

Needleman HL, Bellinger D (1991) The health effects of low level exposure to lead. Annu Rev Public Health 12:111-140

Neelands TR, King AP, MacDonald RL (2000) Functional expression of L-, N-, P/Q-, and R-type calcium channels in the human NT2-N cell line. J Neurophysiol 84(6):2933-2944

New York State department of health (1993) New York State Poison control network. Annual report to the legistative 1-9

Nihei MK, McGlothan JL, Toscano CD, Guilarte TR (2001) Low level Pb^{2+} exposure affects hippocampal protein kinase C gamma gene and protein expression in rats. Neurosci Lett 298(3):212-216

Nishimura M (1988) Zn^{2+} stimulates spontaneous transmitter release at mouse neuromuscular junctions. Br J Pharmacol 93:430-436

Nriagu JO (1979) Global inventory of natural and anthropogenic emissions of trace metals to the atmosphere. Nature 279:409-411

Nriagu JO (1992) Saturine drugs and medical exposure to lead: an historical outline. In: Needleman HL (ed.), Human lead exposure, pp 3-22. Boca Ration (Florida) CRC Press Inc

Ootgiesen M, van Kleef RG, Bajnath RB, Vijverberg HP (1990a) Nanomolar concentrations of lead selectively block neuronal nicotinic acetylcholine responses in mouse neuroblastoma cells. Toxicol Appl Pharmacol 103:165-174.

Oortgiesen M, Lewis BK, Bierkamper GG, Vjiverberg HP (1990b) Are postsynaptic nicotinic end-plate receptors involved in lead toxicity? Neurotoxicology 11:87-92

Pacyna JM (1986) Atmospheric trace elements from natural and anthropogenic sources. In: Nriagu JO, Davidson CI (eds) Toxic Metals in the Atmosphere, pp 33-52. New York: John Wiley & Sons

Padich RA, Dietrich KN, Pearson DT (1985) Attention, activity level, and lead exposure at 18 months. Environ Res 38:137-143

Pekel M, Platt B, Büsselberg D (1993) Mercury (Hg^{2+}) decreases voltage gated calcium channel currents in rat DRG and Aplysia neurons. Brain Res 632:121-126

Pellmar TC, Carpenter DO (1980) Serotonin induces a voltage-sensitive calcium current in neurons of Aplysia californica. J Neurophysiol 44:423-439

Perl DP, Brody AR (1980) Alzheimer's disease: X-ray spectrometric evidence of aluminum accumulation in neurofibrillary tangle-bearing neurons. Science 208:297-299

Peters S, Koh J, Choi DW (1987) Zinc selectively blocks the action of N-methyl-D-aspartate on cortical neurons. Science 236:589-593

Piomelli S, Seaman C, Zullow D, Curran A, Davinow B (1982) Threshold for lead damage to heme sythesis in urban children. Proc Natl Acad Sci USA 79:3335-3339

Platt B, Haas HL, Büsselberg D (1993) Extracellular pH modulates aluminum-blockade of mammalian voltage-activated calcium channel currents. Neuroreport 4:1251-1254

Platt B, Büsselberg D (1994a) Combined actions of Pb^{2+}, Zn^{2+} and Al^{3+} on voltage activated calcium channel currents. Cell Molec Neurobiol 14:831-840

Platt B, Büsselberg D (1994b) Actions of aluminum on voltage activated calcium channel currents. Cell Molec Neurobiol 14:819-829

Platt B, Haas H, Büsselberg D (1994c) Aluminum reduces glutamate-activated currents of rat hippocampal neurons. NeuroReport 5:2329-2332

Platt B, Carpenter DO, Büsselberg D, Reymann KG, Riedel G (1995) Aluminum impairs hippocampal long-term potentiation in rats in vitro and in vivo. Exp Neurol 134:1-14

Pocock SJ, Smith M, Baghurst P (1994) Environmental lead and children's intelligence: a systematic review of the epidemiological evidence. BMJ 309: 1189-1197

Razani-Boroujerdi S, Partridge LD, Sopori ML (1994) Intracellular calcium signaling induced by thapsigargin in excitable and inexcitable cells. Cell Calcium 16:467-474

Reymann KG, Brodemann R, Kase H, Matthies H (1988) Inhibitors of calmodulin and protein kinase C block different phases of hippocampal long-term potentiation. Brain Res 461:388-392

Rice DC, Gilbert SG (1987) Effects of developmental exposure to methylmercury on spatial and temporal visual function in monkeys. Toxicol Appl Pharmacol 102(1):151-163

Rosen JF, Chesney RW, Hamstra AJ, Delucca HF, Mahaffey KR (1980) Reduction in 1,25-dihydroxyvitamin D in children with increased lead absorption. N Engl J Med 302:1128-1131

Sakaki Y, Sugioka M, Fukuda Y, Yamashita M (1997) Capacitative Ca^{2+} influx in the neural retina of chick embryo. J Neurobiol 32:62-68

Sarafian TA (1993) Methylmercury increases intracellular Ca^{2+} and inositol phosphate levels in cultured cerebellar granule neurons. J Neurochem 61(2):648-657

Schirrmacher K, Nonhoff D, Wiemann M, Peterson-Grine E, Brink PR, Bingmann D (1996) Effects of calcium on gap junctions between osteoblast-like cells in culture. Calcif Tissue Int 59:259-264

Schirrmacher K, Bingmann D (1998) Effects of vitamin D3, 17beta-estradiol, vasoactive intestinal peptide, and glutamate on electric coupling between rat osteoblast-like cells in vitro. Bone 23:521-526

Schirrmancher K, Wiemann M, Bingmann D, Büsselberg D (1998) Effects of lead, mercury and methylmercury on gap junctions and $[Ca^{2+}]_i$ in bone cells. Calcif Tiss Int 63:134-139

Schwartzwelder HS (1986) Central neurotoxicity after exposure to organic lead: susceptibility to seizures. Neurosci Lett 58:225-228

Settle DM, Patterson CC (1980). Lead in albacore: guide to lead pollution in Americans. Science 207:1167-1176

Shafer TJ (1998) Effects of Cd^{2+}, Pb^{2+} and CH_3Hg^+ on high voltage-activated calcium currents in pheochromocytoma (PC12) cells: potency, reversibility, interactions with extracellular Ca^{2+} and mechanisms of block. Toxicol Lett 99:207-221

Shore D, Wyatt RJ (1983) Aluminum and Alzheimer's disease. J Nerv Ment Dis 171:553-558

Sideman S, Manor D (1982) The dialysis dementia syndrome and aluminum intoxication. Nephron 31:1-10

Simons TJ, Pocock G (1987) Lead enters bovine adrenal medullary cells through calcium channels. J Neurochem 48:383-389

Simons TJ (1993) Lead transport and binding by human erythrocytes in vitro. Pflügers Arch 423:307-313

Smith DB, Lewis JA, Burks JS, Alfrey AC (1980) Dialysis encephalopathy in peritoneal dialysis. JAMA 244:365-366

Strong MJ, Garruto RM (1991 a) Neuron-specific thresholds of aluminum toxicity in vitro. A comparative analysis of dissociated fetal rabbit hippocampal and motor neuron-enriched cultures. Lab Invest 65:243-249

Strong MJ, Garruto RM (1991b) Chronic aluminum-induced motor neuron degeneration: clinical, neuropathological and molecular biological aspects. Can J Neurol Sci 18:428-431

Sun X, Tian X, Tomsig JL, Suszkiw JB (1999) Analysis of differential effects of Pb^{2+} on protein kinase C isozymes. Toxicol Appl Pharmacol 156:40-45

Suszkiw J, Toth G, Murawsky M, Cooper GP (1984) Effects of Pb^{2+} and Cd^{2+} on acetylcholine release and Ca^{2+} movements in synaptosomes and subcellular fractions from rat brain and Torpedo electric organ. Brain Res 323:31-46

Teyler TJ, Discenna P (1987) Long term potentiation. Ann Rev Neurosci 10:131-161

Trapp GA, Miner GD, Zimmerman RL, Mastri AR, Heston LL (1978) Aluminum levels in brain in Alzheimer's disease. Biol Psychiatry 13:709-718

Trinkwasserverordnung (1990) Verordnung über Trinkwasser und über Wasser für Lebensmittelbereiche vom 12. Dezember 1990, Bundesgesetzblatt I:2613

Tsien RW, Fox AP, Hess P, McCleskey EW, Nilius B, Nowychy MC, Rosenberg RL (1987) Multiple types of calcium channel in excitable cells. Soc Gen Physiol Ser 41:167-187

Tsubaki T, Krukuyama K (1977) Minamata Disease. Amsterdam, Elsevier Scientific Publishing Co

Ujihara H, Albuquerque EX (1992a) Developmental change of the inhibition by lead of NMDA-activated currents in cultured hippocampal neurons. J Pharmacol Exp Ther 263:868-875

Ujihara H, Albuquerque EX (1992b) Ontogeny of N-methyl-D-aspartate-induced current in cultured hippocampal neurons. J Pharmacol Exp Ther 263:859-867

Umweltbundesamt (2002) Ausgewählte Elemente und Verbindungen im Blut der 18- bis 69-jährigen Bevölkerung in Deutschland 1998. 20-51

Uteshev V, Büsselberg D, Haas HL (1993) Pb^{2+} modulates the NMDA-receptor-channel complex. Naunyn Schmiedebergs Arch Pharmacol 347:209-213

Villalobos C, Gracia-Sancho J (1995) Capacitative Ca^{2+} entry contributes to the Ca^{2+} influx induced by thyrotropin-releasing hormone (TRH) in GH3 pituitary cells. Pflügers Arch 430(6):923-935

Walsh TJ, Schulz DW, Tilson HA, Dehaven DL (1986) Acute exposure to triethyl lead enhances the behavioral effects of dopaminergic agonists: involvement of brain dopamine in organolead neurotoxicity. Brain Res 363:222-229

Weahl T, Thomson AM (eds) (1991) Excitatory Amino acids and synaptic transmission. Academic Press, London

Wedeen RP (1984) Poison in the pot: The legacy of lead. Southern Illinois University Press, Carbondale, Illinois

Wehner JM, Sleight S, Upchurch M (1990) Hippocampal protein kinase C is reduced in poor spatial learners. Brain Res 523:181-187

Weiss JH, Hartley DM, Koh JY, Choi DW (1993) AMPA receptor activation potentiates zinc neurotoxicity. Neuron 10:43-49

Westbrool GL, Mayer ML (1987) Micromolar concentrations of Zn^{2+} antagonize NMDA and GABA responses of hippocampal neurons. Nature 328:640-643

WHO (1987) Toxicological evaluation of certain food additives and contaminants. WHO Food Additives Series: 21, prepared by the 30th Meeting of the Joint FAO/WHO Expert Committee on Food Additives. Cambridge University Press, Cambridge

Wiemann M, Schirrmacher K, Büsselberg D (1999) Interference of lead with the calcium release activated calcium flux of osteoblast-like cells. Calcif Tiss Int 65:479-485

Xia SL, Ferrier J (1992) Propagation of a calcium pulse between osteoblastic cells. Biochem Biophys Res Commun 186:1212-1219

Yokoyama M, Koh J, Choi DW (1986) Brief exposure to zinc is toxic to cortical neurons. Neurosci Lett 71:351-355

Yuan B, Klein MH, Contiguglia RS, Mishell JL, Seligman PA, Miller NL, Molitoris BA, Alfrey AC, Shapirp JI (1989) The role of aluminum in the pathogenesis of anemia in an outpatient hemodialysis population. Ren Fail 11:91-96

Zalutsky RA, Nicoll RA (1990) Comparison of two forms of long-term potentiation in single hippocampal neurons. Science 248:1619-1624

Zietz B, Vergara JD, Kevekordes S, Dunkelberg H (2001a) Lead contamination in tap water of households with children in Lower Saxony, Germany. Sci Total Environ 275:19-26

Zietz B, Paufler P, Kessler-Gaedtke B, Dunkelberg H (2001b) Bleiverunreinigung von Trinkwasser durch Leitungssysteme in Berlin. Z Umweltchem Ökotox 13:153-157

Zimmermann HP, Doenges KH, Roderer G (1985) Interaction of triethyl lead chloride with microtubules in vitro and in mammalian cells. Exp Cell Res 156:140-152

Zorumski CF, Yang J (1988) AMPA, kainate, and quisqualate activate a common receptor-channel complex on embryonic chick motoneurons. J Neurosci 8: 4277-4286

Zwart R, van Kleef RG, van Hooft JA, Oortgiesen M, Vijverberg HP (1997) Cellular aspects of persistent neurotoxicants: effects of Pb^{2+} on neuronal nicotinic acetylcholine receptors. Neurotoxicology 18:709-717

Chapter 16

Effects of organometal(loid) compounds on neuronal ion channels: possible sites for neurotoxicity

J. Gruner[1]*, K. Krüger*[1]*, N. Binding, M. Madeja, U. Mußhoff*

Toxicity of organometal(loid)s

Organometal(loid)s from anthropogenic or geogenic sources represent an important group of environmentally hazardous substances. The use of a variety of compounds such as preservatives, antifouling paints and biocides, as well as the biomethylation of inorganic substrates leads to widespread contamination of water, air, sediments etc. (e.g. Thayer 1995; Appel et al. 2000).

The group of metal(loid) compounds contains numerous highly toxic substances, with the organometal(loid)s often exhibiting a higher toxicity than the respective inorganic compounds. This was attributed to the lipophilic properties of organometal(loid)s resulting in an increased resorption rate (Craig 1986).

After resorption, most organometal(loid)s are metabolized to monovalent cations through the loss of an organic substituent. These cations have both lipophilic and hydrophilic properties allowing for efficient transportation in body fluids as well as for easy penetration of cell membranes. Thus, the main toxicity is often attributed to these monovalent cations, since high effective concentrations in target organs may be attained easily and rapidly (Craig 1986).

Neurotoxic aspects of organometal(loid)s

Monovalent organometal(loid) cations may exhibit their toxicity by coordination to the basic regions of enzymes, proteins, hemoglobin or cytochrom P450, for example (e.g. Craig 1986; Denny et al. 1993; Komulainen et al. 1995). Interferences with DNA and protein synthesis as well as with the immune system have been described (Cheung and Verity 1985; Kuznetsov and Richter 1987). But it is often adverse effects on the nervous system that are predominant in organometal(loid) intoxication. In general, neurotoxicity may be defined as adverse functional or structural changes in the nervous system induced by chemical or biological agents. Depending on the neurotoxic agent, the central or the peripheral nervous system or both are affected. Clinical signs of neurotoxic diseases range from impairment of muscle functions, paralysis, sensory dysfunctions, impairment of the autonomic

[1] Both authors contributed equally to the paper

nervous system, and seizures to affections of mood, cognitive functions, learning, memory, speech and behavior. Though quite a number of neurotoxic diseases caused by chemicals are clinically well described, little is known in most cases about the specific target structures in the nervous system or about the mechanisms involved. This is true especially of the organometal(loid)s. Only organomercury and organotin compounds form an exception. This is mainly due to epidemic intoxication with methylmercury in Japan ("Minamata disease") and to the widespread environmental occurrence of organotin compounds which are used in antifouling paints and preservatives.

In Minamata disease, consumption of fish contaminated with methylmercury was responsible for complex neurological/psychiatric disorders in people from Minamata Bay/Japan. The clinical signs varied with the severity of intoxication, sex and age (Takeuchi 1968; Takeuchi 1982): Visual (tunnel vision) and sensory (paresthesia, numbness of the extremities) disturbances, tremor, ataxia, impairment of speech, hearing, gait and mental disturbances developed in almost all patients. Hypersalivation, muscular rigidity, chorea, athetosis and contractures were less frequent.

Different pathomechanisms are involved in Minamata disease. Pathological brain lesions correlating well with the clinical signs were found in the visual cortex, the dorsal root ganglia and the cerebellum (Takeuchi 1982). Inhibition of brain protein synthesis (Cheung and Verity 1985) and mRNA metabolism (Kuznetsov and Richter 1987), decreased neuromuscular transmission (Atchison and Narahashi 1982), decreased Ca^{2+} uptake in synaptosomes (Shafer and Atchison 1989), and interferences with presynaptic Na^{+} and Ca^{2+} ion channels of neuromuscular junctions (Shafer and Atchison 1992) are outcomes of the neurotoxicity of methylmercury. Blockades of voltage-operated Ca^{2+}- (Shafer 1998), K^{+}- and Na^{+}-channels (Büsselberg 1995) as well as of GABA-receptor-mediated synaptic transmission (Yuan and Atchison 1997) have been observed.

Clinical symptoms after accidental intoxications with organotin compounds are described in two case reports. Delirium, disorientation, and disturbances of memory occurred shortly after intoxication with trimethyltin. Seizures occurred a few months later; disturbances of memory and cognitive functions, seizures and dysphoria persisted for years (Feldmann et al. 1993). After intoxication with triphenyltin, the patient presented symptoms like spontaneous involuntary movement of the hands, facial twitching, silly smile, crying, diplopia, drowsiness, giddiness, vertigo, bidirectional nystagmus, impairment of calculation ability, and disorientation to time, people and place (Lin et al. 1998). Compared to methylmercury, only few investigations have been published on pathomechanisms of organotin neurotoxicity: trimethyltin reduces the permeability of neuronal membranes in hippocampal slices (Harkins and Armstrong 1992); triphenyltin increases the intracellular Ca^{2+} concentration in CNS neurons of the rat (Oyama et al. 1992); in cultured neurons, tributyl- and triphenyltin decrease the intracellular Ca^{2+} concentration and induce apoptosis. Di- and tributyl-, triphenyl- and trimethyltin do not interfere with potassium channels of neuroblastoma cells. However, with the exception of trimethyltin they all inhibit potassium currents in lymphocytes (Oortgiesen et al. 1996). In isolated snail neurons, inorganic divalent tin as well as the divalent dimethyl compound decreases acetylcholin-induced inward currents (Salanki et al. 1998).

Despite the clinical experience and investigations using experimental animals, brain tissue and cultured neurons, the pathomechanisms of organic mercury and tin neurotoxicity are not yet completely understood and for other organometal(loid)s are not even known. The main reason for this inadequate knowledge is certainly the complexity of the nervous system with its multiple target structures for neurotoxic agents.

Neurotoxicity of inorganic and organic arsenicals

For inorganic arsenicals, clinical neurotoxic signs are known from occupational or environmental exposure, attempted homicides, deliberate long-term poisoning or accidental intoxication. The symptoms are not very different from those of mercury and tin compounds. Subclinical nerve injuries associated with long-term occupational exposure were found in Swedish copper smelter workers (Lagerkvist and Zetterlund 1994). Delirium and encephalopathy were described as an unusual manifestation of occupational arsenic exposure (Morton and Caron 1989). Peripheral neuropathies were found in persons chronically exposed to arsenic-contaminated drinking water in south Calcutta (Mazumder et al. 1992). Environmental exposure to arsenic-containing dusts resulted in peripheral neuropathy for about 15% of the exposed (Gerr et al. 2000). Chronic intoxication resulting from burning of coal containing high concentrations of arsenic is reported from China: frequently occurring symptoms were loss of hearing, loss of taste, blurred vision, tingling and numbness of the limbs (Liu et al. 2002). Numbness and tingling of fingertips and toes and a decrease in muscle strength was reported after chronic intoxication from contaminated bird's nest soup (Luong and Nguyen 1999). Severe neurotoxic manifestations were reported in a case of chronic pesticide intoxication in Surinam. Encephalopathy, mental deterioration, epileptic seizures, motor deficits and a demyelinating polyneuropathy were diagnosed in a 16-year-old girl who had had continuous contact with an open bag of the pesticide copper acetate arsenite at home (Brouwer et al. 1992). The severity of the symptoms was attributed to reduced detoxification caused by a methylenetetrahydrofolate reductase deficiency.

A dose-independent but exposure time-dependent increase in arsenic brain concentration and alterations in the monoamine content in the midbrain and cortex were found in experimental animals (rats) (Rodriguez et al. 2001). The rats showed (reversible) reduced locomotor activity. In mice, arsenic exposure led to a decrease in brain monoamines (Delgado et al. 2000). Arsenite (Namgung and Xia 2000) and arsenic (Chattopadhyay et al. 2002) induced apoptosis in cultured neuronal cells.

Methylation to tri- and pentavalent organoarsenicals is considered a main detoxification pathway for inorganic arsenicals. Pentavalent monomethylarsonic acid, dimethylarsinic acid and trimethylarsine oxide are the major metabolites detected in urine. Trivalent compounds such as methylarsonous acid and a complex of methylarsinous acid with glutathione, which are considered as intermediates in enzymatic detoxification, have also been identified in urine samples (Cullen et al. 1984;Vahter 1994; Aposhian 1997).

Recently, the question of whether the generally accepted metabolism of inorganic arsenicals really results in detoxification or whether the methylated species also have deleterious biological effects was raised (e.g. Styblo et al. 2000; Petrick et al. 2000; Vega et al. 2001). In cytotoxicity studies with cultured human cells, trivalent arsenicals were found to be much more toxic than pentavalent compounds (Styblo et al. 2000). The high toxicity of trivalent arsenicals, especially of monomethylarsonic acid, has been demonstrated in a cytotoxicity assay with Chang human hepatocytes (Petrick et al. 2000). Viability testing in human epidermal keratinocytes showed increasing relative toxicities from inorganic trivalent arsenicals to methylated trivalent species, then to methylated pentavalent compounds and to inorganic pentavalent arsenate. The trivalent arsenicals stimulated secretion of growth-promoting cytokines, granulocyte macrophage colony stimulating factor and tumor necrosis factor α (Vega et al. 2001).

Target structures of neurotoxicity

The papers referred to above focused on the cytotoxicity of organoarsenicals and their effect on cell proliferation and cytokine secretion. Considering the neurotoxicity of arsenicals, the need for experimental investigations on the targets of neurotoxicity of methylated arsenicals and on the mechanisms involved is obvious. Neurotoxic substances may be classified according to their target structures in the nervous system (Valciukas 1991):

1. substances causing neuropathological changes of the gross morphology of nervous tissue
2. substances causing neuropathological alterations at the cellular level
3. substances inducing neuropharmacological action at the molecular level

Cytotoxicity, the induction of apoptosis as well as cerebral edema, may contribute to the morphological alterations observed in neurotoxicity. At the cellular level, neurotoxin-induced changes are seen as a breakdown of the neuron's cell body (neuronopathy), its axon (axonopathy) or its nerve sheath (myelinopathy). At the molecular level, neurotoxins may interfere with the genetic material of the neuron, protein synthesis, and the basic mechanisms of the generation, transmission and processing of neuronal signals. For example, disturbances of ion channels induced by neurotoxins often cause dysfunction of the nervous system.

Function of ion channels in the nervous system

The fundamental task of nerve cells is to receive, conduct and transmit bioelectrical signals. Electrical signals are usually received on the dendrites or the soma of neurons, sent out along the axon in the form of action potentials, and passed on to other neurons at chemical synapses. Action potentials trigger in the presynaptic axon terminal the secretion of neurotransmitters, which induce membrane potential changes in the postsynaptic cell in the form of excitatory (neurotransmitter: glutamate) or inhibitory (neurotransmitter: γ-aminobutyric acid, GABA) postsynaptic potentials.

Neuronal signaling depends on rapid changes in electrical potential differences across nerve cell membranes. Responsible for changes of this membrane potential in neurons are movements of ions (Na^+, K^+, Ca^{2+}, Cl^-) across the membrane through activated channel-forming proteins. These ion channels are characterized by a high selectivity for one or more types of ions. Two classes of ion channels are of crucial importance for electrical signaling:

(1) Voltage-operated ion channels mediate rapid, voltage-dependent changes in ion permeability during action potentials (Hille 1991). The voltage-operated sodium channels have only one known function: they are responsible for the initial inward current of sodium ions during the depolarization phase of action potentials. The voltage-operated potassium channels, by contrast, have a variety of functions that include speeding repolarization after the action potential, controlling the firing pattern of neurons, maintaining the resting potential, and hyperpolarizing the cell. As a result, potassium channels play an important role in regulating the excitability of neurons. According to their different functional roles, potassium channels constitute the most diverse class of ion channels with respect to kinetic properties, regulation, pharmacology, and structure.

(2) Transmitter-operated ion channels are located in the membrane of the postsynaptic neuron. These channels convert the chemical signal, e.g. the cell-specific neurotransmitter, of the presynaptic cell back into an electrical one. Receptors for the neurotransmitters cause ion channels to open and thereby depolarize (*via* glutamate-activated receptor channel complexes) or hyperpolarize (*via* GABA-activated receptor channel complexes) the postsynaptic area of the neuron. The neurotransmitter receptors fall into two classes based on whether they activate ion channels directly or indirectly. Directly activating receptors consist of multimeric protein complexes that form both the receptor element and the ion channel (ionotropic receptor channel complex). This type produces relatively fast changes of the postsynaptic potential. Receptors that activate ion channels indirectly involve the separated receptor element and the ion channel. The receptor structure activates through GTP-binding proteins intracellular second messenger molecules, which can activate the ion channel (metabotropic receptor). This type results in slow and long-lasting changes of the postsynaptic potential.

The most important excitatory neurotransmitter is glutamate. Excitatory synaptic transmission is mediated by glutamate-operated receptor channels, which represent the most abundant excitatory receptor type in the central nervous system.

Disturbances of the glutamatergic synaptic transmission can lead to pathological conditions such as epileptic activity or neuronal damage. Several classes of glutamate-activated receptors have been identified and colocalized in the postsynaptic membrane (Dingledine et al. 1999). Their names are based on the pharmacological agonist that binds to the specific receptor subtype and selectively opens the associated ion channels: the ionotropic α-amino-3-hydroxy-5-methyl-4-isoxazole propionate receptor (AMPA receptor), the ionotropic N-methyl-D-aspartate receptor (NMDA receptor), and the quisqualate receptor, which has been called metabotropic glutamate receptor (mGluR).

Prediction of neurotoxic potency of hazardous substances with an in vivo expression system: *Xenopus laevis* oocytes

The mechanisms of neurotoxicity may be investigated with different experimental approaches. Alterations in neuronal signaling and neurotransmission by neurotoxic agents may be analyzed using electrophysiological techniques on cultured nervous tissue or on isolated neurons. However, experiments on the detailed effects of neurotoxic substances on certain ion channels are often hampered in intact brain tissue by the complexity of the central nervous system. Compared to in vivo experiments, brain slices or cultured neurons, electrophysiological experiments using mRNA-injected oocytes have some practical advantages: oocytes are easy to handle and known concentrations of agents can be easily applied either by perfusion or by intracellular injection. Furthermore, confounding factors such as interference of other ion channels and pumps, transmitter release or uptake from neighboring nerve and glia cells can be excluded in this single-cell system. Thus, the Xenopus oocyte expression system permits a direct assessment of drug effects on specific neuronal ion channels. Therefore, the oocytes can be used as a tool for neurotoxicological investigations, since any change in ion channel function may be taken as proof of the neurotoxic potency of the tested substance.

The use of *Xenopus laevis* oocytes for electrophysiological studies of ion channels

Historical aspects

The functional expression of foreign proteins in *Xenopus laevis* oocytes following injection of genetic information into the cell was pioneered by Gurdon and coworkers. Xenopus oocytes translate injected messenger RNA for hemoglobin efficiently, assemble the proteins correctly, process the nascent polypeptides, and target them to the proper compartment (Gurdon et al. 1971). Oocytes were first used for the heterologous expression of neuronal membrane proteins in 1981 (Sumikawa et al. 1981) and then quickly became a commonly used in vivo expression system for the functional characterization of voltage- and ligand-gated ion channels (Dascal 1987; Madeja and Mußhoff 1992; Sigel 1990; Wagner et al. 2000). Various experimental groups have shown that the Xenopus oocyte expression system is able to assemble multi-subunit ion channels properly and to incorporate them in a functional form into the plasma membrane of the oocytes (c.f. Sigel 1990). The properties of these heterologously expressed ion channels are similar if not identical to those found in the cells of origin. Therefore, the Xenopus oocyte became the most common transient expression system of message encoding ion channels. In recent years, Xenopus oocytes have taken over a new role in a wide number of different applications to investigate ion channels, including structure-function-relationships, translation, assembly, posttranslational modification and sorting, modulation by second messengers, and effects of drugs and toxicants.

Preparation and maintenance of oocytes

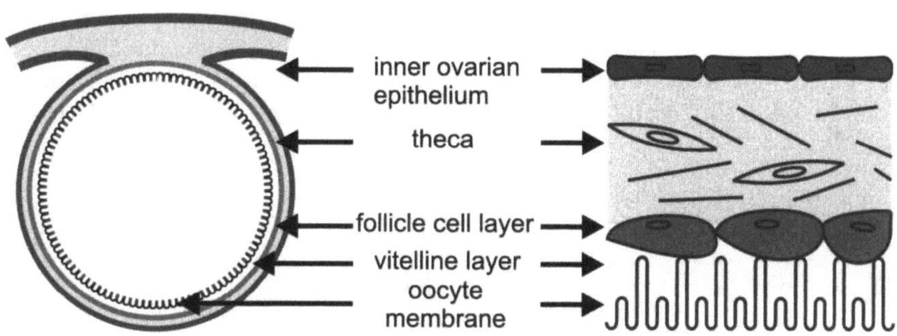

Fig. 1. Schematic diagram of an oocyte of *Xenopus laevis* and surrounding tissues. The oocytes surrounded by a number of cellular and non-cellular layers. The oocyte plasma membrane is covered by a non-cellular vitelline layer. Several additional layers flank the vitelline layer. These include a monolayer of follicle cells, the theca, which is a fibrous layer containing blood vessels, nerve fibers and fibroblasts, and the inner ovarian epithelium

Various protocols have been described by different laboratories. Here we describe the method routinely used in our laboratories (Madeja et al. 1997c), which differs only in a few points from others. Female *Xenopus laevis* frogs are anesthetized by submersion in an ethyl-m-aminobenzoate solution. Pieces of ovary are removed through a small abdominal incision, and the incision is subsequently closed with a resorbable suture. Full grown oocytes, measuring 1-1.2mm in diameter, are surrounded by the noncellular, fibrous vitellin layer and by several layers of follicular cells (Fig. 1; Dumont 1972). The follicular tissues are particularly suitable for experimental work, since they stabilize electrophysiological recordings and increase the life-time of the oocyte (Madeja et al. 1997c). However, follicular tissues may affect the access of substances to the oocyte plasma membrane, thus possibly inducing errors in pharmacological measurements on ion channels. To avoid this, control experiments were performed in which the follicular cells were removed mechanically to give defolliculated oocytes. The oocytes were maintained under tissue culture conditions at 20°C in a solution composed of (in mmol/l) NaCl 88, $CaCl_2$ 1.5, KCl 1, $NaHCO_3$ 2.4, $MgSO_4$ 0.8, HEPES 5, pH 7.4, supplemented with penicillin (100 IU/ml) and streptomycin (100 µg/ml). Every second day, the sterile culture solution was exchanged and damaged oocytes were removed. Under these conditions oocytes may be stored for 7-12 days after injection.

Injection of genetic information into Xenopus oocytes

A precondition for electrophysiological measurements on ion channels in oocytes is the injection of a sufficient amount of intact mRNA (Fig. 2A). The mRNA can be derived from various sources: either total RNA or poly(A)$^+$ RNA extracted from nervous tissue samples (rat brain) or cDNA-derived mRNA (cRNA). Extraction procedures have been described for total RNA and poly(A) RNA (Blumenau et al. 2001). The cRNA was synthesized as described by Stühmer and coworkers (Stühmer et al. 1988). The mRNA was dissolved in distilled water and the solution was loaded into a glass capillary (tip diameter: 10 µm) by suction. The mRNA was injected in aliquots of about 50 nl (50 ng poly(A)$^+$ RNA; 1 ng cRNA) per oocyte.

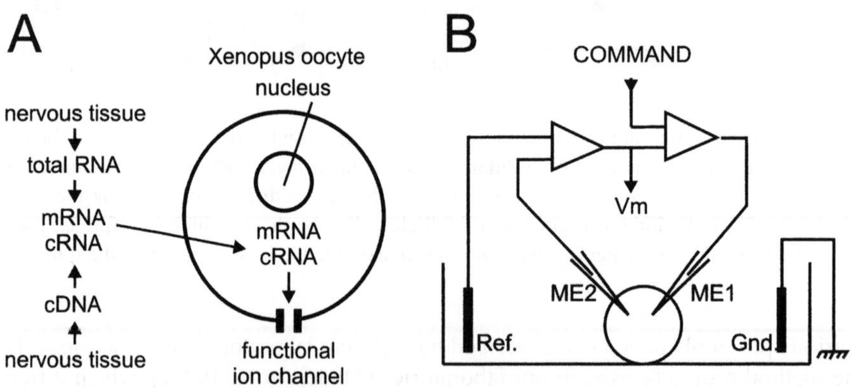

Fig. 2. A: Schematic diagram illustrating the routes for expression of foreign plasma membrane proteins in *Xenopus laevis* oocytes. B: Schematic diagram of the main components of a two-electrode voltage clamp. The microelectrode (ME1) delivered the current needed to clamp the oocyte to the desired membrane potential (command). The resting membrane potential is measured through the microelectrode 2 (ME2). Vm, the output of the left amplifier-comparator, is the difference in membrane potential between the bath electrode (Ref) and ME2. The current injected into the oocyte flows to ground (Gnd) through the bath ground electrode (modified after Stühmer 1992)

Electrophysiological measurements from mRNA-injected oocytes

For electrophysiological recordings the membrane potential of the oocytes has to be under control. This is achieved by "clamping" the oocyte to predefined membrane potentials using the two-electrode-voltage clamp technique (Fig 2B). For this purpose, two glass microelectrodes are impaled into the oocyte. Both electrodes are filled with 2 M KCl and have resistances of 0.4 to 2 MΩ. The membrane potential electrode connects to a feedback amplifier where the signal is compared to the command potential given by a generator. The difference in these signals is applied as a current through the current-delivering electrode, across the plasma membrane of the oocyte, and to the bath-grounding electrode. Ion fluxes through channels are measured as a deflection from the baseline current. The experimental setup has to be carefully grounded and shielded against external currents by a Faraday cage.

For the pharmacological investigations an application setup was used, which allows rapid solution exchange during electrophysiological measurements (Madeja et al. 1991a, 1995a). This system is based on the principle of the concentration clamp technique after Akaike and coworkers (Akaike et al. 1986). Briefly, the setup consists of two rectangular intercrossing tubes. The oocyte, fixed by suction to a flamed micropipette, is positioned in the intersection of the horizontal and vertical tubes. The horizontal tubes serve as guide rails for the microelectrodes. For rapid exchange of the solution the lower end of the vertical tube is immersed into the respective solution. The solution surrounding the oocyte is exchanged by applying a negative pressure of 10 kPa. About 90% of the solution is exchanged in less than 10 ms with a fluid stream velocity of about 0.2 m/s. There is no fluid stream of the solution during the remaining time of the experiment. The membrane potential and resistance of the oocyte do not change with the rapid solution exchanges.

Testing the Xenopus expression system for neurotoxicological investigations: Effects of lead on voltage- and transmitter-operated ion channels

As mentioned above, the Xenopus oocytes are excellently suited to the investigation of drug actions on heterologously expressed neuronal ion channels. In a number of investigations we used this tool to study the actions of various convulsants (Bloms-Funke et al. 1992, 1994, 1996; Madeja et al. 1991b, 1994, 1996, 1997b; Mußhoff et al. 1994, 1995a), antiepileptic drugs (Madeja et al. 2000), antiarrhythmic drugs (Mergenthaler et al. 2001; Rolf et al. 2000), neurotoxins (Binding et al. 1996; Flott-Rahmel et al. 1997, 1998; Madeja et al. 1995b; Mußhoff et al. 1995b; Ullrich et al. 1999) and substances with a neurotoxic potency (Madeja et al. 1997a; Mußhoff et al. 1999) on voltage- and transmitter-operated ion channels. The following section will describe the effects of the neurotoxin lead (Pb^{2+}) on ion channels expressed in Xenopus oocytes (c.f. Madeja et al. 1995b; Mußhoff et al. 1995b).

Lead-induced neurological symptoms are characteristic of decreased neuronal activity, e.g. coma and paralysis, as well as of enhanced neuronal activity, e.g. agitation and seizures. Furthermore, lead intoxication causes disturbances in cognitive functions, especially in developing animals and man. The effects of lead were tested on heterologously expressed ion channels: (i) the delayed rectifier potassium channel Kv1.1, which is a very common and possibly the most abundant potassium channel and (ii) glutamate-operated ion channels of the AMPA-type, which are responsible for the fast postsynaptic responses at glutamatergic synapses. The experiments showed that lead decreased ion currents through both the delayed rectifier potassium channels and the AMPA-receptor channels at a threshold concentration of about 0.1 μmol/l (Fig. 3).

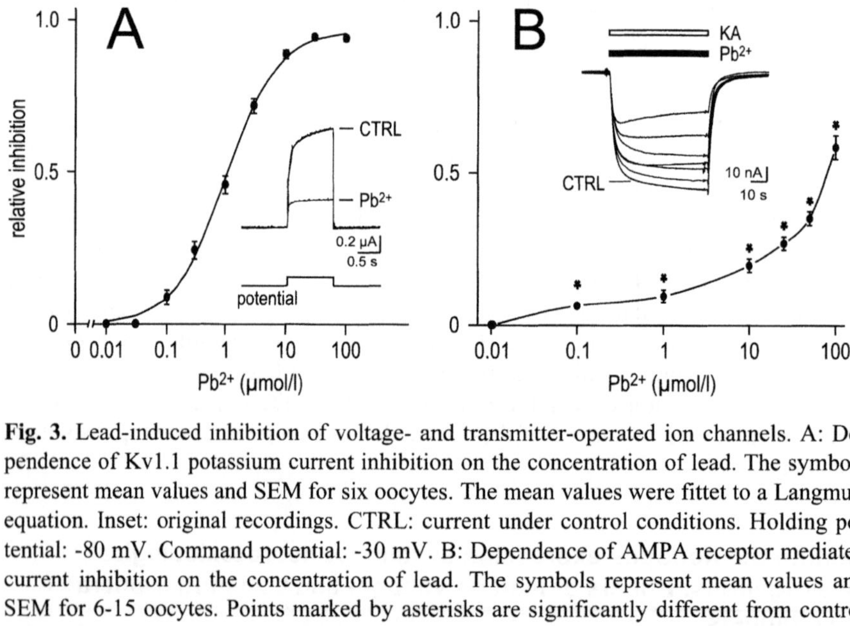

Fig. 3. Lead-induced inhibition of voltage- and transmitter-operated ion channels. A: Dependence of Kv1.1 potassium current inhibition on the concentration of lead. The symbols represent mean values and SEM for six oocytes. The mean values were fittet to a Langmuir equation. Inset: original recordings. CTRL: current under control conditions. Holding potential: -80 mV. Command potential: -30 mV. B: Dependence of AMPA receptor mediated current inhibition on the concentration of lead. The symbols represent mean values and SEM for 6-15 oocytes. Points marked by asterisks are significantly different from control values. Inset: original recordings. Concentration of the agonist kainate (KA): 50 μmol/l. CTRL: current under control conditions. Holding potential: -70 mV. Inward current: downward deflection (modified after Madeja et al. 1995 and Mußhoff et al. 1995)

The depressant effects of lead were dependent on the membrane potential: at a potential of -30 mV, the concentration needed for a 50% reduction of the potassium currents was approximately 1 μmol/l (Fig. 3A), while at a potential of -70 mV, the concentration needed for a 50% reduction of ion currents through the AMPA-receptor channels was approximately 50 μmol/l (Fig. 3B). The site of action is most likely the extracellular region of the channels since patch clamp experiments showed that application of lead to the intracellular side of the channels (inside-out membrane patch) was ineffective, whereas application to the extracellular side (outside-out membrane patch) exerted a significant effect (data not shown; Madeja et al. 1995b). In additional experiments, the depressive effects of lead on native delayed rectifier potassium channels were evaluated on hippocampal pyramidal cells (Madeja et al. 1997d). Furthermore, the effect of lead was tested on other cloned potassium channels which showed different inactivation types (Kv1.2, Kv1.4, Kv2.1, Kv3.4). In these experiments the depressant effect of lead was obtained with all potassium channel types.

It is generally accepted that important mechanisms of lead neurotoxicity are disturbances of calcium channel functions and impairment of the glutamate receptor channels of the NMDA-type (Alkondon et al. 1990; Audesirk 1993). However, the experiments on Xenopus oocytes showed that the mechanisms underlying the lead neurotoxicity are far more complex and heterogeneous than was previously understood. The action of lead on potassium channels and AMPA-operated glutamate receptor channels may also contribute to malfunctions of the nervous system after lead intoxication. Furthermore, the investigations corroborate the view that the oocyte expression system is a promising tool for the investigation of neurotoxic effects on elementary mechanisms of the generation and processing of information in the nervous system at a molecular level.

Effects of arsenicals on transmitter-operated ion channels

The effects of the two organic arsenic compounds, monomethylarsonic acid ($MeAsO(OH)_2$) and dimethylarsinic acid (Me_2AsOOH), and the inorganic arsenite on the function of transmitter-operated ion channels sensitive to glutamate were tested.

Transmembraneous ion currents through the different glutamate-operated ion channels

Glutamate-operated ion channels were heterologously expressed in Xenopus oocytes after injection of poly $(A)^+$-RNA from rat brain. The receptor activity was examined with the two-electrode voltage clamp technique by measuring transmembraneous ion currents through the different channels types, triggered by bath application of various specific agonists. Either uninjected oocytes or oocytes that had been injected with an equivalent volume of distilled water were used as controls for heterologous expression. These control oocytes did not show any measurable transmembraneous ion currents after application of the receptor agonists (n=30).

The glutamate-operated ion channels consist of different subtypes with distinct functional, molecular and pharmacological properties. These subtypes are commonly termed the NMDA-, AMPA- and metabotrope glutamate (mGlu) receptors (see Introduction). The different subtypes could be selectively activated in Xenopus oocytes by the reasonably specific (synthetic) agonists N-methyl-D-aspartate (NMDA; activates the NMDA-receptor), kainate (KA; activates the AMPA-

AMPA-receptor) and quisqualate (QUIS; activates the mGlu receptor). Typical agonist-induced responses are shown in Fig 4. Application of NMDA and KA induced smooth ion currents with different kinetics. The transmembraneous currents induced by NMDA are carried by the influx of sodium and calcium ions, whereas those induced by KA are carried only by the influx of sodium. The administration of QUIS evoked oscillatory currents carried by chloride ions. This response is mediated by the mGlu receptor type, which activates *via* the phospholipase C signaling pathway calcium-dependent chloride channels in the oocyte membrane (Blumenau et al. 2001). The specificity of the receptor responses was confirmed by different antagonists, which selectively blocked the receptor activity (data not shown).

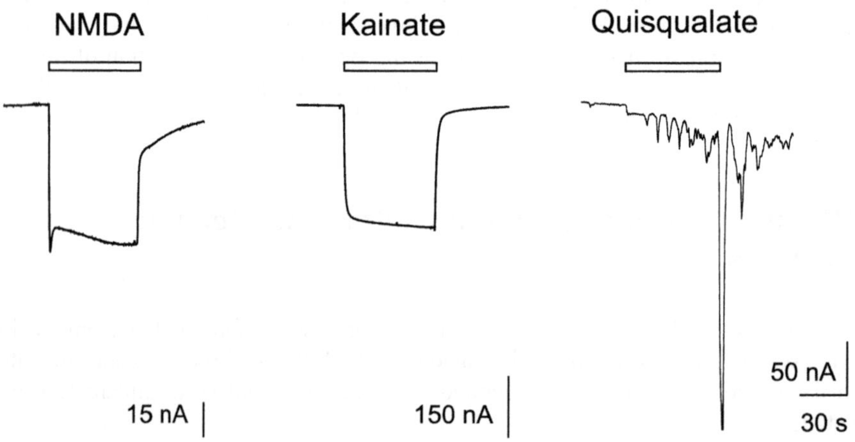

Fig. 4. Membrane currents of mRNA-injected oocytes of *Xenopus laevis*. Currents were elicited by the glutamate receptor channel agonists N-methyl-D-aspartate (NMDA; 50 µmol/l), kainate (50 µmol/l) and quisqualate (10 µmol/l). Recordings 5 days after injection of mRNA from rat brain. Application of the ligands is marked by horizontal bars. Holding potential: -70 mV. Inward current: downward deflection

Electrophysiological techniques

The electrophysiological investigations were performed from day 3 to day 12 after injection of poly(A)$^+$-RNA. The results were obtained from 97 oocytes of 35 donors. Membrane currents were recorded with the two-electrode voltage-clamp technique. The membrane currents were measured at a holding potential of -50 mV or -70 mV. The oocytes had membrane resistances of 1.01 ± 0.073 MΩ (mean ± SEM). In the majority of oocytes the membrane potential and resistance did not change with solution exchanges and remained constant for at least several hours.

The control bath fluid was a Ringer solution composed of (in mmol/l): NaCl 115, KCl 2, $CaCl_2$ 1.8, HEPES 10, pH 7.2. The concentrations of the tested ligands were (in µmol/l): KA 50 or 100, NMDA 100 + Glycin 10, QUIS 10. The different arsenicals in concentrations of 0.1 to 100 µmol/l were added to the bath solution and applied for 60 s. The agonists of the glutamate-operating ion channels and the arsenicals were 1 mmol/l in standard solution and were kept at 4 °C in glass bottles.

Reproducibility of the receptor-mediated ion currents (control experiments)

A prerequisite for testing pharmacological effects of substances on the function of transmitter operated ion channels is the stability and reproducibility of the agonist-induced membrane currents. To test these, series of control experiments were performed. The results showed that the amplitude of the ion current induced by a given concentration of an agonist could vary between individual oocytes and, in particular, between batches of oocytes from different frogs, but was relatively stable and reproducible in a given oocyte. These observations were taken to indicate that the efficiency of heterologous mRNA translation varied between oocytes from different donors, which affects the number of receptors synthesized, but the channel properties of each receptor type were essentially constant. However, repeated prolonged application of agonists with short recovery periods led to a gradual decline in the responsiveness of the oocytes. To avoid or minimize this, desensitization control experiments were performed for each receptor type.

The *AMPA receptor* was activated by 6 successive applications of KA for 60 s each and left for 10 min between applications (n=15). Using this protocol there was almost no desensitization, as demonstrated in 3 consecutive responses from the same cell in Fig. 5A. The amplitudes of the membrane currents were measured 2 s before termination of substance application. The first activation of the receptor was taken as 100% and the following ones were normalized to this control reaction. The mean and SEM of the membrane currents were determined separately for each of the 6 successive applications (represented by the white bars in Figs. 6-12).

The *NMDA receptor* was also activated by 6 successive applications of NMDA for 60 s each and left for 15-20 min between applications (n=19). However, activation of the NMDA receptor showed a desensitization of the second response (a reduction of the amplitude of approximately 20-25%), whereas the following responses were relatively stable compared to the second response, as demonstrated in 3 consecutive responses from the same cell (Fig. 5B). Therefore the amplitude of the first application was discarded, the second one was taken as 100%, and the following ones were normalized to this second reaction.

Activation of the *mGlu receptor* by QUIS (n=5) showed a constant desensitization of the responses, even when the delay between the consecutive application was prolonged to 30 min (Fig. 5C). For this reason, the respective arsenicals were applied during the first activation of the mGlu receptor by QUIS.

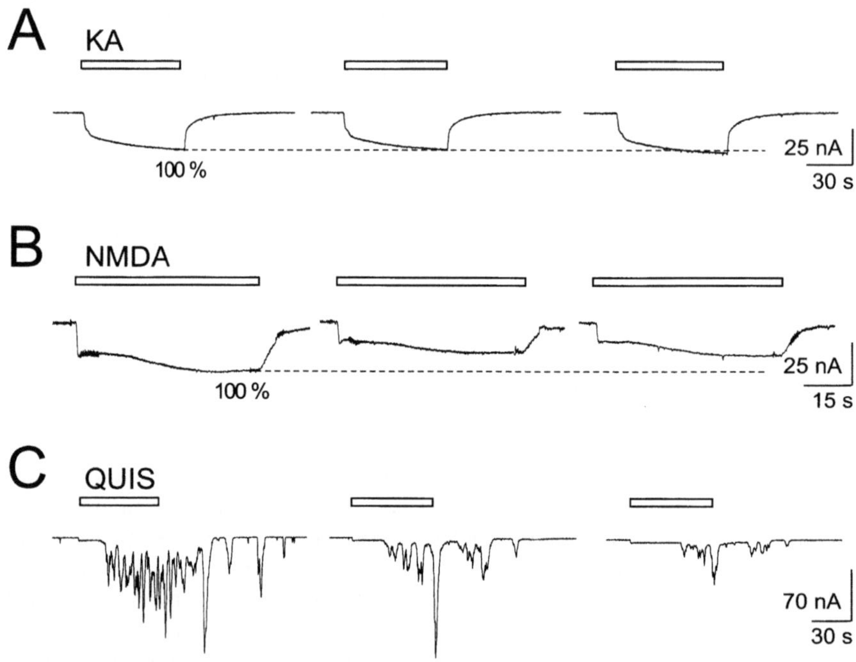

Fig. 5. Original recordings of the first 3 out of 6 control reactions from a single oocyte after application of 50 µmol/l kainate **(A)**, 50 µmol/l NMDA **(B)** and 10 µmol/l QUIS **(C)**. Delay between the applications: 10 min (A), 15 min (B, C). Application of the ligands is marked by horizontal bars. Holding potential: -70 mV. Inward current: downward deflection

Application of the arsenicals

To investigate the effect of arsenite, $MeAsO(OH)_2$ or Me_2AsOOH on the membrane potential of the oocytes these substances were administered to the oocytes in a concentration of 100 µmol/l without co-application of receptor agonists. However, none of substances showed any effect.

Testing of described effects of the arsenicals on the receptor channels was done analogously to the control experiments. KA or NMDA were administered to the oocytes first separately and then simultaneously with the arsenical, beginning with the lowest concentration (0.1 µmol/l) and continuing with increasing concentrations in the sequence 1, 10 and 100 µmol/l. Finally, a control application of KA or NMDA was performed to prove the reversibility of the effects. Corresponding to the control experiments, the first response of the AMPA receptor and the second response of the NMDA receptor was taken as 100%. The responses with co-application of arsenical and agonist were normalized to these control reactions.

For analysis of possible effects of the arsenicals on the receptor channels the amplitudes of all membrane currents were measured 2 s before termination of substance application. The mean value and SEM of the membrane currents were then determined separately for each of the 5-6 successive applications (represented by the dark bars in Figs. 7-14) and compared to the mean and SEM of the control experiments. The values obtained under control and test conditions were compared using a t-test or the Mann-Whitney-rank-sum test. Values of $p<0.05$ were taken as statistically significant.

Effects of inorganic arsenite

The interaction of arsenite with KA, NMDA- and QUIS-induced membrane currents was tested by simultaneous application of the agonists with arsenite in different concentrations.

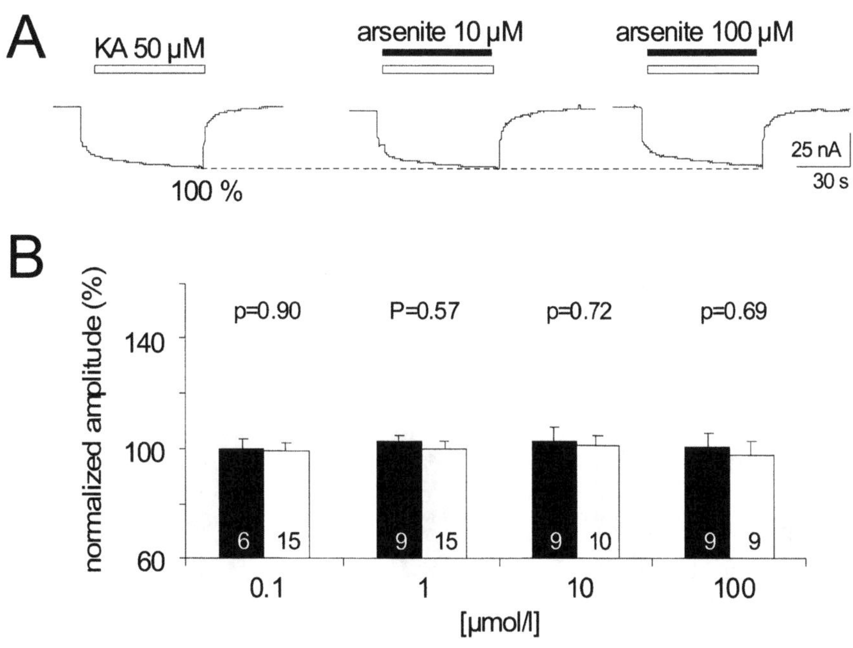

Fig. 6. Effects of inorganic arsenite on AMPA receptor channels. **A:** Original recordings of kainate (KA) induced membrane currents from a single oocyte after application of 50 µmol/l KA and after co-application of 50 µmol/l KA and 10 and 100 µmol/l arsenite. Applications of the ligands and the drugs are marked by horizontal bars. Holding potential: -70 mV. Inward current: downward deflection. **B:** Comparison of the results for KA-induced membrane currents after co-application of arsenite in comparison to the control measurements. Each bar represents mean value and SEM of the normalized current amplitudes determined for the number of oocytes inserted at the bottom of the column. White columns, control experiments; dark columns, experiments with co-application of arsenite.

AMPA-type receptors: Three original recordings of KA-induced membrane currents from a single oocyte are shown in Fig. 6A. The application of 50 μmol/l KA elicited an inward current which was not changed by co-application of 10 and 100 μmol/l arsenite. Fig. 6B summarizes the results for KA-induced membrane currents in comparison with the control measurements. Each column represents the mean value and SEM of the normalized current amplitudes determined for the number of experiments inserted at the bottom of the column. We found no significant effect of arsenite on the glutamate receptor of the AMPA-type.

NMDA-type receptors: Corresponding recordings and the summary for the NMDA-induced membrane currents are shown in Fig. 7. In this example, the application of 100 μmol/l NMDA elicited an inward current which was slightly reduced by co-application of 10 and 100 μmol/l arsenite. However, we found a high variability of the effects of arsenite on the glutamate receptor of the NMDA-type, which in summary also showed no significance.

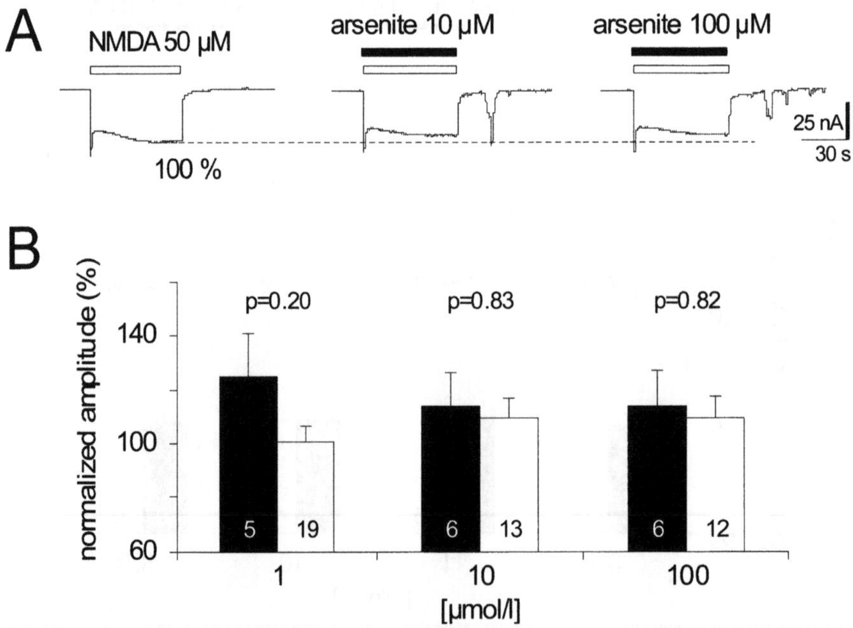

Fig. 7. Effects of inorganic arsenite on NMDA receptor channels. **A:** Original recordings of NMDA-induced membrane currents from a single oocyte after application of 50 μmol/l NMDA and after co-application of 50 μmol/l NMDA and 10 and 100 μmol/l arsenite. Applications of the ligands and the drugs are marked by horizontal bars. Holding potential: -70 mV. Inward current: downward deflection. **B:** Comparison of the results for NMDA-induced membrane currents after co-application of arsenite in comparison to the control measurements. Each bar represents mean value and SEM of the normalized current amplitudes determined for the number of oocytes inserted at the bottom of the column. White columns, control experiments; dark columns, experiments with co-application of arsenite

mGluR-type receptors: Original recording of QUIS-induced membrane currents from a single oocyte are shown in Fig. 8A. The application of 10 µmol/l QUIS elicited oscillatory inward currents which were not changed by co-application of 10 and 100 µmol/l arsenite, indicating that arsenite has no effect on glutamate receptors of the mGluR-type (n=5).

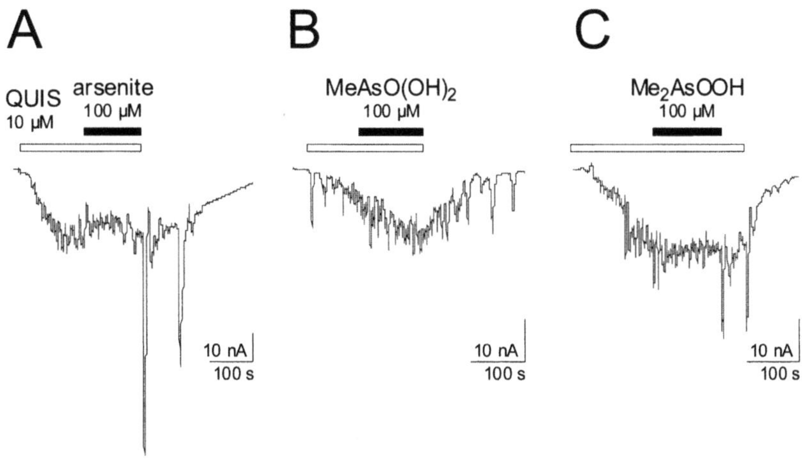

Fig. 8. Effects of arsenicals on mGluR-type receptor channels. Original recordings of QUIS-induced membrane currents from a single oocyte after application of 10 µmol/l QUIS and co-application of 100 µmol/l inorganic arsenite (**A**), 100 µmol/l monomethylarsonic acid ($MeAsO(OH)_2$) (**B**) and 100 µmol/l dimethylarsinic acid (Me_2AsOOH) (**C**). Applications of the ligands and the drugs are marked by horizontal bars. Holding potential: -60 mV. Inward current: downward deflection

Effects of monomethylarsonic acid ($MeAsO(OH)_2$)

AMPA-type receptors: Original recordings from a single oocyte after application of KA and after the simultaneous application of KA and $MeAsO(OH)_2$ are shown in Fig. 9A. The application of 50 µmol/l KA elicited an inward current which was not significantly changed by co-application of KA with 10 and 100 µmol/l $MeAsO(OH)_2$. Fig. 9B summarizes the experiments with different concentrations of $MeAsO(OH)_2$ in comparison with the control reactions. There was no significant effect of $MeAsO(OH)_2$ on the glutamate receptor of the AMPA-type.

NMDA-type receptors: Fig. 10A shows original recordings of NMDA-induced membrane currents from a single oocyte. In the example the application of 100 µmol/l NMDA elicited an inward current which was seriously enhanced by co-application of 10 and 100 µmol/l $MeAsO(OH)_2$. This enhancement of the NMDA-induced response was detected for all concentrations of $MeAsO(OH)_2$ tested and was significant for concentrations of 10 and 100 µmol/l. Fig. 10B summarizes the experiments with different concentrations of $MeAsO(OH)_2$ in comparison with the control measurements. The NMDA-induced inward current was significantly enhanced to 11%-32% in a concentration dependent manner when $MeAsO(OH)_2$ was applied additionally.

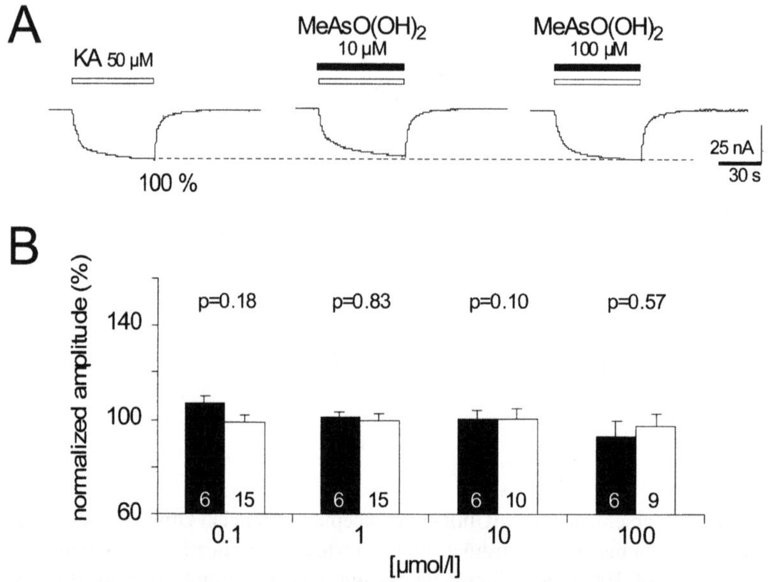

Fig. 9. Effects of monomethylarsonic acid ($MeAsO(OH)_2$) on AMPA receptor channels. **A:** Original recordings of kainate (KA) induced membrane currents from a single oocyte after application of 50 µmol/l KA and after co-application of 50 µmol/l KA and 10 and 100 µmol/l $MeAsO(OH)_2$. Applications of the ligands and the drugs are marked by horizontal bars. Holding potential: -70 mV. Inward current: downward deflection. **B:** Comparison of the results for KA-induced membrane currents after co-application of $MeAsO(OH)_2$ in comparison to the control measurements. Each bar represents mean value and SEM of the normalized current amplitudes determined for the number of oocytes inserted at the bottom of the column. White columns, control experiments; dark columns, experiments with co-application of $MeAsO(OH)_2$.

mGluR-type receptors: Original recordings of QUIS-induced membrane currents from a single oocyte are shown in Fig. 8B. The application of 10 µmol/l QUIS elicited inward currents which were not changed by co-application of 10 and 100 µmol/l $MeAsO(OH)_2$, indicating that $MeAsO(OH)_2$ has no effect on glutamate receptors of the mGluR-type (n=7).

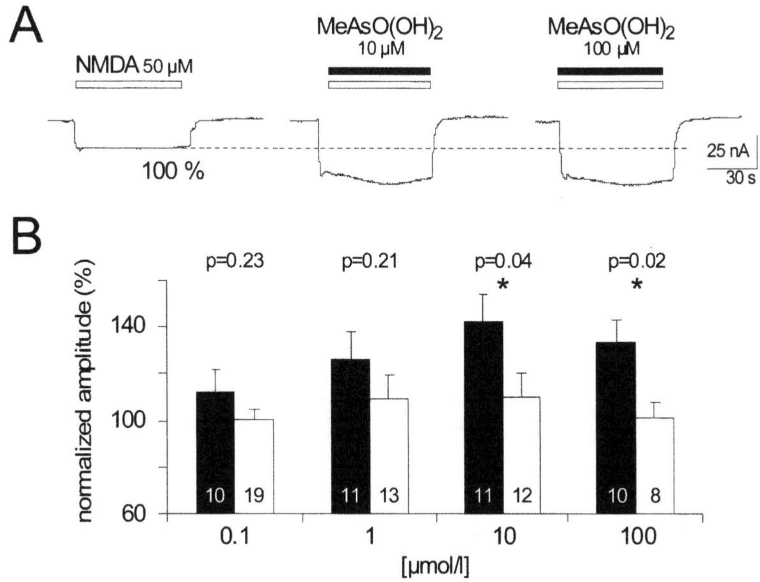

Fig. 10. Effects of monomethylarsonic acid ($MeAsO(OH)_2$) on NMDA receptor channels. **A:** Original recordings of NMDA-induced membrane currents from a single oocyte after application of 50 μmol/l NMDA and after co-application of 50 μmol/l NMDA and 10 and 100 μmol/l $MeAsO(OH)_2$. Applications of the ligands and the drugs are marked by horizontal bars. Holding potential: -70 mV. Inward current: downward deflection. **B:** Comparison of the results for NMDA-induced membrane currents after co-application of $MeAsO(OH)_2$ in comparison to the control measurements. Each bar represents mean value and SEM of the normalized current amplitudes determined for the number of oocytes inserted at the bottom of the column. White columns, control experiments; dark columns, experiments with co-application of $MeAsO(OH)_2$. Bars marked by asterisks are significantly different ($p<0.05$) from control values

Effects of dimethylarsinic acid (Me_2AsOOH)

AMPA-type receptors: Original recordings from a single oocyte after application of 50 μmol/l KA and after simultaneous application of KA with Me_2AsOOH are shown in Fig. 11A. The KA-induced inward current was reduced by co-application of 10 and 100 μmol/l Me_2AsOOH. The experiments with different concentrations of Me_2AsOOH in comparison with the control reactions are summarized in Fig. 11B. A reduction of the KA-induced membrane currents of 6%-13% was found for all concentrations of Me_2AsOOH tested. It was significant, however, only for concentrations of 1 - 100 μmol/l Me_2AsOOH.

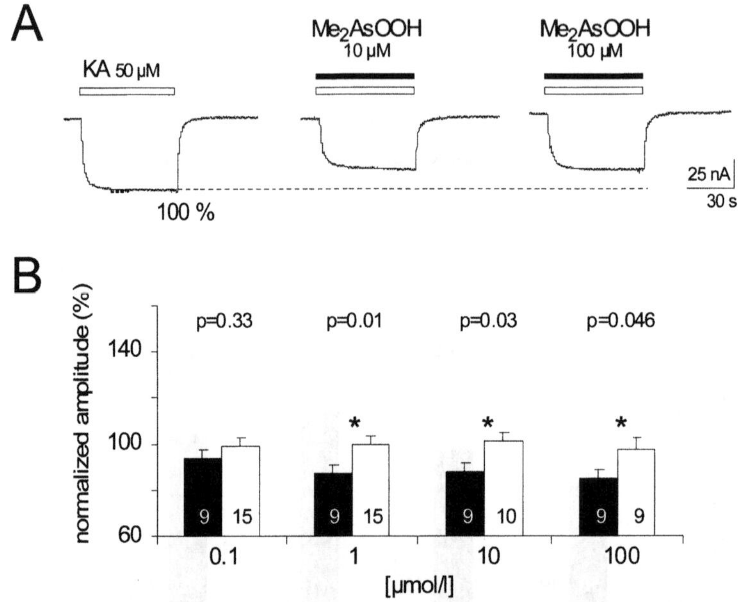

Fig. 11. Effects of dimethylarsinic acid (Me_2AsOOH) on AMPA receptor channels. **A:** Original recordings of kainate (KA) induced membrane currents from a single oocyte after application of 50 µmol/l KA and after co-application of 50 µmol/l KA and 10 and 100 µmol/l Me_2AsOOH. Applications of the ligands and the drugs are marked by horizontal bars. Holding potential: -70 mV. Inward current: downward deflection. **B:** Comparison of the results for KA-induced membrane currents after co-application of Me_2AsOOH in comparison to the control measurements. Each bar represents mean value and SEM of the normalized current amplitudes determined for the number of oocytes inserted at the bottom of the column. White columns, control experiments; dark columns, experiments with co-application of Me_2AsOOH. Bars marked by asterisks are significantly different ($p<0.05$) from control values

NMDA-type receptors: Original recordings from a single oocyte after application of NMDA and after co-application of Me_2AsOOH are shown in Fig. 12A . The application of 100 µmol/l NMDA elicited an inward current which was significantly reduced by co-application of 10 and 100 µmol/l Me_2AsOOH. Fig. 12B summarizes the experiments with different concentrations of Me_2AsOOH to the NMDA-induced membrane currents in comparison with the control experiments. The NMDA-induced inward current was reduced by 19%-39% in a concentration dependent manner with a maximum at 10 µmol/l Me_2AsOOH when Me_2AsOOH was applied additionally. However, this was significant only for concentrations of 0.1 - 10 µmol/l Me_2AsOOH.

mGluR-type receptors: Original recording of QUIS-induced membrane currents from a single oocyte are shown in Fig. 8C. The application of 10 µmol/l QUIS elicited inward currents which were not changed by co-application of 10 and 100 µmol/l Me_2AsOOH, indicating that this organic arsenic compound has no effect on glutamate receptors of the mGluR-type (n=5).

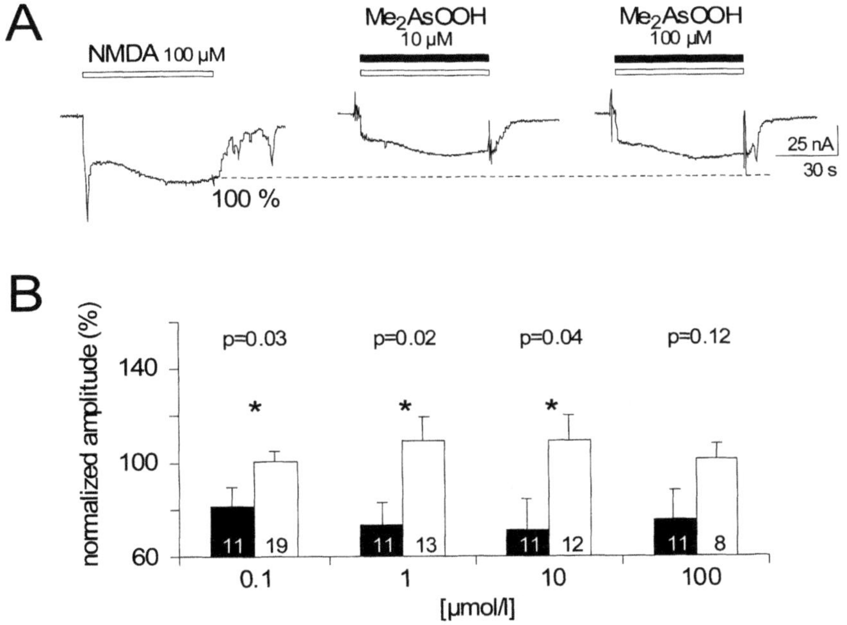

Fig. 12. Effects of dimethylarsinic acid (Me_2AsOOH) on NMDA receptor channels. **A:** Original recordings of NMDA-induced membrane currents from a single oocyte after application of 100 µmol/l NMDA and after co-application of 100 µmol/l NMDA and 10 and 100 µmol/l Me_2AsOOH. Applications of the ligands and the drugs are marked by horizontal bars. Holding potential: -70 mV. Inward current: downward deflection. **B:** Comparison of the results for NMDA-induced membrane currents after co-application of Me_2AsOOH in comparison to the control measurements. Each bar represents mean value and SEM of the normalized current amplitudes determined for the number of oocytes inserted at the bottom of the column. White columns, control experiments; dark columns, experiments with co-application of Me_2AsOOH. Bars marked by asterisks are significantly different ($p<0.05$) from control values

Effects of arsenicals on voltage-operated potassium channels

The effect of the two organic arsenicals $MeAsO(OH)_2$ and Me_2AsOOH and the inorganic arsenite on the function of different voltage-operated potassium channels was tested.

Transmembraneous ion currents through different voltage-operated potassium channels

Cloned voltage-operated potassium channels were expressed in Xenopus oocytes after injection of cRNA for the channels Kv1.1, Kv1.2, Kv2.1, Kv3.1 and the sodium channel rbII. The cDNA encoding for the potassium channel subunits was transcribed to cRNA using a commercial kit (mMessge, mMachine, Ambion, Austin, TX) and T7 RNA polymerase. Denaturating agarose gel electrophoresis was used to check the quality of the cRNA product of each reaction and to quantify the yield.

Voltage-operated potassium channels exist in a large number of distinct molecules with different physiological and pharmacological properties. Thus, injection of brain mRNA would lead to the expression of a large variety of voltage-operated channels and would involve the risk that potential antagonistic effects of arsenic compounds might offset each other, thus giving rise to negative results. Therefore, cRNA was injected, giving expression of only one type of potassium channel in the respective oocyte. The channels Kv1.1, Kv1.2, Kv2.1 and Kv3.1 were chosen since they are common in brain and represent different Kv subfamilies, and since a large set of chimeric channels is available which might be useful in detecting possible sites of action.

Electrophysiological techniques

For the electrophysiological recordings the outer part of the follicular tissues was stripped off some hours after injection of cRNA in order to prevent reduced access to the cell membrane (Madeja et al. 2001). Currents were recorded with the two-electrode voltage-clamp technique. The holding potential was -80 mV and command potentials were applied up to a potential of +60 mV. The control bath fluid was a Ringer solution (in mmol/l): NaCl 115, KCl 2, CaCl2 1.8, HEPES 10; pH 7.2. Arsenic compounds in concentrations of 1 to 1000 µmol/l were added to the bath solution and applied at least 30 s before eliciting currents. All experiments were performed on days 3 and 4 after injection of cRNA and were carried out at room temperature (22 ± 1°C).

The potassium currents obtained in two-electrode recordings were low-pass filtered at 1 kHz and were transferred to a computer system (pClamp program, Axon instruments, Foster City). The amplitudes of the total outward currents were corrected for leakage. Leakage currents and capacitive transients were subtracted online using a p/-4 pulse protocol. The potassium current amplitude was measured at the peak of current obtained during the depolarizing voltage step. Conductance-voltage relations were obtained by normalizing the conductance data to the maximal value under control conditions and by fitting the data to the Boltzmann equation $y = G_{max} [1 + \exp((V_{1/2} - V) / b)]$ where y is the normalized conductance, G_{max} is the normalized maximal conductance, $V_{1/2}$ is the potential of the half-maximal conductance, V is the voltage, and b is the slope factor. The measured values are given as mean or mean ± SEM. Statistical significance was tested using a t-test or a Mann-Whitney-rank-sum test. Values of $p<0.05$ were taken as statistically significant. Curve fitting and all statistical procedures were performed using the program SigmaPlot (Jandel Scientific, Erkrath, Germany).

Effects of arsenite, $MeAsO(OH)_2$ and Me_2AsOOH on the different potassium channels

In agreement with previous results, outward currents mediated by the Kv channels expressed in Xenopus oocytes appeared with positive going potential steps. For example, potassium currents of the Kv2.1 channel were found positive to -30 mV (Fig. 13A). The currents increased relatively slowly and did not inactivate significantly during the 500 ms test pulses. Depending on the current amplitudes (ΔI) and voltage steps (ΔMP), conductance values were calculated for the entire potential range and were normalized to the maximal value at +60 mV under control conditions (G_{rel}; Fig. 13B). Fitting the data to a Boltzmann equation yielded the open probability curve representing the voltage dependence of conductance of the respective channel (Fig. 13B). Eliciting the currents also in the presence of the arsenicals allowed the changes in the parameters of the open probability curves i.e. the changes in potential of half-maximal conductance ($V_{1/2CTRL}$-$V_{1/2TEST}$), in maximal conductance ($G_{relTEST}/G_{relCTRL}$) and slope factor ($slope_{TEST}$ / $slope_{CTRL}$) induced by the arsenicals to be compared.

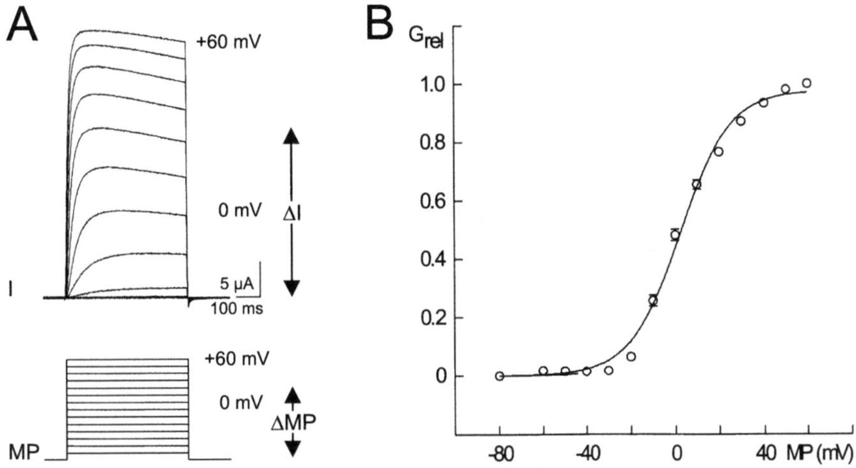

Fig. 13. Conductance-voltage relation of a voltage-operated potassium channel. **A:** Original recordings of currents (I) of Kv2.1 potassium channels at potentials (MP) from -80 mV up to +60 mV. The arrows show the voltage step (ΔMP) to +20 mV and the resulting current amplitude (ΔI). **B:** Conductance-voltage relation of Kv2.1 channels. For each potential, conductance values were calculated (ΔI/ΔMP) and normalized to the conductance value at +60 mV (Grel). Values are given as mean ± SEM of 10 experiments. The data points were fitted to a Boltzmann equation

Table 1: Effects of arsenicals on Kv potassium channels.
The arsenic compound was applied in a concentration of 1 mmol/l each. Channel: channel name; GmaxTEST/GmaxCTR: relative maximal conductance with the arsenic compound in relation to control. V1/2CTRL-V1/2TEST: difference of potential of half-maximal conductance under control and with the arsenic compound. slopeTEST/slopeCTRL: relative slope factor with the arsenic compound in relation to control. n: number of experiments

Substance	Channel	$G_{relTEST}/G_{relCTRL}$	$V_{1/2CTRL} - V_{1/2TEST}$	$slope_{TEST}/slope_{CTRL}$	n
inorganic arsenite	Kv1.1	0.99 ± 0.01	+0.5 ± 0.9 mV	0.99 ± 0.01	4
	Kv1.2	0.92 ± 0.01	+3.5 ± 1.4 mV	1.06 ± 0.02	5
	Kv2.1	1.03 ± 0.01	-0.5 ± 0.2 mV	0.99 ± 0.01	5
monomethyl-arsonic acid $MeAsO(OH)_2$	Kv1.1	0.94 ± 0.02	+0.5 ± 0.8 mV	0.97 ± 0.02	5
	Kv1.2	0.96 ± 0.01	-0.5 ± 0.6 mV	1.00 ± 0.02	5
	Kv2.1	0.99 ± 0.01	+2.1 ± 0.8 mV	1.00 ± 0.02	5
	Kv3.1	0.99 ± 0.02	-0.6 ± 0.9 mV	1.01 ± 0.01	5
dimethylarsinic acid Me_2ASOOH	Kv1.1	0.97 ± 0.01	+1.8 ± 1.0 mV	0.98 ± 0.01	5
	Kv1.2	0.96 ± 0.01	-0.3 ± 0.8 mV	0.96 ± 0.02	4
	Kv2.1	0.97 ± 0.01	+0.2 ± 0.3 mV	1.01 ± 0.00	5
	Kv3.1	0.97 ± 0.02	-0.4 ± 0.3 mV	1.00 ± 0.01	4

The results of the effects of $MeAsO(OH)_2$, Me_2AsOOH and arsenite on the parameters of open probability for the channels Kv1.1, Kv1.2, Kv2.1 and Kv3.1 are shown in the table above (Table 1). Only the data of the extremely high concentration of 1 mmol/l are shown; further experiments were made with 10 µmol/l which, however, showed smaller effects. In general, none of the arsenicals showed significant effects at any channel. The change in the maximal conductance of the channels was less than 10%, with the largest effect found with arsenite at the Kv1.2 channel (decrease of 8%). Shifts of the potential of half-maximal conductance were in the range of single millivolts only; the largest shift was again found for arsenite at the Kv1.2 channel (shift of +3.5 mV). In line with these missing or only negligible effects, the slope factors of the open probability curves were virtually unaffected by the arsenicals (changes ≤6%).

For sodium channels, the largest outward currents were obtained with voltage steps to around –10 mV in *Xenopus* oocytes. Currents appeared and declined within the first 50 ms of the voltage steps. Application of the arsenicals, however, showed no significant effect. At every potential and with every arsenic compound tested changes in current amplitude were less than 10 % of the control value.

Neuronal ion channels as targets for organic arsenicals

Many investigations have shown that ion channels are targets for neurotoxins (e.g. Madeja et al. 1995b, 1997a, 1997b; Mußhoff et al. 1995b, 1999, 2000). These investigations revealed that dysfunction of ion channels may be one important source of neurotoxic symptoms. The present investigation analyzed marked effects of organic arsenicals on transmitter-operated ion channels. These effects have to be discussed under different aspects.

Perspectives and limitation of Xenopus oocytes

Xenopus laevis oocytes have proved to be a reliable expression system for neuronal ion channels. This expression system offers several advantages in the investigation of the effects of arsenicals on ion channels:

- This in vivo expression system permits a direct assessment of drug effects on different neuronal ion channels, thus allowing for the analysis of basic mechanisms of drug action. Furthermore, factors (e.g. transmitter release or uptake from neighboring cells) which would inevitably impede investigations in nervous tissues can be excluded in this single-cell system.
- The Xenopus oocyte expresses ion channels in a common membrane environment, thus removing one source of variability. This consistent and defined background allows stable and reproducible ion currents of the receptors and channels.
- This system allows characterization of the mechanism of drug action on single defined ion channels without contamination by others. Furthermore, the injection of mRNA for different subtypes of specific ion channels allows drug action to be analyzed at a molecular level.
- Oocytes can be studied with the two-electrode voltage-clamp technique which perturbs the intracellular environment less than most whole-cell patch-clamp approaches.
- Xenopus oocytes allow investigations of intracellular effects of the substances.

Concerning pharmacological investigations, however, the Xenopus oocyte has one serious disadvantage: For several substances the pharmacological sensitivity of channels is reduced in comparison with mammalian cell lines. Thus it is obvious that the oocyte system is not suitable for measuring the "absolute" pharmacological values, although it is a useful tool in studies determining "relative" values, e.g. basic mechanisms of drug actions (Madeja et al. 2000). Despite this restriction, however, the oocyte expression system is a promising tool for investigating neurotoxic effects on elementary mechanisms of the generation and processing of bioelectrical information in the nervous system at a molecular level.

Effects of arsenicals on glutamate-operated ion channels

Various organic arsenicals were tested in the oocyte expression system for their ability to disturb the functions of the excitatory ionotropic and metabotropic glutamate receptors from the AMPA-, NMDA- and mGluR-type. The principal finding of this study is that monomethylarsonic acid ($MeAsO(OH)_2$) and dimethylarsinic acid (Me_2AsOOH) affect the ionotropic glutamate receptors in a different and concentration-dependent manner (Fig. 14). On the one hand, Me_2AsOOH reduces both the ion currents through AMPA- and NMDA-receptors. On the other hand, $MeAsO(OH)_2$ induces a marked increase of the NMDA-mediated response, whereas the AMPA receptor-mediated responses are unaffected by this substance. Furthermore, arsenite obviously does not interact with any of the glutamate receptor channels.

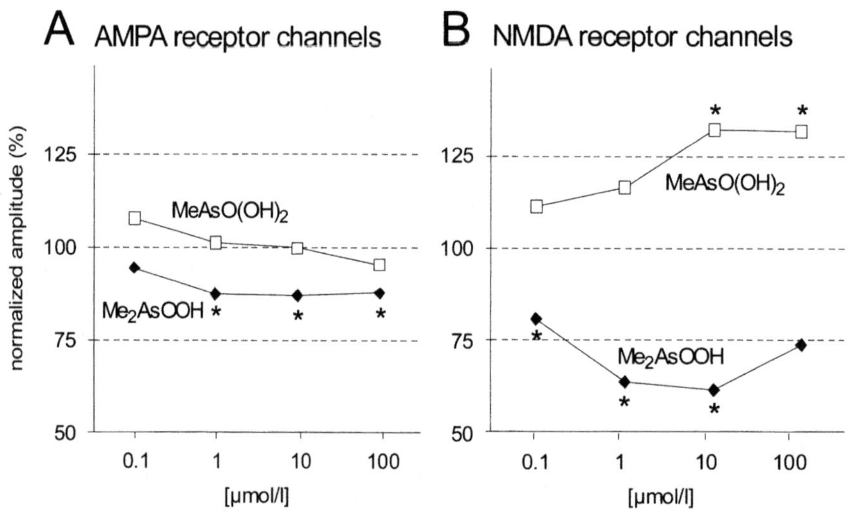

Fig. 14. Summary of the effects of $MeAsO(OH)_2$ and Me_2AsOOH on glutamate receptor channels. **A**: AMPA receptor channels, **B:** NMDA receptor channels. Each point represents the difference between the normalized mean amplitude after co-application of the arsenic compound and the control measurements. Points marked by asterisks are significantly different ($p<0.05$) from control values

The different and partially opposing effects of $MeAsO(OH)_2$ and Me_2AsOOH on the glutamate receptor channels are somewhat surprising considering the structural similarity of the organic arsenicals. They show that small changes in the chemical structure, in the case of Me_2AsOOH an additional methyl group, are obviously responsible for selective effects on receptor function. Since the effects of both organic arsenicals are restricted to the ionotropic glutamate receptor channels, whereas the metabotropic glutamate receptor and the voltage-dependent potassium channels are not impaired by this substance, it is likely that these compounds interact specifically with the receptor channels and not *via* a nonspecific interaction, for example with the surrounding plasma membrane. This is further supported by the finding that $MeAsO(OH)_2$ and Me_2AsOOH have no effects on the membrane potential of the oocytes (data not shown). From the presented experiments, however, it cannot be concluded whether the primary target site of the substances is located directly on the extra- or intracellular part of the channel proteins or on other intracellular factors that regulate the channel activities. To identify the molecular target sites, further and extended investigations are needed.

Although it is difficult to extrapolate directly from oocyte experiments to the situation in the nervous tissue, it can be speculated that the organic arsenicals would affect the excitatory transmission in a different manner. On the one hand, the blocking effects of Me_2AsOOH on the AMPA- and NMDA receptors may contribute to a general decrease in the neuronal activity. On the other hand, the enhancement of the NMDA responses by $MeAsO(OH)_2$ is an extraordinary effect. Overactivity of NMDA receptors, which causes alterations in Ca^{2+} homeostasis of the cells, plays an important role in the pathogenesis of certain brain dysfunctions such as epilepsy, stroke and neuronal degeneration.

In summary, the oocytes can be used as a tool for neurotoxicological investigations, since any change in ion channel function may be taken as proof of the neurotoxic potency of the tested substance. Taken together, the effects of monomethylarsonic acid and dimethylarsinic acid on NMDA- and AMPA- receptors of the glutamate receptor family point to a neurotoxic potential of these substances. Compared to the (absent) effect of arsenite on these ion channels, the organic arsenicals possess a considerably higher neurotoxic potential.

Effects of arsenicals on voltage-operated potassium channels

In the present investigation, the Kv channels and the rat brain sodium channel were found to be insensitive to arsenite, $MeAsO(OH)_2$, and Me_2AsOOH. The effects on parameters of the open probability curves were less than 10% in all cases with the concentration of 1 mmol/l each. In contrast, significant effects have been found for a variety of divalent cations. Metal ions like gadolinium, zinc and lanthanum belong to this group (e.g. Armstrong and Cota, 1990; Elinder and Århem, 1994; Tytgat and Daenens, 1997). These cations are assumed both to act as a screen for the surface charges and to bind to the channel, affecting the voltage sensor directly. Whether the arsenic ion also belongs to this group cannot be tested since tri- and pentavalent arsenic ions are not stable in aqueous solutions. Thus, it can only be assumed that the introduction of methyl groups and/or the oxidation of the ion is preventing the effect.

Although it has not been clarified up to now how and where metal ions affect the potassium currents, it is likely that the metal ions interact with superficially located charges of the channel molecule. An introduction of side chains or more complex compounds might shield the charge of the ion or might prevent the metal ion from gaining access to its proper binding site. Besides speculations on the mechanism of the missing effect, however, it has to be stated that the greater toxicity of arsenic metalloid compounds is most probably not due to an effect on voltage-operated potassium channels.

Significance in arsenic poisoning

Though the understanding of neurotoxic diseases has grown during recent years, there still is inadequate knowledge about the mechanisms involved. This is especially true of the association of experimentally demonstrated alterations of nervous tissue and neuronal functions with neurological, psychiatric or behavioral disorders. The results of the investigations presented here prove that organic arsenicals interfere with the basic mechanisms of neuronal signal generation, transmission and processing. Though the inhibiting effect of Me_2AsOOH on AMPA- and NMDA-receptors and the enhancing effect of $MeAsO(OH)_2$ on the NMDA-mediated response are in accordance with some of the clinical symptoms found in arsenic poisoning (e.g. seizures, loss of hearing, loss of taste, blurred vision), there is no concluding evidence that all or some of the symptoms are causally determined by interferences with ion channel functions. Alterations of other target structures may co-contribute to arsenical neurotoxicity. Nevertheless, interferences with ion channel functions may lead to severe neurological malfunctions and therefore should at least be regarded as important indicators of the neurotoxic potency. Thus it has to be concluded that, compared with the noneffective inorganic arsenate, the organic arsenicals $MeAsO(OH)_2$ and Me_2AsOOH possess a considerably higher neurotoxic potential with respect to the ionotropic glutamate receptors.

Literature

Akaike N, Inoue M, Krishtal OA (1986) Concentration clamp`study of γ-aminobutyric acid-induced chloride current kinetics in frog sensory neurones. J Physiol 379:171-185

Alkondon M, Costa ACS, Radhakrishnan V, Aronstam RS, Albuquerque EX (1990) Selective blockade of NMDA-activated channel currents may be implicated in learning deficits caused by lead. FEBS Lett 261:124-130

Aposhian HV (1997) Enzymatic methylation of arsenic species and other new approaches to arsenic toxicity. Annu Rev Pharmacol Toxicol 37:397-419

Appel KE, Böhme C, Platzek T, Schmidt E, Stinchcombe S (2000) Organozinnverbindungen in verbrauchernahen Produkten und Lebensmitteln. Umweltmed Forsch Prax 5:67-77

Armstrong CM, Cota G (1990) Modification of sodium channel gating by lanthanum. Some effects that cannot be explained by surface charge theory. J Gen Physiol 96:1129-1140

Atchison WD, Narahashi T (1982) Methylmercury-induced depression of neuromuscular transmission. Neurotox 3:37-50

Audesirk G (1993) Electrophysiology of lead intoxication: Effects on voltage sensitive ion channels. Neurotox 14:137-148

Binding N, Madeja M, Mußhoff U, Neid U, Altrup U, Speckmann E-J, Witting U (1996) Prediction of neurotoxic poteny of harzardous substances with a modular in vitro test battery. Toxicol Lett 88:115-120

Bloms P, Mußhoff U, Madeja M, Müsch-Nittel K, Kuhlmann D, Spener F, Speckmann E-J (1992) Suppression of a ligand operated membrane current by the epileptogenic agent pentylenetetrazol in oocytes of *Xenopus laevis* after injection of rat brain RNA. Neurosci Lett 147:155-158

Bloms-Funke P, Mußhoff U, Madeja M, Spener F, Speckmann E-J (1994) Decrease and increase of responses to glutamate receptor agonists in RNA-injected Xenopus oocytes by the epileptogenic agent pentylenetetrazol: dependence on the agonist concentration. Neurosci Lett 181:161-164

Bloms-Funke P, Madeja M, Mußhoff U, Speckmann E-J (1996) Effects of pentylenetetrazol on GABA receptors expressed in Xenopus oocytes: extra- and intracellular sites of action. Neurosci Lett 205:115-118

Blumenau C, Berger E, Fauteck J-D, Madeja M, Wittkowski W, Speckmann E-J, Mußhoff U (2001) Expression and functional characterization of the mt_1 melatonin receptor from rat brain in Xenopus oocytes: Evidence for coupling to the phosphoinositol pathway. J Pineal Res 30:139-146

Brouwer OF, Onkenhout W, Edelbroek PM, de Kom JFM, de Wolff FA, Perters ACB (1992) Increased neurotoxicity of arsenic in methylenetetrahydrofolate reductase deficiency. Clin Neurol Neurosurg 94:307-310

Büsselberg D (1995) Calcium channels as target sites of heavy metals. Toxicol Lett 82:255-261

Chattopadhyay S, Bhaumik S, Chaudhury AN, Das Gupta S (2002) Arsenic induced changes in growth development and apotosis in neonatal and adult brain cells in vivo and in tissue cultures. Toxicol Lett 128:73-84

Cheung MK, Verity MA (1985) Experimental methyl mercury neurotoxicity: locus of mercurial inhibition of brain protein synthesis in vivo and in vitro. J Neurochem 44:1799-1808

Craig PJ (1986) General comments on the toxicities of organometallic compounds. In: Organometallic compounds in the environment. Longmans Group, Harlow, 30-31pp

Cullen WR, McBride BC, Reglinski J (1984) The reaction of methylarsenicals with thiols: some biological implications. J Inorg Biochem 21:179-194

Dascal N (1987) The use of Xenopus oocytes for the study of ion channels. CRC Crit Rev Biochem 22:317-387

Delgado JM, Dufour L, Grimaldo JI, Carrizales L, Rodriguez VM, Jiminez-Capdeville ME (2000) Effects of arsenite on central monoamines and plasmatic levels of adrenocorticotropic hormone (ACTH) in mice. Toxicol Lett 117 :61-67

Denny MF, Hare MF, Atchison WD (1993) Methylmercury alters intrasynaptosomal concentrations of endogeneous polyvalent cations. Toxicol Appl Pharmacol 122:222-232

Dingledine R, Borges K, Bowie D, Traynelis SF (1999) The glutamate receptor ion channels. Pharm Rev 51:7-61

Dumont JN (1972) Oogenesi in *Xenopus laevis* (Daudin) I. Stages of oocyte development in laboratory maintained animals. J Morphol 136:153-180

Elinder F, Århem P (1994) Effects of gadolinium on ion channels in the myelinated axon of *Xenopus laevis*: four sites of action. Biophys J 67:71-83.

Feldmann RG, White RF, Eriator II (1993) Trimethyltin encephalopathy. Arch Neurol 50:1320-1324

Flott-Rahmel B, Falter C, Schluff P,Fingerhur R, Christensen E, Jacobs C, Mußhoff U, Fauteck J-D, Deufel T, Ludolph A, Ullrich K (1997) Nerve cell lesions caused by 3-hydroxyglutaric acid: A possible mechanism for neurodegeneration in glutaric acidaemia I J Inher Metab Dis 20:387-390

Flott-Rahmel B, Schürmann C, Schluff P, Fingerhut R, Musshoff U, Fowler B, Ullrich K (1998) Homocysteic and homocysteine sulphinic acid exhibit excitotoxicity in organotypic cultures from rat brain. Eur J Pediatr 157:112-117

Gerr F, Letz R, Ryan PB, Green RC (2000) Neurological effects of environmental exposure to arsenic in dust and soil among humans. Neurotoxicol 21:475-488

Gurdon JB, Lane CD, Woodland HR, Marbaix G (1971) Use of frog eggs and oocytes for the study of messenger RNA and its translation in living cells. Nature 233:177-181

Harkins AB, Armstrong DL (1992) Trimethyltin alters membrane properties of CA1 hippocampal neurons. Neurotox 13:569-582

Hille B (1991) Ionic channels of excitable membranes. Sinauer associates, Sunderland, USA.

Komulainen H, Keränen A, Saano V (1995) Methylmercury modulates $GABA_A$ receptor comlex differentially in rat cortical and cerebellar membranes in vitro. Neurochem Res 20:659-662

Kuznetsov DA, Richter V (1987) Modulation of messenger RNA metabolism in experimental methyl mercury neurotoxicity. Int J Neurosci 34:1-17

Lagerkvist BJ, Zetterlund B (1994) Assessment of exposure to arsenic among smelter workers: a 5-year follow-up. Am J Ind Med 25:477-488

Lin T-J, Hung D-Z, Kao C-H, Hu W-H, Yang D-Y (1998) Unique cerebral dysfunction following triphenylacetate poisoning. Human Exp Toxicol 17:403-405

Liu J, Zheng B, Aposhian HV, Zhou Y, Chen M-L, Zhang A, Waalkes MP (2002) Chronic arsenic poisoning from burning high-arsenic-containing coal in Guizhou, China. Environ Health Perspect 110:119-122

Luong KVQ, Nguyen LTH (1999) Organic arsenic intoxication from bird's nest soup. Am J Med Sci 317:269-271

Madeja M, Mußhoff U (1992) Die Eizellen des Krallenfrosches als Modell in der Neurophysiologie. EEG-LABOR 14:25-37

Madeja M, Mußhoff U, Speckmann E-J (1991a) A concentration-clamp system allowing two-electrode voltage-clamp investigations in oocytes of *Xenopus laevis*. J Neurosci Meth 38:267-269

Madeja M, Mußhoff U, Kuhlmann D, Speckmann E-J (1991b) Membrane currents elicited by the epileptogenic drug pentylenetetrazol in the native oocyte of *Xenopus laevis*. Brain Res 553: 27-32

Madeja M, Stocker M, Mußhoff U, Pongs O, Speckmann E-J (1994) Potassium currents in epilepsy: effects of the epileptogenic agent pentylenetetrazol on a cloned potasssium channel. Brain Res 656:287-294

Madeja M, Mußhoff U, Speckmann E-J (1995a) Improvement and testing of a concentration-clamp system for oocytes of *Xenopus laevis*. J Neurosci Meth 63:211-213

Madeja M, Binding N, Mußhoff U, Pongs O, Witting U, Speckmann E-J (1995b) Effects of lead on cloned neuronal voltage-operated potassium channels. Nauyn-Schmiedeberg's Archives of Pharmacology 351:320-327

Madeja M, Mußhoff U, Lorra C, Pongs O, Speckmann E-J (1996) Mechanism of action of the epileptogenic drug pentylenetetrazol on a cloned neuronal potassium channel. Brain Res 722:59-70

Madeja M, Binding N, Mußhoff U, Witting U, Speckmann E-J (1997a) Effects of n-hexane and its metabolites on cloned voltage-operated potassium channels. Arch Toxicol 71:238-242

Madeja M, Mußhoff U, Speckmann E-J (1997b) Diversity of potassium channels contributing to differences in brain-area-specific seizure susceptibility: sensitivity of different potassium channels to the epileptogenic agent pentylenetetrazol. Eur J Neurosci 9:390-395

Madeja M, Mußhoff U, Speckmann E-J (1997c) Follicular tissues reduce drug effects on ion channels in oocytes of Xenopus leavis. Eur J Neurosci 9:599-604

Madeja M, Mußhoff U, Binding N, Witting U, Speckmann E-J (1997d) Effects of Pb^{2+} on delayed-rectifier potassium channels in acutely isolated hippocampal neurons. J Neurophysiol 78:2649-2654

Madeja M, Müller V, Mußhoff U, Speckmann E-J (2000) Sensitivity of native and cloned hippocampal delayed-rectifier potassium channels to verapamil. Neuropharmacol 39:202-210

Mazumder D, Das Gupta J, Chakraborty AK, Chatterjee A, Das D, Chakraborti D (1992) Environmental pollution and chronic arsenicosis in south Calcutta. Bull World Health Org 70:481-485

Mergenthaler J, Haverkamp W, Hüttenhofer A, Skryabin BV, Mußhoff U, Borggrefe M, Speckmann E-J, Breithardt G, Madeja M (2001) Blocking effects of the antiarrhythmic drug propafenone on the HERG potassium channel. Naunyn-Schmiedeberg`s Archiv Pharmacology 363:472-480

Morton WE, Caron GA (1989) Encephalopathy: an uncommon manifestation of workplace arsenic poisoning. Am J Ind Med 15:1-5

Mußhoff U, Madeja M, Bloms P , Speckmann E-J (1994) Effects of the epileptogenic agent bicuculline methiodide on membrane currents induced by N-methyl-D-aspartate and kainate (Oocyte; *Xenopus laevis*). Brain Res 639:135-138

Mußhoff U, Madeja M, Bloms-Funke P, Speckmann E-J (1995a) Effects of the epileptogenic agent strychnine on membrane currents elicited by agonist of the NMDA and non-NMDA receptors in Xenopus oocytes. Comp Biochem Physiol 111A:65-71

Mußhoff U, Madeja M, Binding N, Witting U, Speckmann E-J (1995b) Lead-induced blockade of kainate-sensitive receptor channels. Naunyn-Schmiedeberg's Archives of Pharmacol 353:42-45

Mußhoff U, Madeja M, Binding N, Witting U, Speckmann E-J (1999) Effects of 2-phenoxyethanol on N-methyl-D-aspartate (NMDA) receptor-mediated ion currents. Arch Toxicol 73:55-59

Mußhoff U, Madeja M, Binding N, Witting U, Speckmann E-J (2000) 2-phenoxyethanol: a neurotoxicant? Arch Toxicol 74:284-287

Namgung U, Xia Z (2000) Arsenite-induced apoptosis in cortical neurons is mediated by c-jun N-terminal protein kinase 3 and p38 mitogen-activated protein kinase. J Neurosci 20:6442-6451

Oortgiesen M, Visser E, Vijverberg HPM, Seinen W (1996) Differential effects of organotin compounds on voltage-gated potassium currents in lymphocytes and neuroblastoma cells. Naunyn-Schmiedebergs Arch Pharmacol 353:136-143

Oyama Y, Chikahisa L, Hayashi A, Ueha T, Sato M, Hideki M (1992) Triphenyltin-induced increase in the intracellular Ca^{2+} of dissociated mammalian CNS neuron: its independence from voltage-dependent Ca^{2+} channels. Japan J Pharmacol 58:467-471

Petrick JS, Ayala-Fierro F, Cullen WR, Carter DE, Aposhian HV (2000) Monomethylarsonous acid (MMA^{III}) is more toxic than arsenite in Chang human hepatocytes. Toxicol Appl Pharmacol 163:203-207

Rodriguez VM, Carrizales L, Jiminez-Capdeville ME, Dufour L, Giordano M (2001) The effects of sodium arsenite exposure on behavioral parameters in the rat. Brain Res Bull 55:301-308

Rolf S, Haverkamp W, Borggrefe W, Mußhoff U, Eckhardt L, Mergenthaler J, Snyders DJ, Pongs O, Speckmann E-J, Breithardt G, Madeja M (2000) Effects of antiarrhythmic drugs on cloned cardic voltage-gated potassium channels expressed in Xenopus oocytes. Naunyn-Schmiedeberg`s Archiv Pharmacol 362:22-31

Salanki J, Gyeri J, Platokhin A, Rozsa KS (1998) Neurotoxicity of environmental pollutants: Effects of tin (Sn^{2+}) on Ach induced currents in snail neurons. Russ J Physiol 84:1061-1073

Shafer TJ (1998) Effects of Cd^{2+}, Pb^{2+} and CH_3Hg^+ on high voltage-activated calcium currents in pheochromocytoma (PC12) cells: potency, reversibility, interactions with extracellular Ca^{2+} and mechanisms of block. Toxicol Lett 99:207-221

Shafer TJ, Atchison WD (1989) Block of ^{45}Ca uptake into synaptosomes by methylmercury: Ca++ and Na+-dependence. J Pharmacol Exp Ther 248:696-702

Shafer TJ, Atchison WD (1992) Effects of methylmercury on perineural Na^+ and Ca^{2+}-dependent potentials of neuromuscularv junctions of the mouse. Brain Res 595:215-219

Sigel E (1990) Use of Xenopus oocytes for the functional expression of plasma membrane proteins. J Membrane Biol 117:201-221

Stühmer W, Stocker M, Sakmann B, Seeburg P, Baumann A, Grupe A, Pongs O (1988) Potassium channels expressed from rat brain cDNA have delayed rectifier properties. FEBS Lett 242:199-206

Stühmer W (1992) Electrophysiological recordings from Xenopus oocytes. Methods in Enzymology 207:319-339

Styblo M, del Razo LM, Vega L, Germolec DR, LeCluyse EL, Hamilton GA, Reed W, Wang C, Cullen WR, Thomas DJ (2000) Comparative toxicity of trivalent and pentavalent inorganic and methylated arsenicals in rat and human cells. Arch Toxicol 74:289-299

Sumikawa K, Houghton M, Emtage JS, Richards BM, Barnard EA (1981) Active multisubunit acetylcholine receptor assembled by translation of heterologous mRNA in Xenopus oocytes. Nature 292:862-864

Takeuchi T (1968) Pathology of Minamata disease. In: Minamata disease (Kustuma M, ed), study group of Minamata disease, Kumamoto University, Japan, pp 141-228

Takeuchi T (1982) Pathology of Minamata disease. Acta Pathol Jpn 32:73-99

Thayer JS (1995) Environmental chemistry of the heavy elements: hydrido and organo compounds. VCH Verlagsgesellschaft mbH, Weinheim, pp 29-41

Tytgat J, Daenens P (1997) Effect of lanthanum on voltage-dependent gating of a cloned mammalian neuronal potassium channel. Brain Res 749:232-237.

Ullrich K, Flott-Rahmel B, Schluff P, Mußhoff U, Das A, Lücke T, Steinfeld R, Christensen E, Jacobs C, Ludolph A, Neu A, Röper R (1999) Glutaric aciduria type I: Pathomechanisms of neurodegeneration. J Inher Metab Dis 22:392-403

Vahter M (1994) Species differences in the metabolism of arsenic compounds. Appl Organomet Chem 8:175-182

Valciukas JA (1991) Foundations of environmental and occupational neurotoxicology. Van Nostrand Reinhold, New York

Vega L, Styblo M, Patterson R, Cullen W, Wang C, Germolec D (2001) Differential effects of trivalent and pentavalent arsenicals on cell proliferation and cytokine secretion in normal human epidermal keratinocytes. Toxicol Appl Pharmacol 172:225-232

Wagner CA, Fridrich B, Setiawan I, Lang F, Bröer S (2000) The use of *Xenopus laevis* oocytes for the functional characterization of heterologously expressed membrane Proteins Cellular Physiol Biochem 10:1-12

Yuan YK, Atchison WD (1997) Action of methylmercury on GABA(A) receptor-mediated inhibitory synaptic transmission is primarily responsible for its early stimulatory effects on hippocampal CA1 excitatory synaptic transmission. J Pharmacol Exp Ther 282:64-73

Chapter 17

Panel discussion: Analytical aspects

Discussion Session on "Speciation Analysis of Environmental Samples"

Convener: H. Emons

Summary

The discussion was stimulated by experiences of the participants concerning existing experimental limitations in the areas of sample preparation and various quantification methods for trace metal and mainly metalloid species.

Several participants pointed out that stability studies of pure species in solution provide often only limited information about the stability of the same compounds in real matrices. As an example labile arsenic gluthatheine complexes in diluted solutions can be stabilized by an excess of ligand as expected from classic complexation theory. Therefore, investigations of pure species standards are not useful in such cases. Also the long lifetime of methylmercury in fish tissue (in contrast to the instability of the pure species or its solutions) points to the fact that one has actually to define for matrix-embedded chemical species more than just the "classic" description of chemical bonding to the neighboring atoms or functional groups. Obviously the whole chemical microenvironment of the metal(loid) compound has to be much better characterized and must be more taken into account. The present situation of environmental speciation analysis can be described as adapting the samples to the available analytical instruments and methods. But this should be turned around in the future! Tailoring the whole analytical processes to the problem of interest was felt to be of utmost importance for increasing the significance and acceptance of speciation analysis in environmental sciences. Recent advances of solid-state methods such as EXAFS with which now the first coordination sphere of an element in frozen cells can be investigated open the road for speciation studies without extraction steps. But it was clearly indicated that there will not exist an universal method for speciation analysis also in the future. The present co-existence of "established" approaches (consisting mainly of hyphenated methods with chromatography and spectrometry) and expensive cutting-edge techniques for selected investigations in a few laboratories will remain.

The application of capillary electrophoresis (CE) in environmental speciation studies is still hampered by the strong matrix dependence of this technique. Other limitations such as insufficient limits of quantification in instrumental configurations using optical detection based on Beer's law can be overcome now with MS detectors or laser-induced fluorescence. But the latter requires often a species derivatization. CE will profit from a better understanding of the interfacial processes at the capillary surface which would allow more fundamental improvements of the stability and reproducibility of this critical part of the separation set-up.

Several groups present at this workshop have collected considerable experience with techniques using hydride generation (HG). HG-atomic absorption spectrometry or HG-atomic fluorescence spectrometry are relatively inexpensive approaches for trace analysis of some metalloids and are useful if more sophisticated hyphenated methods are not available. HG works fine for speciation analysis of several As, Sb and Sn compounds. On the contrary mercury speciation is hampered by problems due to 'transmethylation/decomposition' in the hydride generation module.

Another point of discussion was centered upon calibration in speciation analysis. For instance, limited stability has been observed for blood samples and points to the problem of appropriate calibration strategies. Several laboratories have found a species-dependent sensitivity of ICP-MS, mostly in combination with HPLC coupling. But W.R. Cullen reported also such dependencies if As species were directly injected into the ICP. Moreover irreversible effects in the chromatographic columns such as sticking of some species on surfaces of HPLC or even GC columns do not allow to perform the consecutive species separation-measurement process with 100 % yield resulting in a quantification problem. In such cases even post-column isotope dilution does not improve the accuracy of the total species analysis. All participants have recommended to use standard addition as general calibration approach. But questions remain regarding the identification of species based on chromatographic retention times. Different interactions between the standard and the matrix are not unlikely (restricting also the power of isotope dilution techniques because of equilibration problems) and prevent often the identification of new compounds. Therefore, much more qualitative information is necessary for increasing the knowledge in the field of environmental speciation. Combinations of structural methods of analysis (such as ESI-MS) with powerful quantification methods will drive future developments.

Overall it was pointed out that speciation analysis has to be improved with respect to quality assurance (QA) concepts, QA tools including reference materials and achievable QA parameters in particular reproducibility and robustness of the total procedures. This seems to be crucial for its recognition as reliable source of toxicological, environmental and health information.

As a further outlook to future trends of speciation analysis the participants highlighted the necessity to progress the development of "direct methods", i.e. analysis with significantly reduced sample manipulations or even without sample preparation. Such method developments should have highest priority and would allow also real trace species studies in situ in the future. Moreover a better spatial resolution of speciation procedures should permit the recognition of the heterogeneous nature of most of the environmental samples thus pushing this field of analysis closer to the real world situation.

Chapter 18

Panel discussion: Toxicological Aspects

Discussion Session on "Toxicological Aspects of Alkylated Metal(loid) Species"

Convener: A. W. Rettenmeier

Summary

In contrast to the broad knowledge which has accumulated in the field of speciation analysis, substantial toxicological data have only been collected on a few organometal(loid) compounds and only by a small number of research groups. Thus, it was not surprising that the contributions during the panel discussion on toxicological aspects of organometallics were largely of a speculative nature, particularly as far as contributions to mechanistic aspects were concerned. This became already apparent when the first topic of this session, the toxicological relevance of organometal(loid) uptake by the cells, was discussed. While the audience basically agreed upon the assumption that organometal(loid) compounds are generally more toxic than the respectiv inorganic species, contradictory views were expressed with regard to the elements of this increased toxicity. According to the prevailing hypothesis, the high toxicity of many organometal(loid) compounds is due to their high lipophilicity or - in case of the alkylated ionic species - to their amphiphilic properties which both should alleviate the penetration of these compounds through cell membranes. At first sight, this hypothesis seems to be plausible not just from a physico-chemical point of view. It is also supported by the fact that the toxicity of some organic species such as organotins exceeds by far that of the inorganic compounds.

For several participants, however, this view of organometal(loid) toxicity appeared to be too simplistic. As contradictory arguments were advanced:

- The neurotoxicity of inorganic lead is as high as that of organic lead compounds
- Organometal(loid)s may severely damage the neonatal brain, although the lipid content of the neonatal brain is smaller than that of the adult brain.
- Organometal(loid) compounds must not enter cells to exert neurotoxic effects. Such effects can be mediated by the interaction of these compounds with receptors on the cell surface or with ion channels.

It was concluded from the discussion that there is no general mechanism upon which the toxicity of organometal(loid) compounds is based. Every given toxic effect seems to be characterized by a specific mode of action. Proteins suitable as targets for organometalloid attack might be located both on the cell surface and inside the cell. To elucidate the mechanisms responsible for the toxic action of organometal(loid)s, the following tasks were considered particularly important: i) The organometal(loid) species must be identified and quantified in membranes, body fluids, and tissues *in vivo* using modern speciation techniques. Speciation will tell, whether organometal(loid)s must be dealkylated inside the cell to ionic species in order to exert their toxic effects. ii) The stability of organometal(loid) species and the respective equilibria under real conditions has to be assessed. iii) The penetration of organometal(loid) species through cell membranes should be investigated in model systems.

A specific issue addressed in the discussion on mechanisms of organometal(loid) toxicity was the nature of the genotoxic acitivity shown by some organometal(loid) species. Genotoxicity studies of participants demonstrating the induction of micronuclei, chromosomal aberrations, DNA strandbreaks etc. by organoarsenic and organotin species have confirmed the considerable genotoxic potential of these compounds observed by other researchers. Nevertheless, the basic question whether the genotoxicity of these compounds is due to a direct effect of these compounds on DNA or whether it results from an indirect mode of action has not yet been satisfactorily answered. According to the studies of a participant (A. Hartwig), it appears that the genotoxic effects of organic and inorganic arsenicals are caused both by the induction of oxidative DNA damage (mediated by oxygen and arsenic peroxide radicals) and by the interaction with proteins (e. g. by inhibition of repair enzymes). On the contrary, it was considered unlikely that a direct transfer of methyl groups from methylated arsenic species to DNA occurs. Since the intracellular availability of methyl donors such as S-adenosyl methionine is limited, the methylation status of DNA and thus its integrity might still be influenced by a high metal methylation activity.

The genotoxicity/carcinogenicity of organometal(loid)s will remain an important area of research. Nevertheless, it was pointed out in the discussion that other toxic effects of organometal(loid)s might be as well relevant as indicated, for example, by the lethal effect of war gases such as pentafluorophenylarsenic oxide.

While current views on the mechanisms of metal methylation gave little incentive for controversial discussion, the quantitative importance of this metabolic pathway remained unclear. This is particularly true for the bacterial methylation of metal(loid)s in the human gut, a process which seems to occur according to observations made in recent studies. Methylation of inorganic metal compounds mediated by microorganisms mostly takes place in anaerobic environments. Since the bacteria mediating the methylation process in the human gut are not yet known, it is not clear whether the high methane production occuring in a large part of the population provides a favourable environment for metal(loid) methylation. Whether or not biomethylation is important in a quantitative matter, most participants consider the methylation of metals not as a pathway designed primarily to toxify or to detoxify metal compounds but rather as a side-effect of enzymes involved in methylation reactions. Nevertheless, increased research efforts will have to be made to elucidate the quantitative significance of the methylation process in view of the detrimental effects caused by many organometal(loid) compounds on human health.

Resumé

The majority of the contributions in this book deals with metal(loid)s which have already been the objects of studies for several years; the analytical skills acquired during the development and application of methods for the trace analysis of As, Sb, Sn, and Hg species provide a sound basis for the investigation of other organometal(loid) compounds such as Se and Bi. Several authors have emphasized that current challenges in the analysis of organometal(loid)s are rather related to the area of species-conservation during sampling and sample preparation and to the chemical identification of naturally occurring species than to the instrumental quantification of these compounds. In addition, methods ranging from biophysical techniques to procedures used in molecular biology must be applied to enable detailed investigations on the role of organometal(loid) species in complex biological systems. As yet, bioaccumulation and biotransformation processes are not sufficiently understood to predict the fate of anthropogenic emissions in the environment. Therefore, monitoring studies must be designed preferentially to elucidate pathways of consumption (i.e. the 'food chain approach') and not only to assess the concentration of organometal(loid)s in individual environmental compartments and matrices. The fascinating new research possibilities at the microscale will improve our understanding of processes at the molecular level and will give us new insights into the relationship between the structure of organometal(loid) compounds and their reactivity.

Several alkylated organometal(loid) species do not only enter the human body by inhalation, ingestion, or dermal penetration, but are also formed from inorganic precursors by methylation in the liver, kidney, and the colon (Ch. 10). Thus, a critical parameter determining most toxic effects of organometal(loid) compounds is the intracellular dose. Currently, data on human exposure to organometal(loid)s and the biotransformation of these compounds are extremely limited (only methylmercury is discussed in detail). The results of geno- and neurotoxicological investigations presented here however indicate the considerable toxicological potential of these compounds, even at environmental concentrations. One of the most striking examples discussed in this book concerns the element arsenic: While the biomethylation of this element has been considered a detoxification process in the past, methylated tri/pentavalent arsenic species have recently been demonstrated to exhibit strong geno- (Ch. 12) and neurotoxic (Ch. 16) effects, exceeding even those of arsenite or arsenate.

In summary, it is evident from the contributions of this book that only multidisciplinary research efforts will help to better understand the importance of organometal(loid) species both in the environment and in the human body.

Subject Index

algae	11 - 12, 72 - 76, 91, 124, 155 - 165
aluminium	261 - 263, 265 - 267, 269 - 270
amalgam	122 - 126, 131 - 132, 262
AMPA receptor	266, 288, 292 - 296, 307 - 309
antimony	8, 55, 74, 82 - 84, 85, 88, 101 - 104, 106 - 110, 137 - 142, 147, 235
Aplysia	264 - 266
archaea	7 - 8, 138 - 139, 142, 148
arsenate	12, 41 - 67, 155 - 165, 286
arsenic	4 - 7, 17 - 18, 41 - 67, 71 - 92, 97, 102 - 110, 137 - 147, 181, 205 - 215, 221 - 230, 285 - 310
arsenic speciation	42 - 68, 71 - 92, 34 - 35, 164
arsenite	41 - 67, 73, 156, 158 - 165, 221 - 230, 285, 293 - 309
arsenolipids	43 - 46, 64
arsenosugars	44 - 46, 54, 59 - 60, 80 - 81, 155 - 165
Asparagus P	125
axenic growth	148, 156, 162, 164
benzo[a]pyrene	226
bioaccumulation	8, 74, 91, 155 - 165, 263
biohydridisation	187, 138, 140, 147, 149
bioindicator	71 - 91
waste treatment	97 - 98, 103, 110
biomethylation	4 - 12, 97, 108, 137 - 150, 181 - 182, 198 - 200, 221 - 222, 230
biomonitoring	71 - 92
biotransformation	138, 140 - 150, 155 - 165
bismuth	9, 55, 110, 137 - 150, 181 - 201
blood	181 - 201, 238, 243, 247 - 248, 251 - 254, 260 - 263, 270 - 271
chromosomal aberrations (CA)	206 - 215, 221, 235 - 244
cisplatin	167, 175 - 180, 212
compost	97 - 110
copper recycling	236 - 244
diethyltin dichloride (Et_2SnCl_2)	175
dimethylarsinic acid (DMA (V), Me_2AsOOH)	3, 41, 73, 156, 158 - 165, 181, 43 - 44, 49, 55 - 65, 80, 82, 190 - 194, 208 - 213, 222 - 230, 293 - 310
dimethylarsinous acid (DMA III), Me_2AsOH)	3, 41, 181, 208 - 213, 222 - 230

dimethylmercury (Me_2Hg)	2, 8 - 9, 113 - 132, 206 - 207
dimethyltin dichloride (Me_2SnCl_2)	175, 211, 214 - 215
DNA (ligand modelling)	167 - 180
DNA damage	206, 221 - 230
DNA repair inhibition	222, 229
docking	173 - 174
earthworm	66
ecosystem (limnic)	71 - 92
electrospray ionisation mass spetrometry (ESI/MS)	30 - 33, 18, 20, 48 - 52, 57, 60 - 64, 67
environment	1 - 12, 17 - 35, 41 - 68, 71 - 92, 97, 113 - 132
faeces	125 - 126, 130 - 132, 182 - 184, 196 - 200, 205
FISH	148 - 150
food chain	8, 76, 88 - 91, 97, 248, 263
freshwater	8, 11, 55, 71 - 75, 86, 91, 250
Fucus gardneri	155 - 165
GABA	251, 262, 284, 287
garden compost	98, 101 - 103, 108 - 110
gas chromatography	1, 17 - 22, 35, 47, 52 - 54, 84, 114
gas condensate	118 - 122, 130
GC/AFS	50, 114 - 121, 125 - 129
genotoxicity	205 - 215, 221 - 230, 235
glutamate receptors	254, 267, 299 - 302, 307 - 310
HPLC	18 - 19, 29 - 31, 80 - 87, 130, 156 - 162, 184 (also see liquid chromatography)
HPLC/ICP - MS	(*see HPLC*)
hydride generation	27, 54 - 57, 100 - 110, 115 - 118
ICP - MS	18 - 31, 46 - 67, 79 - 87, 100 - 109, 127 - 129
ion channels	264 - 266, 283 - 310
lead	2, 4, 6, 18, 23, 235, 238, 243, 247 - 255, 259 - 261, 267 - 272, 291 - 293
limnic	71 - 95
liquid chromatography	1, 47, 57, 59, 138 (*see HPLC*)
long term potentiation	254, 259, 263, 269
mammalian cells	206, 221, 227, 230
marine biota	5, 18, 34 - 35
mercury	4 - 6, 8 - 11, 99, 104, 113 - 132, 139 - 147, 182 - 201, 205 - 207, 235, 247 - 250, 259, 262 - 269
mercury speciation	25 - 28, 35, 113 - 132, 182 - 201
metabolism	42, 164 - 165, 181, 209, 222 - 226, 250 - 252, 262 - 268, 284, 286
metallothionein	45, 47, 66, 248, 252

Methanobacterium formicicum	8, 7, 139, 146 - 149
methanogens	131, 143 - 147
methylcobalamin	6 - 8
methylmercury	8 - 11, 25 - 26, 114 - 132, 206 - 207, 248 - 250, 263 - 272, 284
Minamata	139, 206, 250, 263, 284
monomethylarsonic acid (MMA (V), $MeAsO(OH)_2$)	2 - 3, 41 - 65, 101, 159 - 162, 192 - 194, 208 - 215, 221 - 230, 285 - 286, 293 - 309
monomethylarsonous acid (MMA (III) , $MeAs(OH)_2$)	3, 41 - 43, 55 - 56, 73, 181, 190, 208 - 213, 222 - 230
NMDA receptor	266 - 270, 288, 293 - 308
occupational exposure	243, 285
organotin	11 - 12, 27, 35, 75 - 92, 179, 211 - 215, 284
oxidative stress	210 - 211, 221 - 255
phosphate	42, 65, 155 - 165, 175, 179
poly(ADP - ribosyl)ation	226 - 230
population monitoring	243
potassium channels	284, 287, 292 - 293, 303 - 309
quality assurance	84 - 86, 91
S - adenosylmethionine	5, 209
sample preparation	17, 21, 25 - 27, 45 - 50, 53, 72, 78
seawater	55, 73 - 74, 155 - 157, 162 - 165
selenium	5, 9, 17 - 18, 31, 51, 55, 110, 139, 147, 181 - 201, 205
sewage gases	139 - 141
sewage sludge	97, 139 - 151
sheep	45, 59, 62
sister chromatid exchange (SCE)	206, 210 - 215, 239
speciation	see arsenic, mercury, tin speciation
stable isotopes	24 - 26
sulfate reducing bacteria	142 - 150
tin	4 - 12, 18, 20, 28, 53, 55, 71 - 92, 104, 107 - 110, 137, 140 - 150, 179, 196, 205, 211, 214, 235, 285
tin speciation	71 - 92, 28
toxicity	5, 11, 43 - 45, 71 - 76, 130, 199, 205 - 215, 221 - 229, 247 - 255, 268 - 271, 283 - 310
triethyllead	259, 261
trimethylantimony (stibine, Me_3Sb)	2, 8, 107 - 109, 139 -145, 215
trimethylarsine	6 - 7, 41, 80, 107 - 108, 141, 209 - 215, 285
trimethylbismuth	107, 139 - 141
urine	45, 125 - 126, 130 - 131, 181 - 201, 205,

	208, 222, 286
volatilisation	8, 132, 137 - 150, 147 - 149, 197
voltage clamp technique	291 - 294, 304, 307
voltage gated channels	265
waste disposal sites	205, 235 - 236, 244
XANES	46, 66 - 67
Xenopus oocytes	288 - 293, 304 - 307
zinc	99, 221, 227 - 228, 259, 262, 269, 309
zinc finger proteins	227 - 228

MIX
Papier aus verantwortungsvollen Quellen
Paper from responsible sources
FSC® C105338

If you have any concerns about our products,
you can contact us on
ProductSafety@springernature.com

In case Publisher is established outside the EU,
the EU authorized representative is:
Springer Nature Customer Service Center GmbH
Europaplatz 3, 69115 Heidelberg, Germany

Printed by Libri Plureos GmbH
in Hamburg, Germany